Practical Business Systems Development Using SSADM

A complete tutorial guide

PEARSON
Education

We work with leading authors to develop the
strongest educational materials in computing,
bringing cutting-edge thinking and best
learning practice to a global market.

Under a range of well-known imprints, including
Financial Times Prentice Hall, we craft high quality
print and electronic publications which help readers to
understand and apply their content, whether studying
or at work.

To find out more about the complete range of our
publishing, please visit us on the World Wide Web at:
www.pearsoneduc.com

Practical
Business Systems
Development
Using SSADM

A complete tutorial guide

Third edition

PHILIP L. WEAVER, NICK LAMBROU

AND MATTHEW WALKLEY

FT Prentice Hall
FINANCIAL TIMES

An imprint of **Pearson Education**

Harlow, England • London • New York • Boston • San Francisco • Toronto
Sydney • Tokyo • Singapore • Hong Kong • Seoul • Taipei • New Delhi
Cape Town • Madrid • Mexico City • Amsterdam • Munich • Paris • Milan

Pearson Education Limited
Edinburgh Gate
Harlow
Essex CM20 2JE
England
and Associated Companies throughout the world

Visit us on the World Wide Web at:
www.pearsoneduc.com

First published 1993
Second edition 1998
Third edition 2002

ISBN 0 273 65575 2

British Library Cataloguing-in-Publication Data
A catalogue record for this book is available from the British Library

Library of Congress Cataloging-in-Publication Data
Weaver, Philip L.
   Practical business systems development using SSADM : a complete tutorial guide /
Philip L. Weaver, Nick Lambrou, and Matthew Walkley.-- 3rd ed.
      p. cm.
   ISBN 0-273-65575-2 (pbk.)
   1. Electronic data processing--Structured techniques. 2. Business--Data processing. I.
Lambrou, Nick, 1956- II. Walkley, Matthew. III. Title.

QA76.9.S84 W438 2002
005.1'2--dc21

                                                          2002072189

10 9 8 7 6 5 4 3
07 06 05 04 03

Typeset in 9.5/12pt Stone Serif by 35
Printed by Ashford Colour Press Ltd., Gosport

Contents

Contents

Part 3
Specification

Part 4
Physical Design considerations

Preface

INTRODUCTION

This book describes how to analyse and design information systems using a structured methodology, in this case the Structured Systems Analysis and Design Method, or SSADM. There has been much talk in recent years about the application of other methods of system design, most notably object-oriented methods and Rapid Application Development or RAD. While the purpose of this book is not to provide an academic justification of structured methods, the facts remain that the techniques of structured methods still form the basis of the majority of real-world information systems developments, and indeed are entirely compatible with approaches such as RAD.

SSADM is without question the structured method for information systems students and practitioners to study. It is the standard information systems development method for UK government projects, and has become a *de facto* standard for the UK private sector. SSADM also forms the core of numerous courses at HND, BSc and MSc levels. The skills acquired in learning SSADM can also be applied to virtually any information systems development, as many of the techniques can be mapped to other standard methods and approaches.

One of the main criticisms of structured methods is that they are bureaucratic. This shows a fundamental misunderstanding of how structured methods should be used in practice. SSADM, in common with many other methods, is *not* a recipe for information systems development. It is more akin to a toolkit, containing a number of tried and tested techniques and products, which are applicable to different circumstances, and which are designed to ensure that information systems are developed to meet user requirements in an efficient and effective manner. It is not intended that every technique is applicable to every project. On the contrary, in the experience of the authors, it is virtually inconceivable that any project will ever need to use all techniques. The real skill in applying SSADM is to recognise which products are suited to the individual circumstances of a project, and to adapt them accordingly. This book aims to give some practical guidance on how to do this.

SSADM is a well-tried and comprehensive method. It takes a top-down approach to system development, in which a high-level picture of system requirements is built and then gradually refined into a detailed and rigorous system design. This is achieved through the careful and staged application of a range of standard techniques, several of which are used more than once during the life of an SSADM project to meet different objectives.

This book is aimed at anyone wishing to learn *how* to apply SSADM in practice. In this latest edition it has been enhanced with even more tips and advice on the practical application of SSADM's techniques and its products. It also includes more references and advice on tools outside the scope of SSADM and other information systems development methods. It has been written to appeal both to students with no previous knowledge of systems analysis and design, and to practising analysts who want to use SSADM and may be thinking of sitting for the Information Systems Examining Board Certificates in SSADM.

▶ THE STRUCTURE OF THIS BOOK

Students frequently complain that SSADM is presented to them in a disjointed manner, one technique at a time. This leaves them with no sense of the purpose, interdependence or context of these techniques, or of how SSADM is used within a real project. The material in this book follows the default structure of the method, which is a framework for showing how the techniques can be linked together, rather than a recommendation for a project structure. It introduces techniques as and when they are needed in the development life-cycle, and uses a comprehensive case study to illustrate their application. In this way the reader learns about techniques in a natural way, rather than being presented with them in isolation. Learning is reinforced through the extensive use of exercises, with selected solutions given in Appendix C, and the remainder on the companion web site.

The book covers the entire method, beginning with Project Initiation and ending with Physical Design. No specific technical environment is assumed in the book, as this would have the effect of alienating a large proportion of readers, and in any case would rather go against the philosophy of SSADM. The book is divided into four parts.

▶ Part 1 – Introduction

Part 1 begins with a chapter that introduces some of the basic concepts of SSADM, and of structured methods in general. It also places SSADM into context of the systems development life-cycle.

The second chapter discusses the principles of project management and project initiation, both of which are essential to the success of any information systems development project. It does not attempt to provide detailed tuition in these non-SSADM areas, but does offer practical guidance, in particular for the initiation and conduct of student projects, and references to more complete sources of information. The chapter closes with a discussion of feasibility studies.

▶ Part 2 – Investigation

Investigation, or requirements definition, provides the foundation for any successful information

systems development project. SSADM is particularly strong in this area, and it is the techniques of Investigation that are the most transferable and widely applicable of the SSADM toolkit.

Part 2 provides comprehensive tuition in the key techniques of activity, data and process modelling, and of requirements definition.

▶ Part 3 – Specification

Part 3 details the techniques used by SSADM to specify the required system in a manner that is understandable to non-technicians, and is largely independent of the precise technical products to be used for system construction.

The 'implementation-independent' specification produced by SSADM is analogous to an architect's drawings and plans for a new building. These plans are intended to represent accurately how the building will look and function. The plans must take account of the constraints imposed by the principles of civil engineering and the properties of the main materials to be used. While they do not detail the precise use of each individual physical product, from concrete mixes to types of wiring and switches, to be used in the construction of the building, the plans must provide sufficient information for technical blueprints to be drawn up.

▶ Part 4 – Physical design considerations

Physical design is the process of translating the system specification into a set of technical construction designs, which by definition are directly dependent on the precise physical products to be used. The range of physical products available for system construction is virtually limitless, and is changing all the time. SSADM can therefore only provide general guidance in the area of physical design.

Part 4 covers techniques used for the more generic areas of physical design, and provides general advice and guidance for the more implementation-specific tasks.

▶ Chapter layout

Each chapter begins with an introduction, a set of learning objectives and links to other chapters.

Throughout the text, hints and tips are given in margin notes and text boxes. These text boxes are also used to provide information on advanced topics and techniques outside of SSADM.

Each chapter ends with a summary and (with one exception) exercises.

▶ Support material

The book includes an extensive range of exercises at the end of most chapters in Parts 2 and 3, which deal with the key techniques of SSADM, and a smaller number of review-style exercises at the end of the more discussion-based chapters of Parts 1 and 4.

Solutions to a selection of the exercises can be found in Appendix C. The full set of solutions is available on the supporting web site.

Comprehensive support materials for the book can be found on the supporting web site (www.booksites.net/weaver), including:

- lecture slides;
- full solutions to all exercises;
- supplementary exercises;
- lecturer's guide containing examination-style questions;
- more advice and guidance in SSADM techniques.

SSADM is an 'open' method in the sense that anyone is free to use it, without paying any fees or infringing copyright laws. As a result many support agencies have been set up to offer advice on its use. The addresses of some of them are given in Appendix D.

▶ ACKNOWLEDGEMENTS

First and foremost we owe thanks to Anna, Philip's wife, without whom the first edition of this book would never have been written. Her expertise and efforts on the PC made the whole thing possible, and her constant encouragement when the going got tough were invaluable.

We also owe many thanks to our university colleague Angelos Stefanidis for his thorough technical review of the book and for many discussions on the finer points of SSADM.

There are many other people who have helped in the writing of this book, some unwittingly (such as the students on our courses) and others more overtly, in particular: Andy MacWilliam for his review of the very first draft; John Cushion at Pitman Publishing for his advice on editorial matters and general approach of the first edition; Penelope Woolf and Michelle Graham for editing the second edition; Joseph Howarth for his meticulous proof-reading; and Alison Kirk, Laura Graham and Anna Herbert from Pearson Education for their support during the drawn-out production of the third edition.

For Euan, Molly and Sam

Introduction to information systems development

Before exploring the development of information systems development using SSADM, it is important to understand some of the underlying principles and background of SSADM and of structured development methods in general.

The aim of Chapter 1 is to provide this background, and to cover some of the key concepts of SSADM.

Chapter 2 discusses a number of the ingredients that need to be in place to ensure that information systems development projects succeed, whatever method is used to deliver the resulting system. Specifically, this chapter deals with the tasks associated with initiating and managing a project effectively, and with assessing its feasibility.

Project management is often mistakenly seen as necessary for complex projects only, while the reality is that the principles of project management are essential for the successful delivery of any project, however small.

Chapter 2 also introduces the central case study used throughout this book.

Basic concepts

INTRODUCTION

The aim of this chapter is to do a little scene-setting. The Structured Systems Analysis and Design Method (SSADM) is a rigorous method of systems development, and newcomers can all too easily find themselves lost in a mass of detail with no sense of where they are heading. Hopefully, by reading this chapter some of these feelings should be avoided. We will begin with a brief look at what structured methods, and SSADM in particular, are all about. We will then discuss some of the major principles and terminology of SSADM before briefly introducing the various business information systems that can be found in organisations today. This will lead us to the identification of the types of system for which SSADM is suitable. We will finish with a short overview of some SSADM features that should help to put the contents of this book into context.

▶ Learning objectives

After reading this chapter, readers will be able to:

◆ understand the basic phases of the systems development life-cycle;

◆ understand the structure of SSADM;

◆ recognise a business information system;

◆ know what projects SSADM is suitable for;

◆ appreciate the underlying principles of SSADM;

◆ appreciate the Systems Development Template and the importance of the Three-schema Architecture.

▶ Links to other chapters

◆ Chapter 2   covers the adaptation of the systems development life-cycle to specific projects.

▶ STRUCTURED METHODS

▶ System development life-cycle (SDLC)

Most information systems share a common life-cycle. This is not to say that they are all developed and operated in the same way, but that they will all pass through the same basic phases in their lifetime.

Figure 1.1 illustrates one version of this life-cycle. There are many variants on this, with some people breaking down these phases and others merging them. The important thing is to form an overall picture of how systems are developed and brought into production.

▶ Strategy planning

In most organisations there will be a formal mechanism for deciding which areas of the business require new or enhanced computer systems. This may be referred to as 'strategy planning' or something similar, but will always involve assessing the relative priorities of different areas, with a view to initiating one or more development projects.

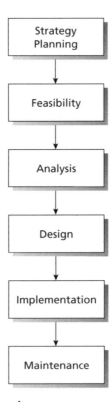

Fig 1.1 System development life-cycle

▶ Feasibility study

Before system development begins in earnest we may need to establish its feasibility, although sometimes there is no choice in the matter – we simply have to provide computer support to remain competitive or to comply with legislation. The ideas that come out of strategy planning are often vague and untried, so some assessment of their feasibility will have to be carried out before we spend too much time and effort on projects that cannot be cost-justified or that are technically impossible.

▶ Systems analysis

Once the project is under way our first task is to establish the requirements of users, and hence of the business. At this point we may have some idea of the eventual shape of the system, but the main thing is to concentrate on what it should deliver, rather than how it should deliver it.

▶ System design

We then translate the user requirements gathered during systems analysis into a computer system design, which details exactly *how* they will be satisfied.

▶ Implementation

The system design provides a blueprint for building, testing and introducing the new system. It is during implementation that programs are finally constructed and hardware is installed. We will also need to provide training for users and assistance in making the transition to the new system.

▶ Maintenance

This is often referred to as the 'production' or 'operational' phase, and covers the period when the system is up and running in support of the business. In the developers' eyes it is definitely the *maintenance* phase, where the system will need to be kept up to date in responding to changing requirements and system errors. In terms of the total development effort required over the life of a system it has been estimated that 70% has been expended in the maintenance phase.

In the early days of systems development (the 1960s and 1970s) each individual developer or project team would devise their own method of moving through the life-cycle, often influenced by hardware and software considerations, but always driven by personal likes and dislikes. This was all very well for the individual analyst or programmer, but did not always lead to the best system design, or to the easiest of systems to maintain. Most of the problems with systems developed in this way were due to poor communication of ideas, between users and developers, or between a system's designers and its maintainers, and to a lack of rigour that led to errors and omissions.

Initial efforts in overcoming these problems were directed at the programming or implementation end of the life-cycle. Conventional programming methods tended to result in code that was as clear as day to the person who wrote it, but was virtually incomprehensible to anyone else (the classic 'spaghetti code'). This meant that it was difficult to maintain or debug. Structured programming methods aimed to overcome this problem by providing a series of steps and diagrammatic program design techniques that ensured that code was well structured and easy to follow, was based on consistent and effective design principles, and was accurately specified. Their structured nature also meant that programming activities could be managed and quality-controlled more effectively.

As a result program quality improved greatly, but this still left a much larger problem: there is no point in producing wonderfully coded programs that fail to provide the facilities that users need to support their business. So attention turned to the earlier phases of the life-cycle and in particular systems analysis and design. The result has been a whole range of structured systems development methods, now more commonly known simply as 'structured methods'.

Structured methods consist of three basic elements:

A brief history of SSADM

SSADM was originally developed by Learmonth and Burchett Management Systems (LBMS), following an investigation by the Central Computing and Telecommunications Agency (CCTA) into adopting a standard information system (IS) development method for use in UK government projects. It was launched in 1981 and by 1983 had become mandatory for all new government developments. This gave SSADM a strong toehold in the structured methods market.

Since 1981 SSADM has been considerably updated, largely in response to practical feedback, leading to the current version (number 4.3), released in 1996. It now occupies a dominant market position and is in many ways the *de facto* standard for information systems development, in both government and industry. In 1988 the Information Systems Examinations Board (ISEB), a subsidiary of the British Computer Society (BCS), introduced an examinable qualification which has now become a requirement for consultants on government projects. Readers who are interested in obtaining the latest SSADM qualification should approach the ISEB to obtain a list of approved course providers. The ISEB website (www.bcs.org.uk/iseb) shows how, through a set of examinations, a professional can amass the necessary qualifications in a modular fashion.

The following dates are significant in the history of SSADM:

1981	Version 1
1983	Version 2
1985	Version 3
1988	ISEB examinations introduced
1991	Version 4
1995	Version 4+ introduced as Version 4.2
1996	Version 4.3
2000	The SSADM manual is published by the Stationery Office in seven small volumes

---

1. a *default structure* of steps and tasks which the project team should consider following;
2. a set of techniques to be applied in each step that provide (largely diagrammatic) *structured* definitions of user requirements and system components;
3. a set of products developed by each of the techniques.

Many claims have been made about the effectiveness and benefits of structured methods as compared with conventional methods. The main benefit is that they result in systems that more closely match the needs of users because user requirements have been more fully understood and communicated from the outset. They do this by applying rigorous techniques that force analysts to examine the business problem thoroughly, and by improving communication through the use of structured diagrams that are easily understood and less ambiguous than text. By providing a firm structure of steps and checkpoints throughout the life-cycle, projects can be carefully managed and personnel can be used effectively.

Structured methods move the project forward through analysis and design, using techniques that gradually transform user requirements into a system specification, and which interrelate and cross-reference each other as they do so. In this way they ensure that information is not lost, which is a danger with conventional methods that leap suddenly from a set of analysis activities ('techniques' is probably too generous a term), using a largely textual specification of requirements, to another set of design activities, possibly carried out by an entirely different team.

It is also claimed that structured methods reduce the costs of a system, but it must be stressed that this is over the entire life-cycle and not just the analysis and design phases. It is precisely because more effort is spent during analysis and design that the resulting system should require less maintenance (due to higher quality, greater flexibility and fewer errors). Bearing in mind that maintenance accounts for 70% of the development cost of a system, any savings in this area are likely to outweigh increased costs in other phases.

Finally, by using a consistent set of techniques and steps for the development of all systems within an organisation, personnel from one project should be able to transfer to another with minimal retraining. Most structured methods are self-documenting in the sense that they produce a comprehensive description of the system and how it operates as part of the analysis and design process. This removes the need for a separate documentation task once the system is in production (a notoriously onerous task), and again ensures consistency.

▶ OVERVIEW OF SSADM

SSADM consists of three main components:

♦ the default structure or framework of an SSADM project

♦ a set of standard analysis and design techniques

♦ the products of each technique

▶ Structure

The SSADM default structural model is based on a series of *modules* that begin with a Feasibility Study and end with Physical Design.

Figure 1.2 shows the modules as recommended by the earlier versions of SSADM, and shown in greater detail in Appendix E. A brief description of each module follows.

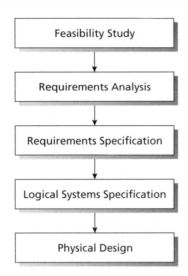

Fig 1.2 The modules of SSADM

Feasibility Study

| Chapter 2 covers the main Feasibility Study principles. |

The scope of the proposed IS project is defined using some of SSADM's core techniques to produce a high-level overview of processing and data. Several options for taking the project forward to a full SSADM study are looked at and a single option is selected by balancing benefits against costs. A decision to abandon the project might be taken if the project is found to be infeasible.

Requirements Analysis

| Chapters 3, 4, 5, 6 and 7 cover the main techniques used in Requirements Analysis. |

In many cases the new computerised information system will be intended as a replacement or extension of existing systems (which may be fully computerised, entirely manual, or a combination of the two).

In this situation we will begin the full analysis of requirements by modelling the current system with a view to drawing out existing problems and new requirements. In early versions SSADM was criticised for concentrating on the examination of existing systems, rather than looking at the requirements of new systems. While in Version 4 the emphasis is very much on the required system, current systems analysis still has a number of points to commend it as a vehicle for uncovering user requirements (if used properly):

◆ **Retained functionality**. Although the current system must have limitations (otherwise we would not be thinking of replacing it), it will usually be providing a large degree of support for the business, which will have to carry forward to the new system. If we analyse this support carefully, then we may well understand the core functionality required of the new system.

◆ **User confidence**. By demonstrating our ability to understand and model the current system, we will increase the confidence of users in our ability to understand the requirements of the new system. By beginning with a scenario that they already know well, users will be encouraged to take an active role in system modelling.

◆ **Identification of requirements**. Many analysts find that the most effective way of uncovering user requirements is by discussing the shortcomings of current systems. By analysing and reviewing current operations closely with users our chances of identifying a complete set of requirements are greatly increased.

◆ **Familiarisation**. Current systems analysis is an effective way for analysts to become familiar with the business area under investigation. In particular it will be of great help in establishing common terms with users.

◆ **Project scoping**. The scope and complexity of the new system can often be deduced quite effectively from those of the current system.

At the end of investigating the current environment we should have a reasonably comprehensive statement of user requirements. We now examine this and put together several options for solving the business problem (or a subsection of the business problem) represented by this statement. Although we will need to take into account some generic physical or technical aspects, our attention will be directed towards defining business (or logical) solutions, and not towards describing any specific technical environment.

A single option will be selected as providing the shape and direction for detailed requirements specification.

Requirements Specification

Requirements Specification lies at the heart of an SSADM project; it is where user requirements are transformed and refined into detailed and precise specification of *what* the system is required to do. Many of SSADM's most powerful modelling techniques are applied in this stage, and we move firmly from analysis into design.

> Chapters 8–11 are dedicated to the techniques used in Requirements Specification.

Logical Systems Specification

During Logical Systems Specification we continue the development of the Logical Design while we also make sure that the technical environment in which the system will finally reside is chosen.

> Chapters 12 and 13 introduce the main concepts of Logical Systems Specification.

The specification resulting from Requirements Specification should provide us with enough information to propose alternative technical environments on which to implement our system design. In many cases we will have no choice of hardware and software, but where we do have some discretion we will draw up several options for hardware, software and development platforms, and help management to select a single option for use in physical design.

In the meantime we will take the system design process as far as is possible without reference to a particular technical environment. The resulting design will be logical in nature and so capable of implementation on a variety of platforms. It will also act as a more or less permanent model of *how* the system satisfies user requirements. The logical nature of the design means that it should reflect underlying business rules and activities rather than physical constraints.

Physical Design

The logical system design is now translated into a physical design based on the selected technical environment. In many areas physical design is dependent directly on technical issues specific to the chosen environment. Consequently SSADM is only able to provide generic guidelines in these areas. In this way it covers most but not all of the systems design phase.

> Chapter 14 explains how to prepare for Physical Design.

Method customisation v. the recipe approach

Methods have developed in response to the need to take control of the design process in order to eliminate project failure. As methods developed and certain techniques started becoming *de facto* standards it was hoped that a recipe-like approach could be somehow discovered to guide new developers. Many methods, including the earliest versions of SSADM, were proposed in the false hope that it would be possible to forge one method that would fit all developments. This method would consist of pre-described steps which, when followed, would generate the right system every time.

Twenty years on we have realised that there is no such silver bullet solution and that each project has its own peculiar set of conditions that require a tailor-made approach.

When studying SSADM, or any other information systems method for that matter, we should be conscious not to follow a set of steps because they happen to have been fruitful in a previous project. Instead, we should be informed by previous successful and unsuccessful approaches and try to understand the reasons for their success or failure. Only then will we be ready to understand the nuances of a system's development and to *customise* the tools on hand to fit the problem.

While we follow a step-by-step development for the case study used to illustrate SSADM here, it should soon become evident that the steps we have chosen fit the conditions of the situation. The reader is expected to reflect on alternative ways of tackling the problems presented by the case study and to continually ask whether any other technique would be more appropriate. Such a reader will then be rewarded by a better understanding of the pieces that constitute a method.

It is hoped that the steps illustrated at the beginning of Parts 2, 3 and 4 of this book will be seen as just one way of working with the Systems Development Template and that any project team should have enough experience to customise an approach whenever a new project arises.

▶ Techniques

Many of the techniques used by SSADM are common to other structured methods, especially those used in the analysis phase. Several of them have been present in most versions of the method and are now well tried and tested. The main feature of SSADM's core techniques is their diagrammatic nature. There is a great deal of truth in the saying 'a picture paints a thousand words'. Diagrams are capable of representing information quickly, in a compact form, and also relatively unambiguously. No one would suggest that the best way of planning or specifying a new building is by using textual reports; they would simply be too vague. The necessary precision is only available from diagrams, plans and models. It would also be very difficult for customers to assess the design of a new building without models and drawings. The same is true of information systems.

A full list of SSADM techniques can be found in Appendix E along with details of where they are used. The three major analysis techniques equate with the three SSADM views: Data Flow Modelling represents system processing; Logical Data Modelling represents system data; and Entity Behaviour Modelling represents the effect of time on data. These are all diagrammatic techniques that closely interrelate. It is a feature of most SSADM techniques that they cross-reference with each other to form a complete, consistent picture of the entire system under investigation.

Logical Data Modelling is applied throughout the life-cycle to provide the foundation of the new system, namely the data model. Entity Behaviour Modelling marks a change in emphasis from analysis of requirements to design of the system. Data Flow Modelling is a powerful technique for analysing processing but does not provide sufficient detail for use in process design. As we progress from analysis to design we introduce a new technique, Function Definition, which transforms the results of Data Flow Modelling into a set of user-defined functions. These functions are initially textual in nature but act as a basic unit of specification to which we add greater detail following the application of further, more rigorous, techniques.

▶ Products

Each step has a number of tasks associated with it, most of which lead to the creation or enhancement of standard SSADM products. At the end of an SSADM project the new system will be described by the sum of these products.

Products can be divided into three basic groups: processing, data and system–user (or human–computer) interface. Clearly, to represent the system as a whole these groups must interlink and complement each other. The ways in which they should interrelate will be explained as each product is introduced in later chapters.

Typically the end of every module should contain an 'assembly' step where all the products (except working documents) from the module are checked for consistency and completeness.

▶ THE SYSTEM DEVELOPMENT TEMPLATE

Any information system consists of three parts: an *external* shell through which users interact with the system; an *internal* or *physical* design from whose presence users are shielded; and a *conceptual* model which represents the business requirements, and upon which the internal design is based. As we progress step-by-step we should always be careful to separate these three concerns and to tackle each one in its own right.

This separation of concerns is known in SSADM as the Three-schema Specification Architecture (3-SSA) – see Figure 1.3. It is hoped that by keeping the specification, and as far as possible the implementation, of these three elements distinct, the final system will be more robust.

The main objective of the systems analyst is to come up with a clear system specification. To do so properly, an investigation has to take place before specification starts. The specification itself has to be followed by the construction of the system if the whole analysis is not to be in vain. An IS project can therefore be seen as the process of moving from *investigation* to *specification* to *construction*.

This process of progressing from investigation to specification to construction is inevitably influenced by the *user organisation* and the prevailing *policies and procedures* of the organisation for which the system is developed. Furthermore, each project has some form of a *decision structure* through which managerial choices are made. One way of representing all this schematically is depicted in Figure 1.4 (note how the Three-schema Specification Architecture is represented in the specification part of the template as the external design, conceptual model and internal design).

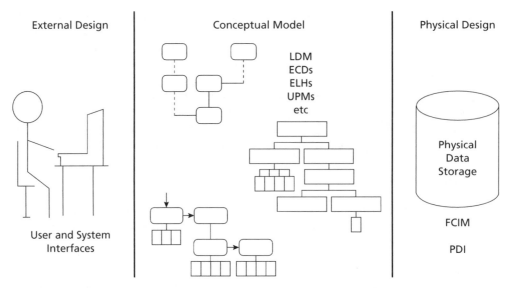

Fig 1.3 A representation of the Three-schema Specification Architecture

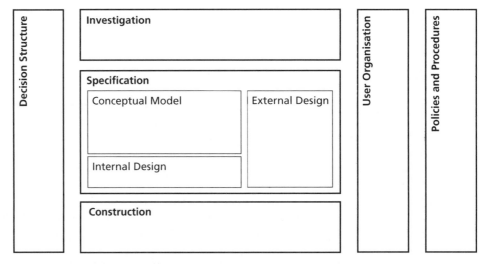

Fig 1.4 The System Development Template

The System Development Template should be used, in conjunction with the default structure, as a guide for project teams to define their own structural model, route map, or approach to systems development, according to the needs of their own project environment.

▶ SEPARATION OF LOGICAL AND PHYSICAL MODELS

One extremely important concept in SSADM (as in most rigorous methods) is the distinction between logical and physical views of system components.

Physical components are those that actually *physically* exist within the real world (or will exist in the future). They represent things as they are, warts and all, complete with constraints imposed by organisational, political or technical factors. For example, in the physical world there may be two processes associated with producing an invoice: filling in the invoice details and calculating the invoice cost. The reason for the separation of the two tasks may be purely organisational, i.e. the job descriptions or policy in the accounts office dictate that one job-holder (e.g. clerk) carries out the description task, and another (e.g. accounts supervisor) carries out the other. The fact is that there are two distinct *physical* processes.

Logical components are those that represent a picture of what underlies the physical components. In a sense they give a picture of what physical components would look like in an ideal world, free from real-world constraints. Using the invoice creation example from above, the logical view would be one of a single activity: creating the invoice. The fact that the process is split in reality is irrelevant – we are only carrying out one 'business' process.

Another example of physical and logical views of a system component might be the invoice itself. In physical terms we have a single object: the invoice. In logical terms we have information about several objects: the customer being invoiced; the items detailed in the invoice; the quantities and costs of each invoice item and of the total invoice.

▶ THE THREE VIEWS OF SSADM

SSADM looks at a system from three different, but highly interdependent perspectives.

The first is that of functionality or processing. This looks at the way in which data is passed around the system and the processes or activities that transform it, i.e. it sets out the functions provided for users by the system.

The second is that of data. An information system (IS) exists only to store and act upon an organisation's data. By understanding the true nature and structure of the data we get to the real heart of the system. Data structures are far more constant than processing or functions, which tend to change fairly frequently; so it is the data view that forms the backbone of SSADM. In this sense SSADM belongs to the family of structured methods referred to as 'data-driven'.

The final view looks at the effects of time and real-world 'events' on the data held within the system. Whereas the function and data views are rather 'snapshot' in nature, the event view is dynamic; it is specifically designed to model system behaviour over time.

▶ BUSINESS INFORMATION SYSTEMS

Organisations tend to be quite complex multi-level edifices that require decisions to be made at various levels. For these decisions to be somehow informed they have to be based on the right information and it is the hope of modern organisations that information technology, with its enormous data-processing power, will be able to supply the right information as and when it is needed.

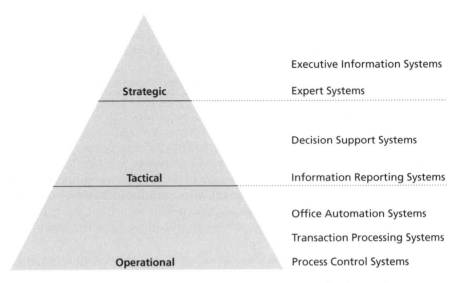

Fig 1.5 Levels of managerial decision-making and the type of information system they require

Most business studies books illustrate the variety of managerial decision-making by using the simple diagram of Figure 1.5 where the levels of decision making are depicted in a pyramid whose base is occupied by 'operational' questions, followed by a middle part of 'tactical' decisions and topped off by the 'strategic' options.

Strategic decisions tend to be unstructured whereas organisational decisions tend to be more structured.

The various levels of decision support that organisations need today are served by a variety of computerised business information systems, such as starting from the more strategically oriented ones and descending towards the operational – executive information systems, expert systems, decision support systems, information support systems, office automation systems, transaction processing systems and process control systems.

Within this huge gamut of systems it would be folly to expect that one method could be used for the design of all of them. It therefore would be useful to identify the type of system for which a method such as SSADM is most suited.

The clue to identifying what type of system we should develop using the method lies in the three views SSADM takes of a system, namely processing, data, and the effects of time on that data. Since most systems contain some form of processing it stands to reason that any method worth its name will be suggesting techniques that capture the nuances of that processing. The distinction therefore must lie in a system's need for data. If a system is not dependent on a significant amount of data then it quickly becomes clear that SSADM should not be used to develop it, since this will mean omitting some of the most powerful data-oriented techniques in its arsenal.

Having decided that SSADM is most suited for data-hungry systems we now need to get a feel about the nature of this data.

What constitutes data?

In normal conversation the word 'data' is used to mean many things. Usually it is confused with information and sometimes even with knowledge. It is relatively recently that we have been forced to make a distinction, relegating data to the bottom of the pile where it forms the base material which when processed can become information.

Within the context of computerised systems we should never forget that computers in their present form cannot think. For example, computers cannot make random associations in the way humans can. Their main advantage is that they can speedily wade through large amounts of data as long as that data is organised in a well-defined manner with clear associations linking each collection of data with the others. For example, this very text is just a string of words for the word processor on which it was first written. One word follows another and any sense that can be derived from this sequence is well beyond its reach, despite the fact that it contains a spell-checker that spells better than the author, a dictionary with more words in it than most humans can master and a thesaurus capable of pointing to many weird and wonderful words. We should therefore not be deceived by the capacity of computers to provide us with material on which we can base informed decisions.

Having decided that a string of text does not constitute organised data we still need to find what this elusive data looks like. The best way to find out is to try and think of the data/information that is held about us by various systems. In a medical system, for example, we would be represented by a forename, a surname, a date of birth, a height measurement, a weight measurement, a blood group indicator, an address, a series of dates representing visits to the doctor, a list of ailments we have been diagnosed with, etc. All the above have a common characteristic, namely that we can *all* give a value to each one of them. Furthermore, the value we give to each will be made of usually one word, e.g. John, Smith, flu, or a date or a number.

We can use the above as a basis for defining a *data item* as:

the smallest piece of information that makes sense to a human within the context of an organisation.

Other examples of data include supplier name, product code, order date, quantity ordered, delivery date, customer address, product retail price. It is hoped that the reader will be able to imagine values for each one of these data items.

Persistent data

The type of data we will be interested in is expected to remain in the system until a decision is taken to delete it. Once the data is entered into the system we should be able to switch off the computer without losing it. When we switch the computer on again the data we have input should be in exactly the same state as when we switched off.

This type of data is known as *persistent data*. It exists in the system as a permanent record.

SSADM is particularly good for the design of systems that contain vast amounts of such persistent data. These systems typically exist at the operational level of organisations where they are used to keep a record of transactions. The data they amass can be used by other systems that can sit on top of these transaction processing systems and use them to extract the data they need.

The data-oriented techniques of SSADM are dedicated to the task of organising the persistent data of a system in such a way that it can become useful to the business for which it is designed.

▶ FURTHER SSADM FEATURES

Although detailed descriptions of SSADM features will come in later chapters, it is worth looking at some of the major ones now, along with a brief overview of SSADM's underlying philosophy.

Top-down approach

In many ways SSADM provides a top-down approach, where a high-level picture is drawn up and subsequently refined into ever-lower levels of detail. This not only applies to individual techniques, but to the way that one technique takes over from another when the earlier one has fulfilled its high-level role.

User involvement

On too many occasions in the past users' views and requirements have been totally lost or overridden by analysts and designers who have marched in and taken over a project. It is a prerequisite for SSADM that user commitment and involvement are agreed right from the start. Most of SSADM's techniques can be taught to users without any prior experience of IS development, who can then participate fully in the project.

Life-cycle coverage

SSADM covers most of the system life-cycle from feasibility study to system design (see Figure 1.6). It assumes that strategy planning will be carried out before an

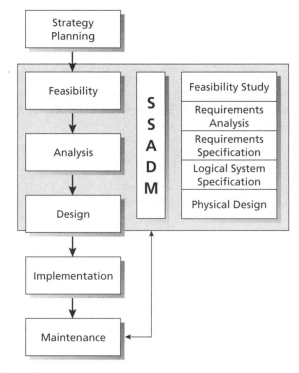

Fig 1.6 SSADM life-cycle

SSADM project is undertaken. Core SSADM is not explicitly intended for use in the maintenance phase. However, it does make maintenance easier by providing accurate documentation of the system's operations, which can be used as the basis for the design of future enhancements using SSADM techniques.

Non-SSADM techniques

There are a number of general systems analysis techniques that SSADM will draw upon during a project, e.g. capacity planning, cost/benefit analysis, fact finding. SSADM does not provide a set of precise step-by-step descriptions of these techniques, but does sometimes provide general guidance on their use or management. For detailed information on these areas the reader is advised to look to the bibliography in Appendix B, as it is not within the scope of this book to discuss them in any depth.

Project management

SSADM provides no specific project management activities of its own, but it does require the presence of effective management procedures. Each stage will involve the completion of a number of key activities in the project, with the end point of the stage acting as an important decision or control point. A structured project management method, such as the government standard PRINCE, is highly desirable but will not be discussed in detail in this book. As part of the project management structure SSADM assumes that a Project Board will exist to control and agree the decisions of the project team. This board should ideally be made up of users and IS management.

> Chapter 2 gives an overview of some of the key project management principles that all projects should follow. For more information see Yeates and Cadle (1996).

Version 4+ of SSADM places more emphasis on tailoring and customisation than earlier versions. The intention is that organisations will adapt the default structural model and standard techniques of SSADM to suit the problem in hand. Clearly this will require careful management, particularly in checking that any modified steps provide the necessary inputs for subsequent development steps.

▶ COMPUTER-AIDED SOFTWARE ENGINEERING (CASE)

Many people regard the use of CASE tools as essential to controlling and documenting large projects. SSADM has been promoted as an open method (i.e. one that can be used by anyone, without the payment of a fee), so there is a reasonable selection of CASE tools available for it.

The claims made in their favour by suppliers are often exaggerated, but most will provide a mix of the following features:

◆ diagramming tools
◆ diagram validation
◆ automatic generation of *first-cut* (initial, rough and ready) low-level diagrams
◆ report production
◆ code generation

The important thing to remember about CASE tools is that they do not carry out analysis and design for you; they merely support it by helping with diagram creation and amendment, and by providing an element of diagram consistency checking.

We can draw an analogy with the production of a book. A word-processor will help you to write and to edit it. It will carry out spelling and grammar checks, generate indexes and contents pages and enable you to develop all manner of attractive layouts. The one thing it will not do is write the book for you, or even teach you *how* to write it. The same is true of CASE tools: you do not need one to understand or apply SSADM, but you will find that they remove many of the administrative headaches associated with a typical information systems development project.

In very small projects (less than one year of total effort) CASE tools may have a negative impact on productivity, as they can involve a lot of hard work in inputting required information. They may also be prohibitively expensive. However, there are on the market a few inexpensive CASE tools of relatively limited functionality, which will be useful in even the smallest of projects in enforcing rigour and providing high-quality 'live' documentation.

▶ SUMMARY

1. SSADM consists of three elements: a default structure of steps and tasks; a set of techniques; and a set of products developed using these techniques.

2. An information system consists of three parts: an *external* shell through which users interact with the system; an *internal* or *physical* design from whose presence users are shielded; and a *conceptual* model which represents the business requirements, and upon which the internal design is based. This is known as the Three-schema Specification Architecture.

3. SSADM makes a clear distinction between logical and physical views of system components.

4. SSADM products are developed to support a system from the perspectives of functions, data and events.

5. SSADM is most suited to the delivery of data-centric business information systems.

6. The focus of SSADM is on the SDLC stages of Feasibility, Analysis, Logical Design and the start of physical design.

▶ EXERCISES

1.1. Consider all the computer applications you have ever used and classify them as executive information systems, expert systems, decision support systems, information support systems, office automation systems, transaction processing systems, or process control systems.

1.2. Think of all the systems that may hold information about you. For each imagine or find the data they hold about you.

1.3. Think of transaction processing systems with which you have come into contact (unless you live on a desert island you should be able to identify many such systems).

Getting started

INTRODUCTION

This chapter deals with the tasks associated with initiating and managing a project, and with assessing its feasibility.

'Project management' is a term used to describe a range of tools, techniques and products that ensure that a project is organised and controlled effectively, and meets its objectives within budget and on time.

Projects fail all too often due to poor project management. Specific reasons for failure include:

◆ lack of business-wide understanding of and commitment to the project;

◆ lack of clearly defined objectives, deliverables and success criteria;

◆ lack of ownership or sponsorship within the business;

◆ problems establishing the right project team;

◆ plans that are too optimistic and lack contingency;

◆ poor day-to-day management of issues and control of project tasks;

◆ lack of awareness of change management and business impact issues;

◆ lack of focus on project goals and milestones;

◆ poor understanding of risks and project dependencies.

Project management and project initiation seek to address these issues.

Project management is often mistakenly seen as necessary for complex projects only, while the reality is that the principles of project management are essential for the successful delivery of any project. What differs from project to project is the extent of the controls we need to apply. For example, the plan for a 1-person student project will clearly be much smaller than that for a 20-person project, and it will not need to use sophisticated resource management techniques. However, the plan is still vital to ensure that all of the tasks to be undertaken within the project are identified and scheduled, their interdependencies are established, and that the project can be completed on time. Once the project is under way, regular monitoring and updating of the plan will ensure that slippages are identified early, and action taken to re-plan and rectify the situation.

▶ **Learning objectives**

After reading this chapter, readers will be able to:

◆ assemble a Project Initiation Document;

◆ understand the need for and application of project management principles;

◆ assess the technical, operational and financial feasibility of a project.

▶ **Links to other chapters**

The project management principles covered by this chapter feed directly into all others within the text. The Feasibility Study has closer links with:

◆ Chapter 4   The requirements established during the Feasibility Study are built upon to form a comprehensive statement of user requirements for the new system.

◆ Chapter 7   Potential solutions for satisfying user requirements are defined, using the outline solution delivered by the Feasibility Study as a start point.

▶ PROJECT INITIATION

The key aims of the initiation phase of a project are to clarify the understanding of the project team and business of what we are to deliver, and to establish a plan for how and when we are to deliver it.

Most projects will begin with a brief statement (usually one or two pages, often in an informal memo) of a business need or systems issue that requires resolution. These 'project briefs' can originate from strategy reviews, examination of systems problems or initiatives within one area of a business.

During project initiation we take these brief statements and, usually following extensive discussions with users and systems development 'experts', develop project objectives and scope to a level where ambiguity is reduced to a minimum in order to:

◆ establish a sound financial business case for the project;

◆ assess costs and business benefits;

◆ agree plans, resources and organisation for the project;

◆ establish key risks and success criteria;

◆ formalise controls and reporting lines.

The product of the project initiation phase is the Project Initiation Document (PID), which will then act as a contract between the project team and users. Typical contents, regardless of project complexity, for a PID are given in Table 2.1.

The level of detail included in the PID will vary with the size of the project, and according to whether a previous detailed study has been undertaken. It is quite common for projects to be split into several smaller sub-projects if they become too large to handle as a single entity. In this case we may have a programme of projects, each with its own PID, but linked together to a parent project or programme. In this

Table 2.1 Typical PID contents

Background	An introduction to the project, describing the main reasons for carrying it out, and placing the project in context.
Project definition	This should explain *what* the project is to deliver. Specifically it should include:
	◆ Project objectives *that can be measured* as to whether we have achieved them. The objectives listed should give the project team a focus to their delivery. It is all too easy in the middle of a project to lose site of its true aims. Woolly statements like 'the system must be easy to use' are meaningless.
	◆ A statement of scope, describing which areas or activities of the business are inside and which are outside the scope of the project
	◆ A description of the deliverables of the project, including intermediate products such as a system design
	◆ Interfaces with other systems
	◆ Change implications for the business, i.e. which functions within the organisation will be affected
Initial business case	This should give the main business benefits (financial and strategic), along with anticipated costs. We should aim to justify *why* we should commit resources to the project. It is quite acceptable to make a case for commencing with the analysis stages of the project, while leaving justification for the remaining stages until requirements and their costs are more fully understood. The costs for analysis are typically low, and therefore businesses are often willing to adopt this approach, as long as a decision milestone is built into the initial project plan before significant development costs are incurred.
Constraints and assumptions	There will usually be some constraints, such as time limits, project budgets, or restrictions over the type of technology to be used.
Project approach and organisation	This should describe the methodology to be adopted and the make-up of the team. An initial project plan and/or list of key milestones should be included. We will also need to establish project-reporting mechanisms, although they are frequently mandated by organisational standards.
Risk management	Every project should produce a log of risks, summarising events that *may* cause issues for the project and the measures that could be taken to counteract the consequences. At the project initiation stage we may already be aware of key risks to the project. If so, we should document them up-front in the PID. Risks should be specific to the project, e.g. software needed from a third party by a specific date may be late. We must avoid generalities, such as 'team members may fall ill' as it will distract from the effectiveness of the risk log. It is the job of a project manager in putting a plan together to build in contingency for such events.

case the PIDs will be quite detailed, and we must be very clear in our description of scope as to which elements our project is to deliver, and which will be delivered elsewhere.

Below is a PID for the central case study used throughout this book. It is not intended to act as a model for what a PID should include, as much of the detail (e.g. business case, project organisation, plan, etc.) has been omitted in order to satisfy the space constraints of this textbook. However, it should give a flavour at least of what a PID might include.

Case Study Project Initiation Document

Background

ZigZag is a music and media distribution company, selling entertainment products in various formats including CDs, DVDs and videos to customers throughout the UK. All of ZigZag's current customers are retail chains or outlets such as shops and petrol filling stations. It does not sell directly to the public at the moment, but has plans to do so in the near future.

Customers order products from ZigZag by telephone, fax and post. One of ZigZag's highest priorities is to expand into on-line selling over the Internet.

ZigZag operates from a large depot, which holds stocks of products from a number of suppliers ranging from music companies to games software importers.

Organisation

ZigZag is divided into four main divisions: Sales and Marketing, Purchasing, Warehousing and Administration. Each division is further subdivided into sections, as shown below:

ZigZag organisation chart

History

ZigZag began trading fifteen years ago, out of a small industrial unit, selling CDs and tapes to independent petrol stations.

Twelve years later, the business has expanded greatly, to sell a range of products, from music and games software to blank tapes and even books to a much wider range of customers, including some national retail chains.

The business moved out of the over-stretched storage unit to a larger depot seven years ago. Around this time computers were introduced to keep records of stock levels and to support the accounts section.

ZigZag is no longer tied to a small number of suppliers. With the growth in its trade ZigZag has been able to negotiate deals with many specialist suppliers, many of whom compete to supply the same or similar products.

Overview of current business operations

ZigZag stock a range of entertainment products, selected by purchasers from within the Purchasing Division.

Customers send orders for items from the ZigZag catalogue to the Sales and Marketing Division, which then passes them to the depot for despatch.

If an item is out of stock, customers are informed of the shortfall, and must order again later if they still require the item.

Information on current stock levels, together with estimates of future customer orders, is assessed by the Purchasing Division, who place purchase orders with suppliers to cover anticipated demand.

Suppliers deliver a complete order at a time to the depot, whose staff check and record the delivery of new stock.

Invoices from suppliers are matched with delivery records by each depot and sent to the Accounts Division where payment is made according to pre-arranged terms (e.g. within 30 days of delivery).

The Accounts Division also deals with the invoicing of customers following the despatch of goods from a depot.

Within the depot, goods are stored in zones, which may consist of areas of floor, shelving, or, for very small items, large plastic bins.

Current systems support

The systems introduced with ZigZag's move to the new depot seven years ago still form the core of current computer support. Many changes and additions have been made and the maintenance of the systems now forms a substantial part of ZigZag's business overheads. There are also a number of manual systems still in existence that are struggling to provide sufficient support for the volumes of business that ZigZag now have.

With recent expansions, particularly into games and books, and with plans to move into e-commerce with the public, ZigZag have decided to review and re-develop the existing systems.

As part of the first phase the Accounts Division has purchased a new package to manage its day-to-day operations.

The second phase of the strategic review covers the stock management and customer ordering systems.

This second phase will consist of a single project to be undertaken by the computer services team from within ZigZag.

Project definition

As part of its high-level strategic review, ZigZag's senior management have identified some major problems with current system support which any new system must address as key objectives:

◆ The current systems provide no facility for direct sales to the public, or internet/e-commerce sales to the existing wholesale customers;

◆ Expansion of trade over the next few years means that the current systems and manual processes will not be able to cope with future volumes of data. For instance, the system will need to cope with up to 10,000 products;

◆ Existing systems are inflexible, and difficult to adapt to new working practices;

◆ Currently suppliers must deliver an entire purchase order at a time. This means that delivery has to wait until all items in the order are available, causing unnecessary delay;

◆ Stock is only recorded as being held in a particular zone. As each zone is fairly large this sometimes causes problems with the location of stock when assembling a customer order;

◆ Inadequate information is available on the performance of suppliers in:

  ◆ promptness;

  ◆ condition of delivered goods;

  ◆ accuracy of order satisfaction.

Clearly many more detailed problems and requirements will be uncovered during the project, but the above points must be addressed.

One of the key aims of the project will be to analyse requirements for e-commerce, and establish whether a true business case exists for its development. While future growth in their current wholesale operation (i.e. sales to shops rather than direct to the public) requires and will justify expenditure on new systems, senior management are concerned that the potential Internet expansion is not just driven by fashion and hype.

Scope

The functional areas to be investigated include the following:

◆ placing of purchase orders with suppliers;

◆ receiving of deliveries from suppliers;

◆ recording and monitoring of levels of stock and its location within the depot;

◆ placement and despatch of customer orders;

◆ monitoring of supplier performance.

The system that monitors customer sales is felt to be fairly effective and will be retained with some small improvements. It is therefore outside the scope of this project.

Constraints and assumptions

Given the success of the accounts package implementation, solutions similar to this should be considered, alongside other alternatives such as upgrades to the existing systems and new bespoke (i.e. 'tailor-made') solutions.

The depot contains a limited amount of recently purchased computer equipment, which must be re-used by any new systems.

The project team must be in a position to implement the new or enhanced system within 18 months of the start of the project.

Project approach and organisation

The project will begin with a feasibility study to investigate the potential for adopting a package system to replace current systems functionality and satisfy the main objectives of the project. A package will only be considered if a positive business case exists, and if modifications to the selected package are minimal. ZigZag does not wish to be restricted in its operations by the limitations of a packaged solution, and is not prepared to finance and manage the development and maintenance of substantial bespoke modifications.

If a suitable package is not found, then the study will also confirm whether a business case exists for proceeding with the analysis of requirements for either a modified package or a bespoke system. A further decision will be taken once analysis is complete and costs and benefits are more fully understood, on whether to proceed with full development.

Although the project will be carried out by the Computer Services team, users from the appropriate areas must be involved, and overall responsibility for the delivery and adoption of the final system will lie with the warehousing director.

The initial project team for the feasibility study will consist of a project manager, a business analyst and a senior user representative (appointed by the warehousing director). This project team will be enlarged to include a systems analyst and a senior analyst programmer if the project proceeds to its analysis phase.

▶ PROJECT MANAGEMENT

Every project should follow a set of predefined project management procedures or a standard method (e.g. PRINCE). A detailed discussion of the activities involved in project planning and management are beyond the scope of this book, but some of the key principles and activities that should be adopted in projects of any size are outlined in this section.

> For a full explanation of project management the reader should consult a text such as Yeates and Cadle (1996).

Project management is a highly skilled discipline, and one in which experience is vital if the project involves large numbers of people or is complex. However, while the 'sixth sense' that warns a project manager in advance of impending issues can only come with experience, a few simple measures can achieve a great deal in keeping a project on track.

▶ Project planning

There are many methods of planning a project, but all share a similar set of steps:

1. **Break the project into a number of stages or phases (e.g. project initiation, feasibility study, analysis, testing).** Each stage will typically represent a key step in the project, with the end point of the stage acting as an important decision or control point. Each stage should also be associated with a major deliverable.

2. **Identify the activities to be undertaken in each phase.** Within each stage there will be a number of key activities that are essential for its delivery. For example, during project initiation, we may have activities such as 'interview key users' and 'prepare Project Initiation Document'. It is also important to identify which activities are dependent on the completion of other activities, as this will determine when and how they are carried out.

3. **Break down the activities in the first stage into a detailed task list.** Most activities will require further breaking down to arrive at self-contained, again interdependent, tasks that can be assigned to team members (see box below for discussion of SSADM structure breakdown). For example, 'interview key users' could be broken down into a number of separate interviews, all of which will need to be undertaken by someone within the team, by a specified date. This can be a very time-consuming process and so is only carried out in detail one stage at a time. Most tasks are dependent on the completion and outcome of other tasks. So detailed plans created too far ahead will always be inaccurate, and involve significant rework as earlier tasks change our detailed understanding of later tasks.

4. **Estimate effort required to complete each task or activity.** Accurately estimating tasks or activities is notoriously difficult, and the key to success is undoubtedly experience. The results of previous similar projects will provide the best guide to the effort required to complete a task. If the project manager is inexperienced in estimating any of the activities, the views of experts or other project managers will be essential. There are a number of estimating tools that can also help with this process (such as function point analysis, covered in Chapter 7 on Business System Options), and package suppliers can provide quite accurate estimates of effort based on previous implementations. It is important to build contingency into activities in order to cover for problems, or unforeseen additional tasks that inevitably emerge as the stage progresses.

5. **Assign resources to each task and activity**. Most projects will be **resource-constrained** (limited by the number of people or technical resources available, the ultimate resource-constrained project being a single-person project) and/or **time-constrained** (must be completed within a specified time limit). As we allocate resources to the project, we must take care to avoid over-allocation. A general rule of thumb is to avoid allocating people for more than 80% of their available time (*after* excluding holidays etc.). This provides us with at least 20% contingency as tasks will tend to overrun, or people will be unexpectedly unavailable. Where a time constraint exists, the only ways to overcome over-allocation of resources are to bring in extra resources, or to reduce the scope of the project. To do otherwise (by reducing task estimates, removing contingency or over-allocating resources) will *always* result in slippage.

6. **Schedule the project.** Once we have identified tasks and their interdependencies, and allocated resources, we can will get an idea of the timings for the project. Invariably, we will find that the plan is not optimised. There will typically be gaps where resources are not being utilised or are over-allocated, and there will be times when activity levels on the project look too high or too low. We need to adjust the plan by reallocating tasks to other people, or by moving tasks to times when they are still feasible but we have spare capacity. For example, in planning the project initiation stage, we may be able to start preparing the PID before the user interviews are completed, even if our initial preference is not to do so. So if we put together a plan for the stage and find some gaps where we are waiting for interviewees to become available, we could begin the PID preparation early into order to save time later.

SSADM structure breakdown

When a project team decides to use SSADM as the basis for the development of an information system, the first thing that has to be decided is the project plan. This project plan should follow a clear structure made of well-defined *steps*. Each step should be split into specific *tasks* that have to be performed, using one of a set of recommended structured *techniques*. For each task to be well defined, the *product* of the task has to be unambiguously specified.

A set of steps can be bunched into a *stage*, and a set of stages can be bunched into a *module*. These modules should roughly map onto the relevant life-cycle 'phases'.

Figure 2.1 illustrates the breakdown of a project's development into a hierarchy of modules, stages, steps and tasks.

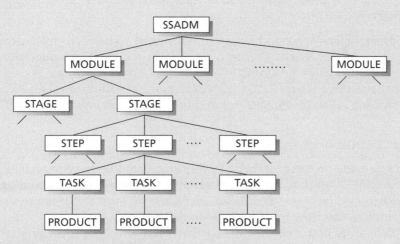

Fig 2.1 SSADM structure breakdown

The SSADM default structure summarised in Appendix F acts as a framework for planning an SSADM-based project. It is not intended as a project plan that any SSADM project should follow. The default structure is a set of stages and steps, with associated techniques and products, which describes all of the tasks that an SSADM project *could* utilise. The process of customisation will tailor this default structure to fit the characteristics of a project, dropping some steps and adding in other non-SSADM activities. Once tailored, the default structure *does* provide an excellent start point for the production of the project plan.

Each stage will typically be made up of between two and eight steps, which provide the framework for applying and controlling the development techniques. The tasks to be carried out within each step define how the techniques should be used, and specify the required standard of the products output from the step. SSADM can be considered to be a product-driven method in that project management will be largely concerned with monitoring the quality and completeness of products, rather than with monitoring the application of the techniques that create them.

Task Name	Effort	Duration	August 2001												Septe
			22	25	28	31	03	06	09	12	15	18	21	24	27	30	02
Investigation	**25 hrs**	**17 days**													
Prepare Stage Plan	6 hrs	3 days													
Schedule Interviews and Workshops	4 hrs	7 days													
Interviews and Workshops	**15 hrs**	**10 days**													
Interview M Portillo	1 hr	1 hr													
Delivery Scheduling Workshop	6 hrs	1 day													
Document DS Workshop	4 hrs	2 days													
Resolve DS Workshop Anomolies	4 hrs	6 days													

Fig 2.2 Gantt chart extract using Microsoft Project

Gantt charts

Project development steps and tasks are not meant to be performed sequentially. As we will see in subsequent chapters, many steps will run in parallel because they inform each other as we go along.

A Gantt chart (see Figure 2.2), which depicts the duration of each task as a horizontal 'time-bar', is an invaluable tool in summarising and visualising a project plan.

▶ Project controls

Once in place, the project plan must be monitored and updated at regular intervals (at least weekly) with progress and details of any slippages or rescheduling. In particular, when new tasks or activities are identified we must add them to the plan in order to understand their impact on delivery timescales.

The plan is the best indicator of whether the project is proceeding effectively. In all projects there will inevitably arise problems that require immediate action or re-planning in order to deliver the project on time. Problems tend to fall into two main groups, both of which should be documented formally, so that the project team and key stakeholders have a full understanding and awareness of them, and can take any necessary actions or decisions:

1. **Project issues.** Most problems will consist of issues that can be solved by a specific set of actions (e.g. a key user is unavailable for interview, where the action might be to find an alternative or request that their manager release them for interview). In these cases there will probably be an effect on the plan, but the overall costs, timings and scope of the project remain unchanged if appropriate action is agreed and taken. The project manager should keep a log of all issues as they arise, define responsibilities and target dates for resolution and track their status.

2. **Requests for change.** Some problems cannot be solved by a simple set of actions. Instead they will force a change of some sort to the 'contract' entered into in the PID (e.g. changes to costs, scope, timescales or estimated benefits), or to agreed deliverables later on in the project, such as the system design. Where problems of this nature arise, a strict change control procedure should be operated (following organisational guidelines), where the problem is reported and permission is sought from the appropriate stakeholders for a change to the project.

Both of these problem types will require a formal report format, which should outline the problem and recommended actions or change, and document the agreement to the recommendation.

While no project is helped by unnecessary bureaucracy, there are two other formal reporting mechanisms that are invaluable in ensuring that a project is effectively managed and controlled, regardless of the size of the project:

1. **Risk Log.** A Risk Log details project-specific events that *may* cause issues as the project progresses. It should be brief and include a description of the event, its consequences, the warning signs that should be monitored, and the actions that should be taken if the event occurs.

2. **Progress reports.** A regular project progress report should be prepared summarising how the project is tracking against the plan and budget. While there may be no specific issues worthy of a Project Issue Report, the progress report should cover adjustments made to the plan and any minor issues that have arisen. The preparation of the report acts as a regular checkpoint, where the project manager can take a little time to assess the state of the plan. Progress reports should be brief and to the point, and if all is going well, a single line reporting 'Project proceeding to plan' may be all that is needed.

▶ Project organisation

A well-defined project structure ensures that the right people are involved in the project and that roles and responsibilities are clearly defined.

The key roles on any project are those of the Project Manager and the Project Sponsor. These are defined below, along with others generally found in a project management structure.

It is quite acceptable for an individual to fulfil more than one role on a project, and for responsibilities and organisational structure to change as the project shifts focus through its life-cycle.

Summary of project roles

Project sponsor

The Project Sponsor is the individual within the organisation who sets or authorises the project scope and objectives. The Project Sponsor will often appoint the Project Manager and is ultimately responsible for the successful completion and implementation of the project.

They are typically senior managers or directors within the organisation, and are essential in providing high-level business backing for the project, i.e. ensuring that policy decisions are taken, funding obtained and resources allocated to the project.

Project manager

The Project Manager is responsible for the day-to-day management of all aspects of the project, ensuring delivery to the agreed timescales, budget and quality.

On major projects this will be a full-time role. On smaller projects it will be just one of the responsibilities of an individual. It is *never* a shared role.

Project board

The Project Board is responsible for monitoring and steering the project under the chairmanship of the Project Sponsor. The Project Board will be made up of high-level representatives of each business area or third party contributing to or affected by the project.

It will normally meet on a regular basis to review progress, as reported by the Project Manager, and sign-off key decisions. Its other major roles are to communicate details of the project and its impact around the organisation, and to ensure that any changes or impacts are understood and carried out fully during implementation of the project.

On very small projects the Project Board may consist of just the Project Sponsor and the Project Manager.

User representatives

In most major projects there will need to be significant direct involvement from user representatives of those functions impacted by or making use of the deliverables of the project.

In small projects this involvement can take the form of interviews or update meetings with users who understand the detail of the project's deliverables. In larger projects user representatives will form part of the project team and be responsible for the delivery of certain components of the project. In all cases they will represent the Project Board in agreeing the details of the proposed solution and signing off deliverables as 'fit for purpose'.

Note

On student projects the role of the Project Sponsor will be undertaken by a representative of the organisation that will be the recipient of the final deliverable(s). In addition there will be an academic supervisor who will act as a co-sponsor, and is responsible for both the mentoring of the Project Manager (student) and agreeing the academic objectives of the project.

▶ FEASIBILITY STUDY

Many of the proposals for new information systems that come out of strategy planning are vague and largely uncosted. Before we commit resources to their development we need to ensure that they are feasible. By 'feasible' we mean justifiable in terms of net benefits compared to overall costs, *and* technically or operationally feasible. Feasibility assessment is an activity that can take many forms, varying from an informal study carried out as part of strategy planning or project initiation, to a high-level systems analysis 'mini-project', depending on the size or complexity of the overall project.

The basic questions to be answered in any kind of feasibility study are:

◆ Is there a computer solution to the given business problem?
◆ Is the solution justifiable in business terms, both organisationally and financially? For example:

- Will benefits outweigh costs?
- Will the proposed solution be politically acceptable?
- Can the solution be developed in time?
- Can the level of change associated with the system be absorbed at this time?
- Are the skills in place and the people available to develop and manage the system?
- Is the risk of project failure acceptable to the organisation?

The approach adopted by SSADM *for larger projects* is to carry out an overview systems analysis exercise in order to answer the first question, and to gather quantitative information as we go along to help in answering the second. A formal SSADM Feasibility Study involves applying several of the core techniques we will encounter later in this text, but applied at a very high level, in order to model data and processing requirements.

Even where SSADM is the chosen method for the main development stages of the project, it is still entirely acceptable to adopt an informal approach to the Feasibility Study. Indeed, one of the issues the Feasibility Study may need to address is the development methodology to be adopted by the project.

Student projects

Student projects can in fact involve the carrying out of just a Feasibility Study if the area under examination is complex or large enough. In these cases it is essential to understand the full range of analysis techniques presented in the following chapters, so that they can be used to present a fully modelled study.

For smaller projects the Feasibility Study can be as little as a two-week exercise, with little in the way of formal modelling. However, the same basic questions laid out below will need to be answered, albeit with less detail.

SSADM, in common with most methodologies, regards the Feasibility Study as an optional stage, as project feasibility may have already been established during strategy planning, or the study may be unnecessary where there is little risk or choice in commencing with a full study straightaway.

There are several points in the life-cycle where a decision to drop the project might be made. The Feasibility Study provides easily the most cost-effective point to do so, as costs will be relatively low, while many of the outputs can form the basis for subsequent detailed analysis work.

Whether an informal approach or an SSADM Feasibility Study is undertaken, the basic steps we go through will be the same:

1. Define the business problem to be solved;

2. Develop high-level alternatives (or 'options') for its solution;

3. Assess the feasibility of the options and select options for discussion with the Project Board;

4. Make recommendations to and document the decision of the Project Board;

5. Develop action plan for further analysis and subsequent development of the chosen option.

This is a very similar process to that undertaken once we have completed our analysis and need to define the precise solution to business requirements (see Chapter 7 Business Systems Options).

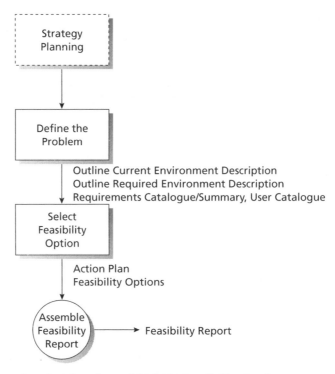

Fig 2.3 Steps undertaken in a formal SSADM Feasibility Study

▶ Defining the problem

Our main aim is to investigate the requirements of the new system to a level that will enable the project team to *outline* possible development solutions. We will also need to establish where and to what extent existing systems fall short of the requirements.

To support the aim of the step we have several specific objectives:

◆ To gain a fuller understanding of the required functional support and information needs of the business area;

◆ To develop a more detailed picture of existing systems in order to identify problems and missing support;

◆ To identify the main users of the new system, together with their current responsibilities.

By the end of this process we should have reasonable high-level descriptions of both the current and required systems. Using these descriptions we will then be able to draw up a Problem Definition Statement, in agreement with management; this summarises and assigns priorities and benefits to the high-level requirements of the new system. By contrasting the required system with the current system, the Problem Definition Statement should then reveal the full extent of the work involved in delivering a solution.

During the Feasibility Study we must be careful to restrict detailed analysis to the most important functions and information requirements only. Lower-priority requirements should be covered at the highest level only, as our purpose is to understand the fundamental requirements, not to specify the complete new system.

It is helpful to impose strict time restrictions on the Feasibility Study in order to limit the amount of effort that goes into it, and prevent it evolving into a full-blown systems analysis exercise (a process known as 'time boxing').

Part of our task in defining the business problem (or opportunity) is to provide a statement of the 'ideal' required system, albeit in outline. Subsequently we may find that it is not feasible to develop this idea, but that it is feasible to develop an intermediate system, some way between the current and ideal systems. We must be careful to leave such decisions until we have a statement or set of models encompassing all high-level user requirements.

Note

In informal Feasibility Studies the Problem Definition Statement may consist of a purely textual description of the key functions and problems that any new system will need to satisfy. What follows is a description of a formal SSADM approach, which requires the reader to have completed Chapters 3 to 7 in order to understand the models and techniques referred to.

SSADM approach

We begin the Feasibility Study with a high-level fact-finding exercise. Given the time constraints of a Feasibility Study the most effective approach is often to use brainstorming workshops, backed up by one or two targeted interviews and examination of any previous documentation. We should document all of the high-level requirements identified during this exercise in a Requirements Summary (a Requirements Catalogue is probably too detailed at this stage for all but the most complex of projects).

The current environment outline will consist of an overview LDS (Logical Data Structure) and a current physical overview DFM (Data Flow Model), while the required environment outline will consist of an overview LDS and a logical overview DFM.

When developing a Current Physical DFD (Data Flow Diagram) we are able to verify its accuracy quite easily by checking that it mirrors what was actually happening on the ground. With the Required System DFD there is nothing 'real' with which to compare. Therefore we are likely to spend significant effort in assessing users' often conflicting and, at this stage, vague wishes, in order to arrive at a sensible and accurate view of business requirements (and into agreeing this view with management).

This gives us the freedom to be creative, as we are not restricted by any physical constraints, enabling us to describe an ideal picture of how the new system should function. On the other hand it means that the success of the model is highly dependent on our ability to ask the right questions of the right users.

We should attempt to be efficient and methodical by using the PID to identify the *major* functional areas that the system will be required to support, and using these as a checklist of high-level processes to be covered by the overview DFD.

For example, using the ZigZag PID we can draw up the following list of functional areas for the new ZigZag system:

◆ placement of purchase orders;

◆ receipt of deliveries;

◆ recording of new stocks;

◆ despatch of customer orders;

◆ invoice matching;

◆ maintenance and monitoring of stock position;

◆ monitoring of supplier performance.

While concentrating on the processing aspects of the system we should not neglect to maintain overview Logical Data Models at the same level of detail as the Data Flow Models we create.

By comparing the overview models of the required and current environments we should be able to identify most of the new high-level functionality, so providing a check that the Requirements Summary is complete.

While updating the Requirements Summary, we can also develop a high-level User Catalogue. In a Feasibility Study the main purpose of the User Catalogue is to support the identification of users in the current environment. If needed, the prevailing work practices can be documented either in a separate Work Practice Model or in an extended Requirements Summary/Catalogue entry.

The current and required environment models, together with the Requirements Summary, provide a formal definition of the business problem. However, to present these products in isolation to users and the Project Board makes it difficult to obtain agreement as to their accuracy.

So a Problem Definition Statement is produced, providing a textual summary of requirements and their relative priority. It should include references to the formal SSADM products (which it does not replace but merely supplements), and should include a list of the minimum requirements.

Once users and the Project Board agree the Problem Definition Statement we can then move on to formulate options for the rest of the project.

▶ SELECTING FEASIBILITY OPTIONS

The main aims of the next step are to identify the best option in business terms for solving the problem defined in the Problem Definition Statement, and to provide an outline plan for development of the chosen option.

A Feasibility Option is really a high-level combination of two standard SSADM products:

◆ **Business System Option (BSO).** A BSO defines the functional scope of a proposed solution. At its most basic level it consists of the set of Requirements Catalogue entries satisfied by the solution. All BSOs must satisfy the minimum requirement as identified by users.

◆ **Technical Systems Option (TSO).** A TSO defines a possible technical environment for the implementation of the system. It will include descriptions of hardware and software, technical support arrangements, distribution of the system and development tools.

The basic steps undertaken in formulating and selecting Feasibility Options are very similar to those used in the selection of fully fledged BSOs and TSOs covered in later chapters, but the overall level of detail is *much* lower.

We begin by outlining several BSOs and TSOs for solving the business problem. We then produce composite options that combine the two, short-list a number of them and flesh them out a little. Finally we present the options to the Project Board who will select one of them as the basis for the rest of the project.

In practice, this whole process is unlikely to proceed in sequence like this. It is far more likely to be an iterative exercise, with new options being proposed throughout.

The organisation will usually have a strategic policy on technical implementations, tied to particular hardware and software platforms, or to specific suppliers. We should use any flexibility in these existing technical policies to provide as wide a range of suitable TSOs as possible. If the proposed environment can be supplied by only one vendor (or if the organisation's policy dictates the choice of vendor), the TSOs might include details of specific suppliers or products.

Each TSO may be capable of supporting more than one BSO and each BSO may be supportable by more than one TSO (see example below). So we could produce a number of Feasibility Options that propose several different physical implementations of a single BSO.

A brief check on the viability of each composite option should prevent the number of options getting out of hand. Feasibility Options are compared with a view to eliminating those options which offer less obvious net benefits or which are impractical given the constraints of the project (maximum cost, timescale, etc.).

In the case of ZigZag, BSOs might vary from an option that delivers improved capacity with little additional functionality, through to an option that delivers new functionality in support of e-commerce and improvements to delivery scheduling etc.

We will assume that there is no existing technical policy, but one of the physical constraints for any new system is that existing PC hardware in the depots should be used as much as possible.

The TSOs for ZigZag might include: a bespoke development using the same tools and infrastructure components as the sales analysis system; a bespoke development using a new technical environment but capable of interfacing with the sales and accounts systems; and a packaged solution using hardware specified by the package vendor.

Composite options could involve combining all of the BSOs and TSOs with each other. In reality, where a package is proposed the option may already be a combined Feasibility Option.

We then expand the Feasibility Options with high-level descriptions of the following:

◆ **Functional support.** Textual descriptions can be supplemented with DFDs and LDSs showing the subset of functional requirements covered by the option.

◆ **Technical overview.** This should extend the outline TSO description and be tailored to the specific needs of the BSO. It may be necessary to produce outline configuration diagrams.

◆ **Costs.** These will be very approximate and must include: hardware; software; human resources; consultancy; training; maintenance and running costs (which are frequently higher than the development costs).

◆ **Benefits.** Include financial benefits (e.g. increased profits or reduced costs), strategic benefits (i.e. the meeting of strategic business objectives), removal of problems (e.g. capacity constraints).

◆ **Organisational Impact Analysis.** Again, this will be at a high level, and will describe the cultural and operational changes associated with the option.

◆ **Approximate timings.**

This information should then enable us to assess the technical, operational and financial feasibility (see box) of each option. Those that we assess as being most feasible will now be short-listed for presentation to the Project Board or management, for assessment and selection.

Financial feasibility

Financial feasibility has two key elements. First, we need to establish whether the funds are available for the solution to be developed and maintained. Even solutions with overwhelming net benefits will be infeasible if the organisation is unable to finance the initial development. For instance, a very small company may find that international expansion of its operations and systems could result in substantial profit increases. But if it could not fund the associated initial development costs (even if it had the human capacity to effect the change), then the expansion is not feasible.

The second element of financial feasibility is the balance of costs and benefits over time.

Cost–benefit analysis

In conducting a cost–benefit analysis (CBA), financial costs are usually easier to estimate than the financial benefits. Systems frequently promise a number of intangible benefits, i.e. benefits with a positive impact on the strategy or operation of a business, but without a quantifiable or measurable financial benefit. For example, a system may claim to improve the decision making of a set of employees, but measuring the increased profits generated directly by that improvement might well prove impossible. Estimates in these cases can be made, but they will not stand up to close financial scrutiny. For the purposes of CBA, we need to establish (and agree with business representatives) reliable estimates of tangible financial benefits. This may result in costs outweighing benefits within the CBA, in which case the organisation will need to assess the feasibility of the project based on its intangible benefits.

There are a number of methods for assessing costs and benefits, including return on investment and payback periods, as outlined below. For a full description of these methods the reader is referred to a text such as Robson (1997). Most organisations will have internal standards for which of the methods should be used in conducting a CBA, and what result will be considered acceptable in assessing feasibility.

Return on investment (RoI)

RoI is the simplest, and one of the most frequently used, measures of financial feasibility. It delivers a percentage figure that can be compared against prevailing interest rates, in order to assess whether the proposed investment is financially worthwhile.

The basic formula is:

$$RoI = (Net\ benefit/Investment) \times 100$$

where Net benefit = the sum of tangible benefits – Total costs, including annual running and development costs.

Standards vary from organisation to organisation regarding the period over which costs and benefits are measured. A common standard is to use the sums of annual costs and benefits over a four-year period; another is to use the costs and benefits over the expected life of the solution.

Standards also vary as to what RoI rate is acceptable, with values such as twice bank base rate, or base rate plus 5% being fairly typical.

Payback period

Another common measure is that of payback period. This is a measure of when sufficient benefits will have accrued to cover both the initial investment costs and the ongoing running costs of the solution.

For example, a project with an investment cost of £120,000, annual running costs of £20,000, and annual benefits of £50,000 will pay back the investment in 4 years.

In assessing overall cost or benefit, measures such as RoI and payback period will frequently be used in combination, and viewed differently by different organisations. For example, some might view an RoI of 20% with a payback period of 2 years as preferable to an RoI of 30% with a payback period of 4 years, depending on their strategic aims and current financial position.

For each of the short-listed options we should draw up a project proposal for its detailed analysis, design and implementation.

When deciding on possible development approaches the following should be considered:

◆ Is SSADM the most suitable method for developing the option? If not, what method should be used?

◆ How many projects are necessary? If the proposed system is large or complex, a phased approach may be best;

◆ Who will develop the option? Possibilities include in-house project teams, contractors, software houses, package vendors, etc.

Once we have decided on the development strategy for each option, we will need to plan all of the necessary projects in outline using the organisation's standard methods.

For tips on presenting options, please refer to Chapter 7 (Business System Options).

Finally, we will present the short-listed options to the Project Board or senior management for assessment and selection. We should also precede the presentation by circulating a summary of the options to be considered.

Our task is to assist the board in assessing the relative merits of each option, so that they can make an informed selection of a single Feasibility Option. We must also ensure that the reasons for selection are fully documented.

It is quite possible that a combination or hybrid of the Feasibility Options will be selected by the board. In this case it may be necessary to take the hybrid option away to investigate further its viability and to prepare it for presentation to management at a second selection meeting. It is also possible that none of the options will appear feasible to management and the project may be halted at this point.

When a choice is made we produce an outline development plan for the chosen Feasibility Option. In most cases the project plans prepared before selection will be carried forward and adjusted to allow for any decisions made by the selection meeting. If management adopts an entirely new option then we may need to produce development plans from scratch.

We will assume that ZigZag's Project Board rejected all package options, as they could not deliver the required functionality and benefits without substantial modifications, which were considered too risky and potentially costly.

The Project Board were unable to make a final decision on the precise functionality of the new system, until substantially more analysis of the detailed requirements and their associated benefits has been undertaken. They were sufficiently confident that a bespoke development of most of the outlined requirements will be financially feasible that they have authorised a full analysis of user requirements, with a view to making a final decision on the BSO to be pursued when that analysis is complete.

The Project Board have decided that the system, whatever its ultimate functionality, should be developed if possible using the same tools as the Sales Analysis system. But a decision on the precise hardware to be used has also been deferred until later in the project when requirements are clearer.

▶ SUMMARY

1. Effective and appropriate project management is an essential ingredient in the success of any project, regardless of its size. Project Management consists of a formal set of plans, controls and organisational structures. The key roles within a project are those of Project Manager and Project Sponsor.

2. A Project Initiation Document (PID) represents a contract for delivery of a project. It is vital that the PID is unambiguous and agreed by all stakeholders in the project.

3. The SSADM default structure provides a good start point in assembling a project plan. It should always be viewed as a checklist of tasks, techniques and products that will require customisation to fit the needs of the project.

4. A Feasibility Study should address technical, organisational and financial feasibility.

▶ EXERCISES

2.1. Consider an organisation of your choice (e.g. a university, a supermarket, an airport) and identify separate departments and sub-departments within it. For each department identify different job titles and for each job title identify job-related activities. Demonstrate your findings in an Organisation Chart.

2.2. Imagine you are required to plan the Feasibility Study for ZigZag. Produce a Gantt chart of your plan and use it to cost the feasibility activities.

Investigation

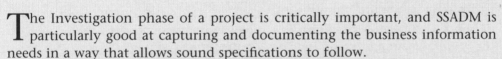

The Investigation phase of a project is critically important, and SSADM is particularly good at capturing and documenting the business information needs in a way that allows sound specifications to follow.

During Investigation several of the core techniques of SSADM are used to model the business activities of the area under study and the data these activities need.

The overall objectives of Investigation are to gain a thorough understanding of the requirements of the new system, and to firmly establish the direction and viability of the rest of the project. With a good project management plan in hand this is also where an analyst can build bridges of trust with the users.

We begin the Investigation by modelling the activities of the people working in the business area. These activities are studied to provide us with a view of what is going on and to help the users visualise our understanding of their job environment. As activities are identified they are studied to see if they lead to the interchange or recording of information. Every such piece of information is captured with care and placed into a data model. The dual search of business processes, and the data they depend upon, allows us to create important holistic models of the whole system.

Investigation comes to an end when we feel we have captured the nuances of the system and created a data model that supports the business.

To model processing we employ the techniques of Business Activity Modelling, Work Practice Modelling and Data Flow Modelling. To model data we employ the most important analysis technique – that of Logical Data Modelling. As we go along we create a list of requirements we will want the future system to satisfy. These requirements will be presented to the Project Board who will decide which of these requirements will be finally taken to full implementation.

In Part 2 we will take the opportunity to introduce each technique gently as we become familiar with some basic concepts. The fledgeling analyst will be tested in many skills ranging from 'people skills' to some interesting conceptual skills. Systems analysis requires that we are able to pay attention to detail without losing sight of the whole picture. Hence most of our models strive to display, as far as this is possible, the whole system in one go.

The entire process is one of accretion where we do not expect to find things in a linear fashion. While the techniques try to dismantle a problem into manageable tasks, there will be moments where good teamwork and some lateral thinking become vital if the project is to succeed.

We begin the Investigation with an analysis of current systems (if there are any), as we can usually get a good idea of a project's scope and complexity by familiarising ourselves with current services in the area under investigation.

As we document the current environment, we should identify points where the new information system can be used to improve efficiency. We may also uncover some of its major problems and shortfalls in functional support. We will document these, along with additional new requirements for the new system, using the technique of Requirements Definition.

During Investigation no attempt is made to model the required environment in a formal way. This will follow after we are sure what system will be finally developed, during Specification. For now we restrict ourselves to documenting the required system textually in the Requirements Catalogue.

▶ Default structure

It is the responsibility of the project manager to understand the nature of the problem in hand, and to devise a structured project plan that fits the environment of the system.

In SSADM there is a recommended default approach that can be used as a starting point by inexperienced project managers. This default approach consists of five steps, three of which are expected to be performed concurrently since they feed ideas into each other. Figures P2.1 and P2.2 depict the default approach pictorially.

Develop Business Activity Model

Any systems development that uses SSADM assumes that a Project Initiation Document (PID) triggers the project. The PID for the case study used throughout the book was discussed in Part 1.

The main purpose of developing the Business Activity Model is to assess the scope of the project and confirm this scope with management. This is done by examining the PID, and producing an understanding of the project environment using the technique of Business Activity Modelling.

If a Feasibility Study or any other relevant studies have preceded the analysis, their results are examined, and overview current environment models created (if they are not already available).

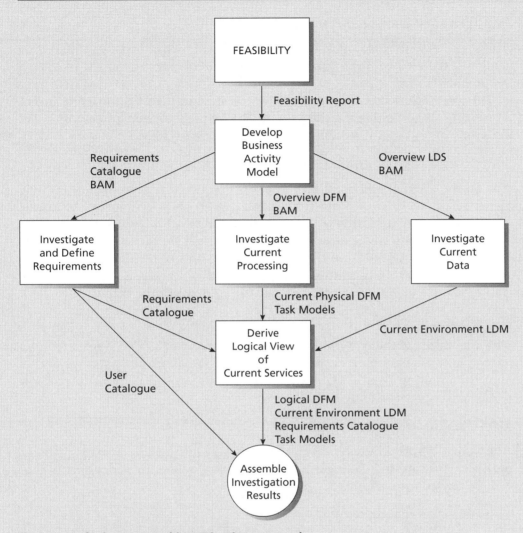

Fig P2.1 Default structured investigation approach

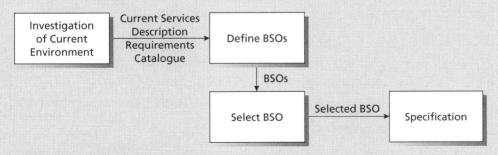

Fig P2.2 Business System Options

Investigate and define requirements

The Requirements Catalogue is created or, if one exists from any preceding studies, expanded to form a detailed and comprehensive statement of system requirements.

This step is carried out in parallel with the steps of Investigating the Current Processing and Investigating the Current Data, which provide inputs to it based on problems or missing functionality in the current system.

Investigate current processing

The Business Activity Model (BAM) from the first step is transformed into a fully fledged Current Physical Data Flow Model (DFM) and decomposed to a level where all Elementary Processes have been identified and described.

During this step we ensure that each Elementary Process is reflected in the Requirements Catalogue and that the data it requires is modelled properly. We also take care to produce textual descriptions for data flows and external entities, and to set up a data catalogue to describe the contents of the data store we find.

Investigate current data

The data needs of the environment are reviewed and extended. We carry out a more rigorous analysis of entity definitions and relationships, and we ensure that the resulting Logical Data Model is cross-referenced with the Data Flow Model we are concurrently constructing.

Derive logical view of current services

The Current Physical Data Flow Model is converted into a logical model of current services. This can then be used in Specification to define the functionality that will be carried forward into the new system.

Assemble Investigation results

The major products from Investigation should provide a complete description of required services in a logical form. These are checked and reviewed with users before we submit them to Business System Options, the technique whereby we take decisions on which requirements to proceed with.

Define Business System Options

A number of Business System Options are outlined, each satisfying at least the minimum system requirements. These are then short-listed to about two or three options to be fleshed out, and supported by cost/benefit and impact analyses.

Select Business System Option

The detailed Business System Options are presented to the Project Board and a single (possibly hybrid) option is selected. This final Business System Option is documented in detail and agreed as the basis for system specification.

Business analysis

INTRODUCTION

A computerised information system does not exist in isolation. It exists to help and support the main functions of an organisation. If we are trying to design a computerised system it stands to reason that we need first to understand the business context within which the system will reside.

In this chapter we will introduce two structured analysis techniques that help ensure that the final systems fits within the mission and objectives of an organisation.

The two techniques are Business Activity Modelling and Work Practice Modelling.

▶ **Learning objectives**

After reading this chapter, readers will be able to:

◆ produce Business Activity Models (BAMs);

◆ understand the importance of performing Business Activity Modelling;

◆ produce resource flow diagrams;

◆ scope a project;

◆ use the results of fact finding to produce Business Activity Models;

◆ undertake user analysis;

◆ produce a Work Practice Model.

▶ **Links to other chapters**

◆ Chapter 4   User requirements identified during Business Activity Modelling are further defined and documented.

◆ Chapter 6   The BAM is used as an input to the development of process models.

◆ Chapter 9   User interfaces (i.e. screens and reports) are designed, using the products of the Work Practice Model.

▶ BUSINESS ACTIVITY MODELLING

Before we embark on an analysis of the activities that may be supported by a computerised information system we first have to understand those activities. We should always remember that the main purpose of performing systems analysis and design before constructing the system itself is to safeguard against delivering the wrong system to our users. There is nothing more unprofessional than delivering a system that is not really what users expected. Business Activity Modelling is the technique SSADM recommends as a start-off technique in our quest for delivering the right system to our users.

Business Activity Modelling is influenced by the Soft Systems Methodology ideas proposed by Checkland (1981; Checkland and Scholes, 1990), but in SSADM we only use the technique to find out *what* is going on in the environment of the system under investigation. Considerations of *who* is involved in each activity, *when* the activity is performed and *why*, are recorded and catered for, but the method assumes that there is no major conflict within the user community. The techniques that are applied after Business Activity Modelling deal in depth with *how* the system works.

▶ Business activities

There are five kinds of activity within any business, as depicted in Figure 3.1. These activities are the actions usually performed by humans, with or without the aid of systems, which deliver the organisation's primary task. We need to understand an organisation's primary task in order to understand the actions of the employees of that organisation. We also need to believe that the employees of the organisation

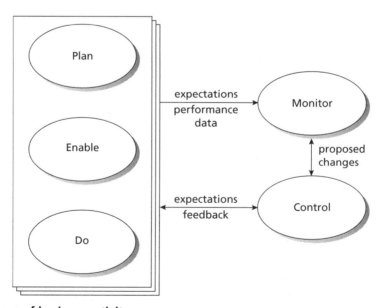

Fig 3.1 Types of business activity

share this understanding of the primary task and that they purposefully participate in the fulfilment of this task. In ZigZag the primary task is to sell entertainment goods to their customers.

At the operational level of an organisation we expect to mostly find *doing* activities. Doing activities are the main activities that deliver the actions for which the business primarily exists. For example, in ZigZag we can identify the doing activities of assembling the customer order and despatching the customer order as activities that clearly address the company's primary task of selling goods.

Doing activities are supported by *enabling* activities that ensure that the resources and facilities needed by the doing activities are available. In ZigZag, 'record purchase order' and 'confirm purchase order' would be such enabling activities. Enabling activities tend to exist at operational and tactical organisational levels.

For the doing and enabling activities to exist we need to recognise *planning* activities such as, in the case of ZigZag, 'suggest products to buy' and 'find appropriate suppliers' which plan how doing and enabling activities will take place.

Planning, enabling and doing activities give rise to *monitoring* activities which collect performance data for comparison with expectations which, if not met, give rise to *controlling* activities which lead to corrective activities' being activated.

▶ Business events, business threads and business rules

A business event is a 'happening' that triggers one or more business activities (see Figure 3.2). For example, the business event of a truck-load arriving at ZigZag's loading bay triggers the activities of checking the delivery, placing the goods in the delivery bay, allocating stock locations, removing goods from the delivery bay, storing goods in the depot and updating the stock levels.

The activities triggered by one business event need not be triggered right away. Sometimes a business event triggers an activity, which then gives rise to another

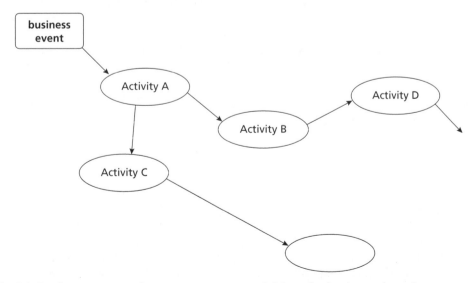

Fig 3.2 Business events trigger one or more activities of a business thread

activity only if and when a new business event happens. For example in ZigZag, the receipt of a supplier's invoice triggers the activity of checking the invoice but does not immediately activate the 'Forward Matched Invoice' event. The latter activity lies dormant until it is triggered by a date around thirty days after the receipt of the invoice.

Business events include inputs from outside the business, decisions made within the business boundary and time-triggered events.

Sets of business activities and business events that are triggered by one business event are known as 'business threads'. The recognition of business threads allows groupings of workflows to be identified, which in turn facilitates the process of Business Process Re-engineering.

To complete an activity, humans perform certain tasks, which may involve sub-tasks. When an activity is to be supported by the new system, task models may be drawn up, as we will see later.

Each business activity has to conform to certain business rules. These rules describe the activity in the form of guidance and possible constraints. The business rules that define each activity form the basis for a bottom-up description of the system.

Business activities indicate either *what is* going on or, in the case of new systems, *what we wish* to go on. Business events are a record of *when* the activities take place and business rules suggest *how* the activities should take place.

▶ The Business Activity Model

The Business Activity Model consists of a diagram depicting business activities and their dependencies, and supported by documentation about business events and business rules. The dependencies between activities are an indication of which activity is responsible for passing a message to which activity.

SSADM is mainly concerned with building information systems where the information is in the form of distinct data items. The purpose then of Business Activity Modelling is to familiarise ourselves with the working environment in which our system will fit, with the distinct aim of isolating those activities that we feel will be aided by a computerised information system.

The future computerised information system should be perceived as a tool to enable the whole or part of an activity, or to provide information to such an activity.

Initial fact-finding results

Initial information gathering activities lead to the following textual summary of the *current* ZigZag depot system:

◆ **Purchase order placement.** Purchasers examine estimates of future customer orders and compare these with the current levels of stock held in depots. Records of stock levels are available in the form of a daily computer-generated report sent by the stock clerk which highlights products which have fallen below their 'standard' order points. The report also includes details of customer orders awaiting despatch. Purchasers decide if new supplies are required to supplement current stock and, if so, details of the required quantities are sent to each depot.

The purchase order clerk at each depot groups these purchase requirements by supplier into purchase orders. The delivery date and time are agreed with the supplier of each order, either at order placement time or later if necessary.

Two copies of the purchase order are filed for later processing when goods are received.

◆ **Goods receiving.** Suppliers deliver a complete order at a time with a delivery note that is checked against the original copies of the purchase order.

If goods are in poor condition or fail to match the order the delivery may be rejected. In this case the purchase order copies are marked as 'Order Cancelled' and passed to the purchase order clerk. The purchase order clerk sends an official cancellation to the supplier, files one copy of the purchase order and sends the other to the purchaser.

If the goods are accepted one copy of the purchase order is sent to the Purchasing department, and the other is placed in a tray for collection by the stock clerk.

◆ **Stock keeping.** The stock clerk enters details of the newly accepted items onto the depot PC system and assigns the goods to a zone depending on their type, e.g. zone 101 is used to store CDs and DVDs, zone 102 is used to store videos, etc.

The stock clerk then passes the purchase order copy to the purchase order clerk for future matching with invoices.

Stock keepers are provided with a printout of where to store each product, and the products are put away.

Stock-takes occur regularly and totals of goods checked with the PC-held records. Any differences are entered as amendments to stock levels. Occasionally stock may pass its sell-by date, resulting in further amendments.

◆ **Despatch.** Customer orders are sent to the despatch clerk by the Sales and Marketing division. Each order may contain details of several products and will refer to quantities required by a single customer. Although a customer may have several food outlets, each outlet is known to the system as a separate customer.

The despatch clerk matches customer orders with the delivery runs timed for the next week and assigns them to specific runs. Details are then entered on the depot PC and the customer order is placed in a pending file.

Every morning a report of despatches for the day is printed and given to the despatch supervisor by the despatch clerk. Delivery runs are put together using this report and a stock listing from the stock clerk. The despatch report is marked with quantities despatched and a copy sent to the stock clerk who updates the stock levels on the PC system.

The despatch report is then passed back to the despatch clerk who checks the quantities against the customer order and sends this back with details of despatched goods to Sales and Marketing. A second copy of the order is sent to accounts for invoicing.

◆ **Invoice processing.** Suppliers send invoices for completed purchase orders to the purchase order clerk, who matches them against the annotated purchase order copy and sends them to accounts for payment.

This textual information forms the basis for the Business Activity Model.

▶ Direct Business Activity Modelling

Like most of systems analysis, Business Activity Modelling is formalised common sense and is simplicity itself. We begin by finding out, at a pretty high level, what the people in the system's environment do. In the case of our case study, for example, a combination of observation, examination of business documents, a study of the terms of reference and, finally, interviews or workshops, leads to the identification of the following activities: products that have to be bought are suggested; appropriate suppliers are identified; proposed purchase orders are recorded; information about product availability is received; alternative products are suggested for purchase; purchase orders are confirmed; invoices are received, checked against deliveries and forwarded to Accounts; deliveries are arranged; delivery schedules are set up; deliveries are checked and, if accepted, placed temporarily in the delivery bay; delivered goods are allocated a stock location, removed from the delivery bay and placed in their proper place; stock records are updated after every delivery and every sale; stock lists are produced; customer orders are received, their despatch organised and forwarded to the despatch bay; ordered goods are assembled and despatched; customer order files are kept up to date; Sales and Marketing are kept informed of despatches.

As the activities increase it becomes convenient to represent each activity as a bubble and to show how these activities interact by joining them with arrows that indicate which activity has to precede which. Figure 3.3 depicts four activities suitably linked.

When an activity has no other activity triggering it, a little incoming thunderbolt can be used to show that information is arriving from somewhere else and we

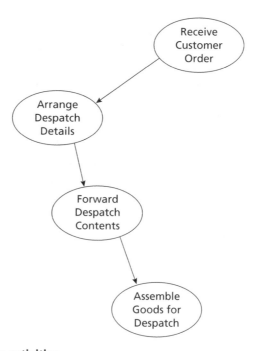

Fig 3.3 Typical ZigZag activities

should be careful to capture it. Similarly, when an activity appears as a dead-end it usually indicates that the activity is probably responsible for sending information out of the system. If this proves to be the case a little outgoing thunderbolt can be used to show this.

Figure 3.4 shows a first Business Activity Model for ZigZag.

Phrasing business activities

Observe how each activity is phrased using a brief statement containing a verb through which we can visualise a person performing the activity. Also note that the verb used is phrased from the point of view of the system. That is why the topmost activity in Figure 3.3 is 'Receive Customer Order' and not 'Send Customer Order'.

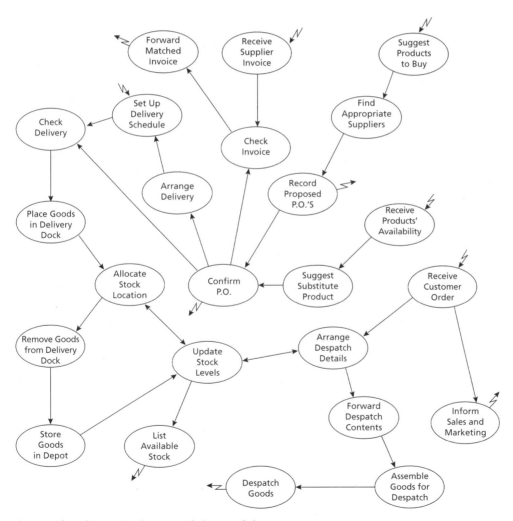

Fig 3.4 The ZigZag Business Activity Model

▶ System events

When a business event occurs, some of the activities it triggers may end up inside the future computer system's boundary and some may not. Those that end up inside and also lead to an update of our records give rise to *system* events. These system events are what experienced analysts look for since they indicate the vital moments when *new* data enters the system. If we don't manage to capture the new data as it arrives our final system will be of no great use and all the reports coming out of it will be incomplete and suspect. 'Doing' activities tend to identify moments when new data come into the business and should enter the computerised system.

Business activity tip

When producing a Business Activity Model there is no need to be bogged down by unnecessary detail. An activity such as 'Find available fork-lift truck' just prior to 'Store goods in depot' is probably superfluous. We should avoid spending time on activity sequences that clearly lie well outside the future information system.

Similarly, common sense dictates that activities such as 'Find pen', 'Open stock file', 'Find appropriate stock', 'Strike out old stock level', 'Write-in new stock level', 'Close stock files' and 'Return pen to holder' should be resisted; 'Update stock levels' suffices.

▶ Resource Flow Diagrams

One way of identifying business activities is to start from a Resource Flow Diagram. A Resource Flow Diagram models the flow of resources as they move around the business.

In any business we can usually identify many different resources. These resources are tangible things, such as goods or pieces of paper, but they may also be people who can be seen as human resources.

Resource Flow Diagrams are drawn because it is easier to track tangible things and to then identify the activities taking place each time the resource moves from one place to another. The idea is that the movement of most of these goods or items will be associated with an activity. In the context of information systems development we also expect the movements of these goods to be accompanied by or associated with a flow of information about them, which can then be used as the basis for identifying the data-oriented activities of the system, thus identifying the computerised system boundary a bit more clearly. For each activity, the need or not for information will be identified through the business rules that explains the activity that moves the goods.

A Resource Flow Diagram consists of two symbols:

1. resource flows (Figure 3.5).

Fig 3.5 Resource flow

2. resource store (Figure 3.6).

Goods In Delivery Dock

Fig 3.6 Resource store

Resource stores refer to areas where goods are stored and not to the goods themselves.

In the ZigZag system resource flows are relatively straightforward. We could begin with a list of sources and recipients of resources, but resource flows and stores are frequently quite obvious, as in this example, and so can be drawn straightaway as in Figure 3.7. Figure 3.7 tells us that deliveries from suppliers arrive at the Loading Bay, then, when accepted, move to the Goods In Delivery Dock from where they move to the Depot Storage Zone as stock.

It now remains for us to note the activities that take place before and after the goods are placed somewhere. As goods arrive they are 'checked'. If the goods are okay they become accepted and are temporarily 'placed' in the Goods In Delivery Dock waiting for a permanent spot in the depot. Once a spot is 'allocated' they are 'removed' and 'stored' in the depot proper.

We can now interlace the Resource Flow Diagram of Figure 3.7 with the activities we have just observed as shown in Figure 3.8. (The reader will spot that the five activities identified in Figure 3.8 have also been discovered through the direct approach to Business Activity Model and reside on the left-hand side of Figure 3.4.)

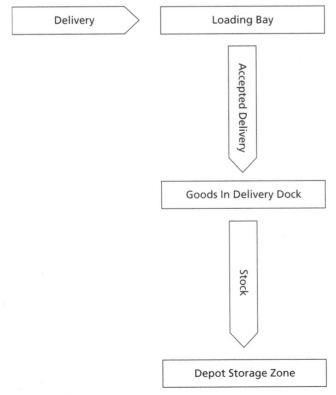

Fig 3.7 A Resource Flow Diagram for goods being delivered by suppliers

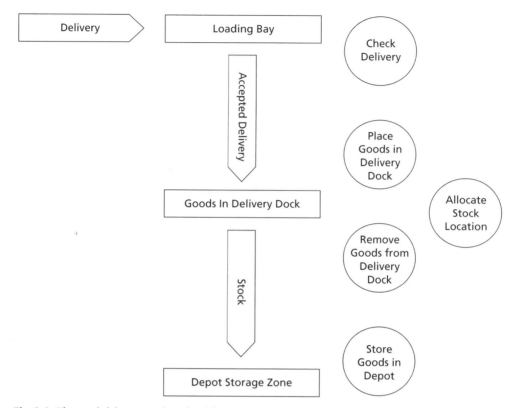

Fig 3.8 The activities associated with the resource flow of Figure 3.7

Identifying activities through Resource Flow Diagrams helps identify *what* takes place in a business without being influenced by *who* performs the activity. This approach, which separates the 'what' from the 'who', allows us to focus on the activities themselves and to think of different communication channels which may better take advantage of new technologies.

Using Resource Flow Diagrams is better at identifying 'doing' and 'enabling' activities and weaker at identifying the more managerial 'planning', 'monitoring' and 'controlling' activities.

▶ Functional Decomposition

Another approach to Business Activity Modelling favoured by business analysts is the one that uses Functional Decomposition, which starts from an organisational chart and drills down to the activities of each department or subsection.

If we look at the ZigZag organisation chart of Figure 3.9 and focus our attention on the Warehousing side for a moment we can ask questions about the activities of the three subsections of Goods Receiving, Stock Keeping and Despatch.

The Goods Receiving subsection employs stock keepers whose job it is to physically move goods into the warehouse. They are principally engaged in the 'doing' activities of checking deliveries, placing goods in the delivery dock, removing them from the delivery dock and storing them in the depot.

Fig 3.9 ZigZag organisation chart

The Stock Keeping subsection employs stock clerks who provide 'doing', 'enabling' and 'planning' activities such as arranging deliveries, setting up delivery schedules, allocating stock locations and updating stock levels. They are also involved in collating monitoring data by producing stock reports and stock storage reports.

The Despatch subsection employs despatch supervisors and despatch clerks who are responsible for receiving customer orders, arranging despatch details, forwarding despatch contents to the despatch bay, assembling goods for despatch, despatching goods and informing Sales and Marketing, through despatch reports and matched despatch reports, of recent despatches. The customer is also sent a despatch note.

Figure 3.10 depicts the results of the above findings.

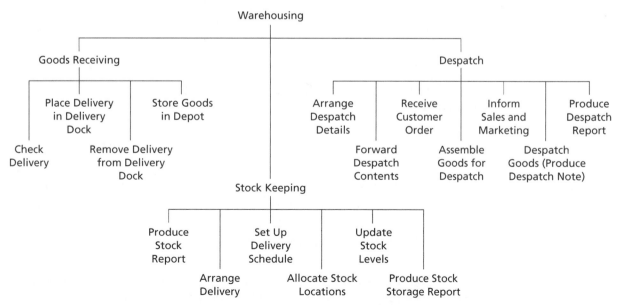

Fig 3.10 Business activities identified through Functional Decomposition

The use of Function Decomposition shows that a Business Activity Model could be quite large, especially if we decide to include, as we should, all activities. It is therefore helpful to 'level' a diagram by focusing on different subsections of a business. Levelling can take the form suggested by the Function Decomposition. If on the other hand we choose to draw the activity model in the form of Figure 3.4 we may soon run out of space. In this case it is suggested that we draw a top-level diagram showing the main departments and then draw an activity model for each of these departments.

When the activities of an organisation are grouped together in functional areas first, the dependencies between activities also allow us to identify communication channels that may require further attention. In Figure 3.11 for example, the dependency between the activities of placing deliveries in the delivery dock, allocating stock locations to these goods and then removing them from the delivery dock identifies the need for communication between the Goods Receiving and the Stock Keeping departments.

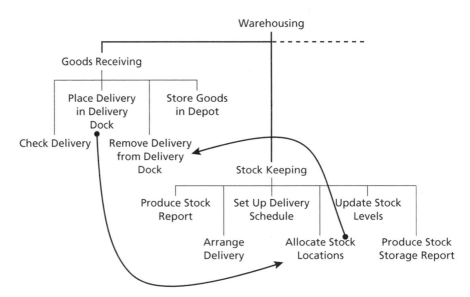

Fig 3.11 Using Functional Decomposition to identify interdepartmental communications

If we were minded to perform some further business analysis we would use Figure 3.11 to ask questions such as:

◆ Can we have the allocation of locations done by the Goods Receiving department?

◆ Can we do the allocation of locations before we place the delivery in the delivery dock?

◆ Can we first store the goods in the depot and then record where these goods were stored?

These questions can lead to reorganising the departments or the procedures of the company. They may also provide insights of how to use the computer system to speed-up the processes.

▶ WORK PRACTICE MODELLING

The computerised information system we are developing will clearly change the way people go about doing their jobs. The development of such a system requires us to design its interface and to study its effects on the working environment itself. It is therefore important to concentrate on the activities that involve direct access to the system and expand our study to include all activities taking place in the environment of the system. This will allow us to understand the whole working practices of our users and so help us integrate the system into their environment.

Until its latest version SSADM took it for granted that the computer system being built would influence the working conditions of the people around it and therefore the project team would deal with any issues as they arose. By not explicitly stating what should be done to study the impact of the system, the impression developed that the SSADM practitioner was meant to look only at the system itself and its direct interaction with its users. Nothing could be further from the truth, and Work Practice Modelling has been introduced to redress the balance.

The Business Activity Model is principally an analysis tool concerned more with *what* the business does, *why* and *when* it does it, and begins the process of finding *how* it does it. The Work Practice Model deals with *who* participates in the activity and is also interested in finding out *where* the activity is performed.

As our analysis progresses the technique of Work Practice Modelling helps us communicate better with the users in order to mould together the future shape of their jobs. This is done by studying the current working practices with a view to not missing them out of the future system.

Let us look, for example, at the 'Forward Despatch Contents' activity in Figure 3.4. At first sight it might appear superfluous, and that we could have decided not to show it on the BAM, linking instead Arrange Despatch Details directly to Assemble Goods for Despatch. The reason we chose to depict it is to remind management to delegate this responsibility if they decide that in future despatch details will be transmitted electronically to the despatch clerk. Should the people arranging the despatch details be responsible for checking that their instructions have been followed, or should the people assembling the customer's order be vigilant lest a despatch has been arranged?

The Work Practice Model begins with the set-up of an organisation chart and the identification of job titles within the area of concern. We have already met ZigZag's organisation chart in Figure 3.9. The job titles that exist within the Warehousing and Purchasing areas are purchaser, stock keeper, goods-in clerk, stock clerk, purchase order clerk, despatch clerk and despatch supervisor.

To start, the Work Practice Model consists of the juxtaposition of business activities with job titles. In this way it becomes the document where the allocation of responsibilities for each activity is recorded. Figure 3.12 contains a Work Practice Model for ZigZag.

As the project develops, the Work Practice Model gains importance, because it becomes the vehicle through which we record changes caused by the new system to the working environment.

Activity	Job Title	Activity Description
Suggest products to buy	Purchaser	The Purchasers' work is outside the scope of this development. Purchasers roam the world in search of new suppliers, study stock lists, and decide what should be bought. They send their suggestions to the P.O. Clerks
Find appropriate suppliers	P.O. Clerk	P.O. Clerks check the product files to find the suppliers of products. They then set-up a purchase order for each supplier
Record proposed purchase orders	P.O. Clerk	Two copies of each purchase order are filed. A copy is sent to the supplier
Receive products' availability	P.O. Clerk	Suppliers respond by informing ZigZag whether they can fulfil the whole order
Suggest substitute product	P.O. Clerk	The files are checked to see if any products can be replaced by similar ones. Currently this activity depends on the clerk's knowledge. It is hoped the new system will facilitate this activity
Confirm purchase order	P.O. Clerk	When it is clear that the supplier can satisfy an order, the order is confirmed
Receive supplier invoice	P.O. Clerk	Suppliers send invoices requesting payment
Check invoice	P.O. Clerk	Invoices are checked against delivered goods
Forward matched invoice	P.O. Clerk	If invoices are correct, they are sent to Accounts for payment
Arrange delivery	Stock Clerk	Delivery dates and times are agreed with suppliers
Set up delivery schedule	Stock Clerk	The Stock Clerk arranges so that ZigZag has enough people in place to unload the delivery trucks
Check delivery	Stock Keeper	Delivered goods are checked against purchase orders. Damaged goods are rejected and the P.O. is annotated accordingly
Place goods in delivery bay	Stock Keeper	Goods are physically placed temporarily in the delivery bay
Allocate stock location	Stock Clerk	A place in the depot is secured for each delivered product
Remove goods from delivery bay	Stock Keeper	Goods are physically removed from the delivery bay
Store goods in depot	Stock Keeper	Goods are physically placed in their allocated spots
Update stock levels	Stock Clerk	The Stock Keeper informs the Stock Clerk of the result of the delivery. Any adjustments to the planned stock allocations is recorded
List available stock	P.O. Clerk, Stock Clerk	At the request of a Purchaser, the P.O. Clerk produces a stock report which is sent to the Purchaser. The Stock Clerk also produces stock reports and stock storage reports to help plan and monitor the stocktaking and allocation process
Receive customer order	Despatch Clerk	Sales and Marketing forward customer orders to the depot. Two copies of each customer order are filed.
Arrange despatch details	Despatch Clerk	The Despatch Clerk checks the stock files and allocates the stock to be given to the customer
Forward despatch details	Despatch Clerk	The Despatch Clerk gives the despatch details to the Despatch Supervisor
Assemble goods for despatch	Despatch Supervisor	The Despatch Supervisor arranges for the physical assembly of the customer's order and the return of the annotated customer order back to the Despatch Clerk
Despatch goods	Despatch Supervisor	The fulfilled customer's order is despatched with a despatch note to the Customer
Inform Sales and Marketing	Despatch Clerk	The Despatch Clerk prepares a 'matched despatch report' and sends a 'matched customer order' copy to Sales and Marketing

Fig 3.12 A first Work Practice Model for ZigZag

The effect of e-commerce on the Business Activity Model for ZigZag

One of the main reasons the ZigZag board of directors decided to investigate the development of a new system is the advent of e-commerce. The board wishes to take advantage of e-commerce to reach retail customers directly. It also wishes to investigate whether the World Wide Web would be a convenient platform to be used by their purchasers who roam the world to communicate their findings.

When it comes to customers' using the Internet to communicate their orders the only real difference is that the input of the order is now to be performed by the customer directly, thus relieving the onus from the ZigZag employee who currently does the input. Such a situation, where the work is shifted from the company to the customer, is just another example in a long-standing shift that manifests itself more clearly in supermarkets where the customer does much of the work we traditionally associate with a shopkeeper.

Internet banking is an industry which has taken advantage of this shift and ZigZag is interested because the banking sector is reporting that a transaction using electronic means is forty times cheaper than a transaction that uses a teller.

The actual information to be stored as a result of expanding into e-commerce is very similar to that which we would have stored in any case. What changes are the users of the system, which now have to encompass bona fide customers, and the activities that have to be added because of the expansion into retailing. There will also be a job shift since now web designers will need to be employed to maintain the new site. We therefore note that B2C (business-to-customer) e-commerce only affects the user organisation and the external design.

With no effect on data, we see that Business Activity Modelling and Work Practice Modelling can deal with the transition to e-commerce quite effectively.

Currently the typical ZigZag customer is a shop that buys wholesale from ZigZag. This wholesale customer sends orders to Sales and Marketing who forward them to the depot where the despatch clerk is responsible for arranging the despatch details and updating the stock levels.

If we wish to allow retail customers direct access to our system we need to first understand the buying activities from their point of view. Only by understanding the customer's buying experience will we be able to design a system to accommodate them.

Up to now we have been performing Business Activity Modelling from the point of view of the business. With the advent of business-to-customer e-commerce we need to also study the activities from the point of view of the customer and to see how the two activity models interact.

To order goods from ZigZag a customer will have to search for goods, place them in a shopping basket, arrange for payment and then forward the order to ZigZag. We can represent these activities quite easily (see Figure 3.13).

We now turn our attention to the activities from ZigZag's point of view. While ZigZag was dealing with wholesale customers, despatches were fairly big and relatively infrequent. With the decision to add retailing to the business we anticipate many smaller despatches. Each retail customer order will be for just a few CDs that will need packing and pricing. Since the packs will be smaller than the ones that ZigZag is used to, we expect these packs to be placed in a despatch dock where they will be assembled once a day (or more often at peak periods) for posting. ZigZag will also need to produce packing lists and despatch reports to be able to tackle an increased number of returns.

ZigZag expects to continue despatching wholesale orders using their own vans but expects to use the post to despatch retail orders. Each retail customer order will incur a delivery charge and every effort will be made to despatch the whole customer order in one go.

We can represent the activities for handling retail orders quite easily (see Figure 3.14).

When the goods leave ZigZag they arrive at the customer's address where they are received, checked and, if found wanting, returned.

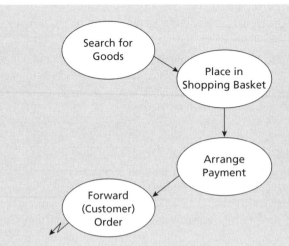

Fig 3.13 E-customer activities

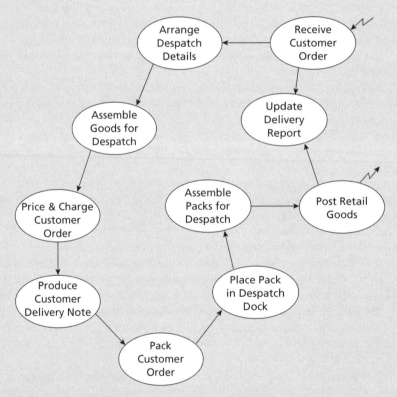

Fig 3.14 ZigZag activities associated with retail customer orders

We can combine the two activity models in one diagram to study how the two activity sets interact with each other (Figure 3.15).

By expanding the area of study to acknowledge the customer as a new user of the system, we immediately spot from Figure 3.15 that we expect the user to search for goods but we have not yet identified a corresponding

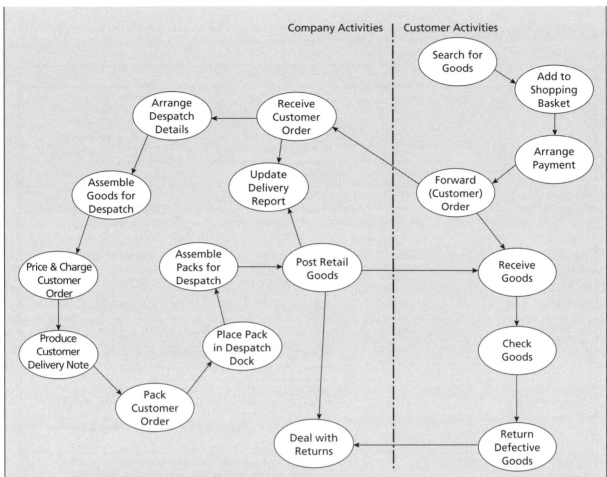

Company Activities | Customer Activities

Fig 3.15 Combining activity models to study e-commerce effects

activity of adding goods on the browser for the customer to find. This realisation immediately raises the question of who is responsible for providing the future web site with product information. Questioning of the ZigZag staff reveals that Sales and Marketing are currently responsible for providing sales brochures and price lists for the wholesale trade. It therefore stands to reason that Sales and Marketing should also be responsible for maintaining the future website.

The above arguments show two things: (a) the power of Business Activity Modelling to identify communication and responsibility gaps within departments, and (b) the need for the IT department of a company to transcend traditional departments in an attempt to integrate as many cross-company operations as possible.

We will have occasion to revisit the creation of the ZigZag website as we progress through the SSADM techniques.

Using the new information system to move from wholesale to retail causes a few upheavals on work practices too. First, the customer has to be studied as a new user, in the same way that e-banks had to study their customers before designing on-line banking systems. Secondly, ZigZag has to acknowledge the need to add web designers to their list of employees. Thirdly, the hustle and bustle in the warehouse will increase with many more pickers needed to accommodate the anticipated increase in the number of transactions that will have to take place.

While investigating the current environment, we record our findings on the Work Practice Model (see Figure 3.16).

Activity	Job Title	Activity Description
Receive customer order	Despatch Clerk	Sales and Marketing forward customer orders to the depot. Two copies of each customer order are filed. It is expected that in the future retail customer orders will be received through the net. These will be dealt with in a similar way to wholesale orders. The only difference is that it seems orders will first come to the warehouse and not Sales and Marketing
Arrange despatch details	Despatch Clerk	The Despatch Clerk checks the stock files and allocates the stock to be given to the customer. Apart from having to deal with an increased volume of smaller orders this activity remains largely unchanged by e-commerce
Forward despatch details	Despatch Clerk	The Despatch Clerk gives the despatch details to the Despatch Supervisor. This activity will probably disappear as it is hoped that the Despatch Clerk and the Despatch Supervisor will be connected to the same computerised system negating the need for a physical movement of the despatch details
Assemble goods for despatch	Despatch Supervisor	The Despatch Supervisor arranges for the physical assembly of the customer's order and the return of the annotated customer order back to the Despatch Clerk. For wholesale customers this activity is probably unchanged
Price & Charge Customer Order	B2C Clerk	This is a new activity for e-commerce customers. While wholesale customers receive their bills from Accounting via Sales & Marketing, retail customers have prepaid and require a receipt with their goods. Accounts should be sent a copy for reconciliation
Produce Customer Delivery Note	B2C Clerk	A 'new' activity for retail customers whose delivery note is slightly different from that of wholesale customers and includes promotional material too
Pack Customer Order	B2C Clerk Packer	The retail customer's order is placed in an appropriate pack, approved by the postal authorities and franked
Place Pack in Despatch Dock	B2C Clerk Packer	The retail customer's pack is placed in a 'new' despatch dock awaiting assembly
Assemble Packs for Despatch	B2C Clerk	The accumulated retail customer packs are assembled and given a batch number for future reference
Post Retail Goods	B2C Clerk	The postman arrives to collect the packs under the supervision of the B2C Clerk
Update Delivery Report	B2C Clerk	The paperwork concerning deliveries is updated for use if there is a dispute over payment or deliveries
Deal with Returns	B2C Clerk	This activity cannot yet be precisely defined so we just acknowledge it here because of its potential significance. It is not yet clear if Sales & Marketing should be responsible for this, in which case we should ensure that they receive despatch information on a regular basis

Fig 3.16 A section of the ZigZag Work Practice Model dealing with e-commerce activities

▶ HIERARCHICAL TASK MODELLING

When an activity is complicated the technique of Hierarchical Task Modelling can be invaluable in understanding its details. Hierarchical Task Modelling is a labour-intensive technique, so we will use it to study only the most complex of activities.

We will use the activity of arranging a delivery for ZigZag to illustrate how to dissect an activity into its constituent tasks and then construct a task model for the activity. As with most analysis techniques, we will first start with a straightforward case and then add potential complications.

Deliveries at ZigZag are quite big events that need to be planned and anticipated. A lorry-load of goods doesn't just appear at the loading bay. It has to be pre-arranged so that ZigZag knows what is expected and can provide the workforce to unload the goods. When a delivery arrives the stock keeper has to have a list of expected goods at hand to record the condition the goods are in. For that list to exist we need to know beforehand what is to arrive, especially since a delivery may include goods from several purchase orders.

The activity of arranging a delivery consists of two main tasks. First, an available time-slot has to be agreed. Once this is done, the supplier has to state the contents of the delivery so that the stock clerk can match them with current purchase orders – ZigZag does not wish to be landed with goods it has not ordered and is therefore quite particular about ensuring that suppliers declare what they wish to deliver *before* the delivery takes place.

A straightforward delivery would therefore consist of the supplier's declaring the date they wish to deliver. The Stock Clerk would then identify the free time-slots for that day. (The free time-slots could be spotted on a diary or, when the activity is computerised, will be output by the system.) From the available time-slots one will be picked if the supplier agrees to it.

Having agreed the time and day, the supplier should then list what is to be delivered and the stock clerk will accept or reject the items. To do so we expect the supplier to state the product, the purchase order reference and the quantity to be delivered. The stock clerk should use this information to identify the relevant order line against which the delivery will be recorded and accepted or rejected accordingly.

We thus see that arranging a straightforward delivery consists of the following tasks: Record Preferred Date, List Available Slots, Pick a Slot, Type Delivery Line Information, and Accept Delivery Line.

We can represent these tasks in a Hierarchical Task Model (HTM) like the one depicted overleaf in Figure 3.17.

The HTM is read from left to right and annotated with 'plans' that show the different ways in which the activity may be fulfilled. The model contains boxes depicting the activity itself, nodes of sub-activities, and finally the elementary tasks. Each 'box' is numbered, preferably in a nested fashion. The 'plan' below the Arrange Delivery box states that either '33' will take place or '33' followed by '36'. This is because task '33' may come to an abrupt end if the supplier and the stock clerk do not find a mutually agreed time-slot for the delivery. (The fact that '33' may not finish successfully is indicated by the plan under it which indicates that task '336' may not take place, i.e. a time-slot may not be picked.)

While the arranging of a straightforward delivery may itself take some time because most deliveries to ZigZag are quite big, there are some added complications

> The primary use of task models will come later in documenting how individual tasks are to be structured in the required system, leading directly to the design of user interfaces.

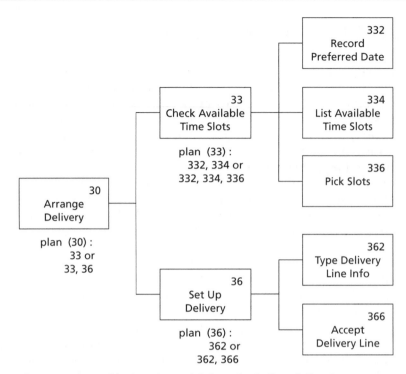

Fig 3.17 A first-cut Hierarchical Task Model for scheduling deliveries

to arranging a delivery. These added complications not only make the job of the stock clerk quite tricky, they would also challenge the design of a usable computerised system.

The complications arise when amendments to currently arranged deliveries take place. These amendments come about when suppliers call after a delivery has been arranged to change something. An amendment may be to the time of the agreed delivery or to the goods to be delivered. Further, an amendment to the goods to be delivered may include the addition of new delivery lines or an amendment on the quantities to be delivered. Figure 3.18 attempts to incorporate these complications into one HTM.

Hierarchical Task Models serve three important purposes: (a) they help us understand the nuances of each job, (b) they can become the basis for a training manual later on, (c) they facilitate the process of our trying to take advantage of graphical user interface (GUI) environments without allowing these environments to overwhelm the design (as we will see in the Function Definition section).

▶ TASK SCENARIOS

One way of understanding and controlling the tasks involved in the successful completion of an activity is to provide concrete examples of real-life situations, which describe from beginning to end the actions needed to complete the activity.

Each business activity is influenced by certain happenings and conditions to which users of the system have to react in order to complete the activity. Each set

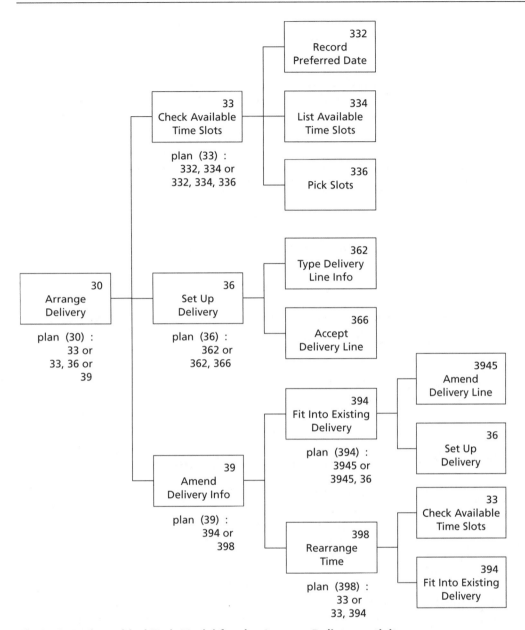

Fig 3.18 A Hierarchical Task Model for the Arrange Delivery activity

of such conditions represents a Task Scenario for the activity. For example, each different plan of the HTM represents a Task Scenario.

Task Scenarios are very helpful in validating Task Models. Usually, the Task Scenarios precede and drive the creation of the Task Model, but, as with all other diagrammatic tools of system analysis, the tables are soon turned and the Task Model generates ideas for Task Scenarios that can be validated with the users.

Task Scenarios become that basis for testing the final system, where they drive the creation of test scenarios.

▶ THE USER CATALOGUE

Users take a central role within SSADM and so the identification of relevant users is quite an important task. Creating a User Catalogue (Figure 3.19) is a formal way of documenting the job titles and the business activities of each user or jobholder. In essence, the User Catalogue is a summary of the Work Practice Model, arranged by job title.

User Catalogue	
Job Title	**Responsibility (Job Activities)**
Purchase Order Clerk	*Placing of Purchase Orders* (Record proposed purchase orders; Find appropriate suppliers; Receive products' availability; Suggest substitute product; Confirm purchase order; List available stock) *Matching Supplier Invoices* (Receive supplier invoice; Check invoice; Forward matched invoice)
Despatch Supervisor	*Despatching of Customer Orders* (Assemble goods for despatch; Despatch goods)

Fig 3.19 A section of the ZigZag User Catalogue

The User Catalogue will later be used to help define the outward appearance of the new system (or at least its interface with users), but to start with its main purpose is to support the identification of users in the current environment.

▶ APPLICATION OF WORK PRACTICE MODELLING

Work Practice Modelling starts during Business Activity Modelling. As the BAM is developed, a User Catalogue is set up. This forms the starting point of User Analysis which will evolve into a complete definition of each user's interface requirements.

During Business System Options, Work Practice Modelling plays an important part in providing a set of alternative outline Work Practice Models that help illustrate the impact of each BSO.

With an agreed boundary in hand, the technique becomes the forum through which the detail of who does what, down to the level of the user interface, is studied using Hierarchical Task Models and Task Scenarios. This is an iterative process which may resort to prototyping to elucidate complicated tasks. Figure 3.20 illustrates the technique.

Work Practice Modelling straddles the region between IT practitioners and human factors specialists. The manner in which it is undertaken will depend largely on the prevailing policies and procedures. These will usually define exactly how Business Activity Modelling and Work Practice Modelling will take place. The approach described here is quite general. As the SSADM manual rightly states, 'The overriding consideration in the approach adopted for Work Practice Modelling is that projects should take account of the complete picture of the business and not just concentrate on the automated system. The automated system must be designed to fit the business needs, not the other way round.'

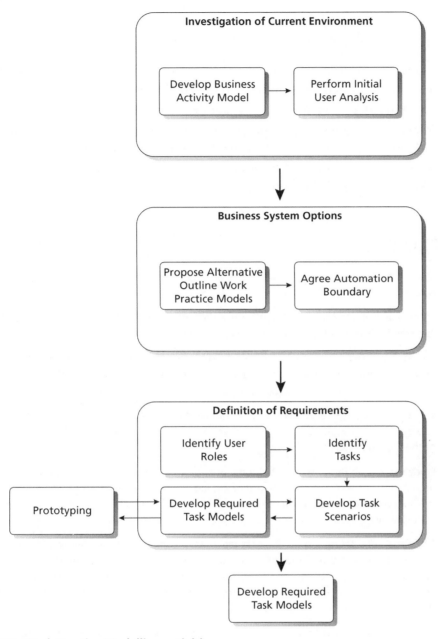

Fig 3.20 Work Practice Modelling activities

▶ SUMMARY

1. A Business Activity Model is a tool for establishing a common understanding of the business area under investigation.

2. There are five types of business activity: planning; enabling; doing; monitoring; and controlling.

3. A business event is something that triggers one or more business activities.

4. Business Activity Models are most often developed from scratch, but can also be developed from Resource Flow Diagrams or Functional Decomposition.

5. Work Practice Models describe who performs which activities, and where. They are used to record changes to the working environment.

6. User Catalogues summarise the Work Practice Model by job.

7. In order to facilitate the development of job specifications and human–computer interfaces, complex activities can be broken down using Hierarchical Task Models.

8. Task Scenarios can be used to illustrate and validate Hierarchical Task Models using real-life situations. Task Scenarios can also be used as the basis for testing the final system.

▶ EXERCISES

3.1. Study the Business Activity Model of ZigZag and identify as many business threads as you can.

3.2. Produce a Hierarchical Task Model for each activity in the ZigZag Business Activity Model. Identify the activities that may need further study.

3.3. Consider yourself a potential web-based retail customer of ZigZag. List what you would expect to find on their website and explain the activities you expect to go through. For each activity consider designing an appropriate screen.

3.4. Investigate websites that may offer a similar functionality to the business-to-customer functionality you expect ZigZag to offer. Write a report to present your findings to the rest of the ZigZag project team.

3.5. Study the ZigZag Business Activity Model and identify areas with business-to-business (B2B) potential. Write a report to the board of ZigZag explaining the effects of B2B commerce on the operations of the area you have identified (it is hoped that your report will not neglect work practice issues).

3.6. ZigZag is considering using an intranet to aid communication between departments. Identify areas where an intranet would be of benefit and write a report explaining the effects of an intranet set-up for ZigZag. (It is hoped that you will work in groups since there are many areas where an intranet opportunity may arise and that you will conduct a brain-storming session to consolidate your findings.)

3.7. Study each ZigZag activity and identify the support it will require from the future comput-erised information system.

3.8. Produce an Organisation Chart for the current manual system described below and use it to identify the main activities involved.

Fresco is a ticket agency, dealing in concert and theatre tickets. All of their business is con-ducted over the telephone, with customers ringing up to request tickets for a wide variety of performances.

Concert and theatre venues provide Fresco with a constant stream of information on forthcoming events, which is then used by Fresco's manager to compile a fixture list for use by the sales staff in responding to customer calls. The manager will also select a number of events for which Fresco will purchase tickets in advance of customer requests (e.g. for popular events). Details of these 'pre-purchase orders' as they are known are passed to the post clerk who is responsible for placing orders with each venue. The post clerk sends out each order with its payment attached. Once tickets are received the post clerk files them in the tickets file.

When customers ring the sales team their ticket requests are checked against the ticket file. If pre-purchased tickets are available they are put in an envelope marked with the customer's name and filed at the back of the ticket file. If not, the sales team fill out a ticket request form and put it in a desk tray for collection by the post clerk, who will then place an order with the appropriate venue in the same way as for pre-purchased tickets.

Details of the customer and the tickets are passed to the payments section for invoicing (by the sales team for pre-purchased tickets and by the post clerk for new tickets). The payments section will then send an invoice to the customer, and await payment if the customer is paying by post or accept payment immediately for credit card holders. A copy of the invoice is placed in the invoice file for matching with payments, and once payment has been made a further copy is placed in a tray labelled 'despatch list' for collection by the post clerk.

The post clerk collects paid invoice copies three times a day, and retrieves the appropriate tickets from the ticket file for despatch to customers.

3.9. By using your Functional Decomposition of the previous exercise, or otherwise, produce a Business Activity Model for the Fresco system above.

3.10. Identify the resources moving within the Fresco ticket agency above and produce a Resource Flow Diagram for each.

3.11. Produce a Work Practice Model for the Fresco ticket agency above.

3.12. Use your Business Activity Model of the Fresco ticket agency to identify activities that could be expanded to take advantage of the World Wide Web. Explain your reasons for choosing each activity and produce a report explaining the costs and benefits that may arise by the activity being offered through the Web.

3.13. Produce a Business Activity Model for Natlib using the following information.

Natlib is a small private library specialising in natural history books. They have a collection of titles available for loan by registered readers, free of charge.

Each reader is allowed to borrow up to eight books at a time. Loans are to be recorded against particular book copies, rather than their title, as Natlib may have several copies of any given book.

When all copies of a book are already out on loan a reader may wish to place a reservation for it. Each reservation is recorded against a book title. When a copy of the title is subsequently returned Natlib will place it on one side, record which copy is to satisfy the reservation, and notify the reader that it is ready for collection

The new system will cover functions in several areas: reader registration; book registration; book loans; book reservations; book returns; and issue of loan reminders.

When readers borrow a book their status is checked to see if they have already reached their loan limit or have any outstanding overdue loans or fines, in which case they will not be allowed to borrow any more books until they have taken appropriate action.

New loans will be recorded and details of their 'due for return' date given to the reader. Overdue loans will be monitored and up to three reminders sent to the reader. Each time a reminder is sent its date will be recorded against the reader, whose status will then be updated.

When readers reserve a book the title will be checked to ensure that Natlib currently holds it, and if so the reservation will be recorded.

When a reader returns a book its return date will be recorded against the loan record and its condition (if it has changed) against its copy record. For overdue returns and damaged copies the librarian will levy a fine against the reader and record its amount (and any subsequent) payment on the loan record. If payment is not made in full, the reader's status will be updated.

Each return will be checked against outstanding reservations. If the title is reserved the book copy will be placed on one side, the reservation will be updated with details of which copy has been returned and the appropriate reader will be notified.

The purchase of new books will not be covered by the new system as it already functions well. However, the system will need to record details of new titles, as sent in by book suppliers and publishers, and of new copies as the librarians at Natlib register them.

3.14. Set up a Business Activity Model for the activities apparent in the following scenario.

The Hergest educational institute runs short courses. Each course is run many times during the year. Once a month the institution's courses are advertised in the local paper. If enough students show an interested in a course, then a lecturer and a substitute lecturer are allocated to it. The role of the substitute lecturer is to replace the main lecturer when that person is unable to teach a particular session. At the end of a course students are presented with a certificate of attendance, which acts as proof of the fact that the student has attended the course. Prior to the start of a course, a class list is printed out and given to the main lecturer.

3.15. Produce a Business Activity Model for the following scenario.

Bodgett & Son is a building firm of medium size, working for a single central office. They carry out most kinds of general building work and have expanded greatly in recent years. They are in the process of developing new systems to handle their administration tasks, and have already computerised the accounts area. The next area of investigation is the estimating and management of building work, or 'jobs'.

All work carried out by Bodgett & Son is preceded by a formal estimate, carried out by an outside surveyor or, for smaller jobs, an employee.

When an estimate request is received from a prospective customer the administration section assesses the likely size of the job and selects either a surveyor from their standard list of acceptable surveyors, or a suitable Bodgett employee. The administration section keeps the surveyor list up to date from changes supplied by Bodgetts' manager. Surveyors are contacted to check that they are available at the time requested by the customer and entered onto a booking sheet. At the end of each day a member of the administration section will use the booking sheet to draw up booking letters to send out to surveyors.

Surveyors (or employees) return the completed estimate to admin, which then prepares a formal estimate letter for the customer. Large pieces of building work may be subdivided into a number of smaller jobs. Details of the estimate and jobs are filed in the estimates file.

Customers will usually send back a slip at the bottom of their estimate detailing their acceptance or rejection of some of or the entire estimate. The estimate file is then updated by admin, and any acceptance forms placed in a tray for processing later in the day by the job administrator.

The job administrator then picks up the acceptance forms and retrieves the relevant estimate details from the estimate file. Details of work already booked for Bodgett are held in the job file. The job administrator checks these details and the newly accepted estimates are scheduled around these existing jobs. Most jobs will at this point be subdivided into a number of much smaller tasks. The job administrator will create a materials order line for any building supplies that are needed for a task, and attach these to the task details for order placement by the order clerk at the start of the next working day. There may be more than one order line for any given task.

All new job and estimate details are then placed at the front of the job file and labelled as ready for ordering. The job administrator then notifies the customer of the provisional start and finish date for their work.

The order clerk checks the beginning of the job file each morning and groups together materials orders for placement with suppliers. Copies of each order are then placed in an orders file at the storeroom entrance. The job file is marked as 'order placed'. As materials are received the order copies are updated until they have been fully satisfied. The job file is then updated to show which tasks have materials ready for them. When the materials for the last task in a job are ready the job itself is marked as 'ready'.

The staff office maintains a list of employees and what skills each one has. This list is constantly updated as the manager hires new employees. Each day the job file is checked and the jobs that are ready are compared with the employee list. Each task record is matched with a single employee and annotated with when the employee is free to carry out the work. One employee will be designated as job supervisor. When all tasks have been allocated the staff office will know the firm start and finish dates for the job and the customer is notified.

The job supervisor will monitor the progress of each job and keep the job file updated. It may be necessary to change the start and finish dates for a job, in which case it is the responsibility of the supervisor to notify the customer.

Once the job is finished, the job supervisor will fill out a completion notice and send copies to the customer and to the accounts section, and will attach a further copy to the job file.

3.16. Identify the resources moving within Bodgett & Son as described above and produce a Resource Flow Diagram for each.

3.17. Produce a Work Practice Model for the Bodgett & Son firm described above.

3.18. Use your Business Activity Model of Bodgett & Son to identify activities that could be expanded to take advantage of the World Wide Web. Explain your reasons for choosing each activity and produce a report explaining the costs and benefits that may arise by the activity being offered through the Web.

Defining requirements

INTRODUCTION

The main purpose of systems analysis is to identify and define the requirements of the new system. In other words we first aim to define precisely *what* the system is required to do before suggesting *how* it will do it.

SSADM provides a number of tools and techniques to assist in analysing, modelling and graphically representing requirements. In any given project the range of techniques used will vary with system complexity, size and the type of solution (e.g. package system versus bespoke development).

Regardless of which analysis tools are used, we will always need to be rigorous in documenting the resulting requirements. Errors made in defining requirements are more expensive to correct than errors made in any other stage of a project, and put the effectiveness and acceptance of the final solution at serious risk. In SSADM, in common with most methods, we record descriptions of requirements in a central Requirements Catalogue which is then cross-referenced with other more detailed products as the project progresses.

The Requirements Catalogue is largely textual in nature, and therefore if written in appropriate business language provides an ideal vehicle for communication with users. Indeed in many projects it is the key document for reviewing and agreeing requirements with users. The catalogue should cover all aspects of the required system in sufficient detail so that users are able to confirm that requirements have been accurately defined, and that none have been overlooked.

The Requirements Catalogue acts as a central reference document for all steps within an SSADM project and so is one of the few products within SSADM that should be regarded as mandatory. Although it provides a definitive list of requirements it is too textual and imprecise on its own to act as a complete specification of those requirements. In order to form a complete picture, a range of further interdependent SSADM products will be developed, all based to some extent directly on Requirements Catalogue entries.

> ▶ Learning objectives
>
> After reading this chapter, readers will be able to:
>
> ◆ understand the process of identifying and developing system requirements;
>
> ◆ define requirements in such a way as to provide a communication tool for all participants and stakeholders within a project.

> ▶ Links to other chapters
>
> This chapter links to virtually all other chapters within the book. There are particularly strong relationships with Chapters 5 and 6, which deal with the analysis and modelling of data and processes, and with Chapter 8, which describes how we begin to translate requirements into the functional design.

▶ REQUIREMENTS DEFINITION

The purpose of the Requirements Definition technique is to identify and document all requirements of the new system.

Requirements Definition is an iterative process carried out in all stages of a project. In the initial investigation phase we will identify largely high-level requirements. As the project progresses we will add further detail and understanding to these requirements, and will identify additional less obvious requirements. Each catalogue entry should be viewed as dynamic in nature and will probably be updated many times.

▶ Functional requirements

Functional requirements deal with *what* the system should do or provide for users. They will detail what facilities are required and what activities the system should carry out. In other words they define the required functional support of the new system. The sort of things we need to record in this area are:

◆ descriptions of actions or processes that create or update information;

◆ outlines of reports or on-line queries. These can be user requested or automatically generated by the system at scheduled times, or when specified events occur;

◆ details of business data to be held and used within the system.

▶ Non-functional requirements

Non-functional requirements detail constraints, targets or control mechanisms for the new system. The kinds of thing covered by this broad categorisation are:

◆ **Required service levels**, e.g. response times for on-line functions, the time that the system should be available in the morning. It is important to establish realistic targets for performance, based on the activities they support. Achieving high levels in all areas would be extremely costly, even if it were feasible. For instance, to specify even a 10-second run time for all queries in a major system would be both impractical and impossible to justify. It is always worth investigating whether processes or enquiries can run in batch overnight (e.g. month-end statement printing) or as background tasks with relatively long run times. System performance and tuning will always involve trade-offs in technical design and infrastructure (hardware and software), so it is crucial to establish early on where the key priorities lie.

◆ **Volumes**. Data volumes and the frequency of updates and enquiries will to a large degree drive the physical design of the database. The choice and sizing of the technical infrastructure will also be driven by questions of volume. We should therefore aim to document volumes as early as possible.

◆ **Transition arrangements**. All system implementations will require user training and changes to working practices. It is important to consider these at the outset and document requirements in this area as soon as possible as the time and effort required to prepare users for the arrival of the system can be very significant. Data conversion requirements are also overlooked in many projects and left to the last minute. Converting data from one system to another is rarely trivial and so again requirements should be documented as early as possible to avoid delays during the implementation phase.

◆ **Technical constraints**, e.g. if the system must make use of existing hardware only.

◆ **Usability**. These requirements need to be carefully handled as no function is ever specified as needing to be difficult to use. It is important to establish levels of usability and required training or experience appropriate to different functions. Some functions may frequently be used by occasional or novice users, while others will be used exclusively by experienced or well-trained users only. In Internet applications those functions to be accessed by customers clearly require higher levels of usability and on-screen help, than those used by the accounts office for example.

◆ **Required interfaces** with other systems.

◆ **Security and access requirements**, e.g. data back-up requirements, which users have access to a function.

◆ **Project constraints**, e.g. system delivery date, cost limits, company policies.

In a sense non-functional requirements detail how well, or within what limits, a functional requirement should be satisfied. Non-functional requirements may apply to the system as a whole or to particular functional requirements.

It is important that requirements have measures of success or quantifiable targets included in their definition, to ensure that the requirements are actually met in full by any systems developed in the future. These may be in the form of related non-functional requirements (e.g. response times) or subjective checks by responsible users (e.g. ease of use). Success measures will also help in reducing ambiguity.

▶ REQUIREMENTS CATALOGUE

As we identify requirements we record them in the Requirements Catalogue (see Figure 4.1).

As well as a brief textual description each catalogue entry should include:

Requirements Catalogue	
Requirement Id.	114
Requirement Name	Overdue Delivery report
Description	Report providing details of overdue deliveries, including the date the delivery was originally due on, number of days late, supplier number and name, purchase order number.
Priority	Desirable.
Source	Delivery scheduling workshop – Stock Clerk (M Patel).
Owner	Warehouse Manager – M Portillo.
Non-Functional Considerations	Required on-line from 8.00 until 18.00, Monday to Saturday. Response time should be under 10 seconds as required up to 20 times per day, giving the latest situation.
Comments/ Suggested Solutions	Should be provided as an on-line report, with an option to print hard copy.
Benefits	Will enable chasing up of late deliveries, monitoring of supplier promptness and help. Target 30% reduction in outstanding late deliveries. This should reduce incidents where customer orders cannot be satisfied due to late delivery of stock from the supplier. Estimated additional sales of 0.1%.
Related Documents	Interview notes No. 3, including suggested layout. Delivery scheduling workshop notes.
Related Requirements	No. 115 Re-schedule delivery. No. 136 Delivery Details Report – will need to be able to access these details in order to establish what items are overdue.
Resolution	

Fig 4.1 Sample Requirements Catalogue entry prior to the development of formal design products and resolution

◆ A priority for the requirement, such as: essential to the operation of the system or business (E), desirable as it will deliver significant business benefit (D), or nice to have (N), i.e. 'luxury' features. Care needs to be taken in establishing real priorities as people have a natural tendency to classify all requirements relevant to themselves as essential;

◆ The source of the requirement, which may be an individual, a document or even a workshop;

◆ The owner or user responsible for agreeing and signing off the requirement;

◆ Associated non-functional considerations, e.g. volumes or frequencies;

◆ Suggested solutions. Any ideas for satisfying the requirement (including retention of the current system or subsystem) should be noted down as they are suggested. Too many good ideas are lost by neglecting to record them;

◆ Benefits. It is not necessary or feasible for every requirement to have an individually identifiable benefit or justification. Many are just needed for the system as a whole to function properly and therefore deliver the overall benefits of the project. Where requirements are merely desirable it is important to provide some justification for their inclusion in the final solution. If possible, benefits should be measurable;

◆ Related project documents or products. These may be SSADM deliverables or general business documents, such as strategy reports or descriptions of company policy and standards;

◆ Related functional requirements. One requirement will often have an impact or dependence on another, or require a common solution, e.g. a report on overdue deliveries will need to be linked to a facility to change delivery due dates;

◆ Resolution. This may provide a textual description as well as reference to the SSADM products that specify the solution. If a requirement is turned down or deferred to another project we should record the reasons for rejection. Analysis may reveal that the requirement would be too costly, take too long to develop (thus risking the project deadline), or is less appropriate than an alternative requirement. Requirements for reports or enquiries are frequently combined or satisfied by the provision of a general querying facility.

Requirements Definition is a text-based technique and therefore is less precise and objective than many of the more rigorous modelling techniques within SSADM. It is not possible to be prescriptive about how detailed the information recorded within the Requirements Catalogue should be.

Where calculations or measures are involved, we should be able to record requirements in precise mathematical terms. In more descriptive areas we need to provide statements which give users confidence that their requirements have been fully understood, and which they are happy to sign off as accurate. The level of detail for new requirements or solutions to current problems should be sufficient to ensure that there is as little ambiguity as possible. In areas where the requirement is to retain the same functionality as the current system, we should state that this is the case, and give enough detail so that the complete requirement can be carried forward into the later analysis and design stages of the project.

In most projects we would also attach project management information to the catalogue entry (as with most SSADM-generated documents). This information (e.g. date of last amendment, version number) will be subject to standards as defined in the organisation's policies and procedures.

The Requirements Catalogue can become very large and may even need to be held in several volumes, each relating to a business area for example. This can make it rather unwieldy, so it is useful to set up a summary, with one row per requirement, giving information such as the requirement ID, name, owner and priority. We can then use the summary as a communication aid in presentations or when reviewing the catalogue with users, sponsors and the project team. The summary also acts as an index to the catalogue (see Figure 4.2).

Requirements Summary			
Id	Name	Priority	Owner
114	Overdue delivery report	D	MP
115	Re-schedule delivery	E	MP
116	Make use of existing PCs in stock office	D	HS

Fig 4.2 Extract from the ZigZag requirements summary

Note that some requirements will be entirely non-functional and will apply to the entire project or system. For example, in the ZigZag requirements summary, requirement 116 is non-functional.

▶ IDENTIFYING AND DEVELOPING REQUIREMENTS

In practice users find little difficulty in generating large numbers of functional requirements. Non-functional requirements then often emerge as we further analyse and debate functional requirements.

Sources of requirements

There are many ways of gathering information on requirements, some of which are summarised below. General fact-finding is outside the scope of this text, but is documented well elsewhere, e.g. Yeates, Shields and Helmy (1994), Chaffey et al. (1999).

◆ **Interviews**. These will almost always form the backbone of any fact-finding exercise. Although the techniques below are useful there is no substitute for thorough detailed discussion with users and managers. It is only through well-prepared, carefully targeted and meticulously documented face-to-face questioning that we can get to the heart of the business problem.

◆ **Workshops**. This way of gathering and clarifying information about business procedures whereby all the key participants of the system are gathered in one room in the presence of two system analysis 'facilitators' is gaining in popularity.

◆ **Examination of business documentation**. The documents that are produced by the day-to-day activities provide an important source of what information will be needed by the business in the future. Sample documents and sample data values should be avidly collected and their content studied. User or procedure guides are also invaluable as a source of information on current practices, and often act as a good catalyst to an interview.

◆ **Questionnaires**. In some organisations these may be politically unacceptable. In any case they are only worthwhile as an initial fact-finding exercise as the results are always fairly limited in scope, owing to the necessarily closed nature of the questions.

◆ **Observation of operations**. This can be very time-consuming and tedious. However, there is occasionally no other way to get at the minute details of operations. At the beginning of the project it is a good idea to visit the 'shop floor' as part of general familiarisation.

◆ **Brainstorming sessions**. There is nothing better for disentangling the issues and ideas raised by members of a system analysis team than putting them all in one room and asking them to articulate their findings by 'thinking aloud'.

All of the above are general information-gathering techniques, which can be applied to many situations outside the field of systems development. There are also a number of SSADM-specific sources of requirements, all of which feed directly into Requirements Definition:

◆ **Project input documents**. The Project Initiation Document, together with copies of company policies and standards will provide a good starting point in identifying requirements. Often these will be high-level and so require a lot of added detail. They may also act as headings, which we will need to break down into a number of constituent requirements.

◆ **Previous or related studies**. Previous feasibility studies may have produced high-level Requirements Catalogues (indeed it is good practice for them to do so) which we can adopt as the starting point for the full catalogue within the project. Detailed requirements will also often be inherited from related projects (particularly if the project is part of a programme of development).

◆ **Current system problem and change logs**. Most systems will have formal logs or files of problems reported by users or the IS department. They may also have logs detailing requests for enhancements to the system.

◆ **Business Activity Modelling**. Discussion of the Business Activity Model will identify requirements for automation and information support for activities, as well as for higher-level management information, including performance monitoring. It will also highlight requirements for external interfaces with other systems, outside agencies, customers, etc.

> For example, discussions of the BAM extract given in Figure 3.19 and the WPM extract in Figure 3.21 highlight new requirements in the areas of customer order despatching.

While the emphasis of Requirements Definition must always be on the new system, the requirements most readily highlighted by users often relate to solving problems associated with the current system support, such as:

◆ **Restrictive operations**. Many existing systems are unresponsive to changing requirements or exceptions to normal processing, often leading to systems' being 'fiddled' in order to achieve the required results.

◆ **Poor system quality**. Systems that have developed over a number of years tend to have done so in a piecemeal fashion, leading to poor performance and clumsy procedures.

◆ **Lack of system availability** due to unreliable software or lengthy overnight batch runs which restrict on-line availability during the day.

◆ **Inadequate or complex user interfaces** requiring significant training for users to be able to operate the system effectively.

◆ **Lack of capacity**, which in extreme examples can restrict the ability of a business to grow.

Where requirements are generated from current problems we must ensure that the future requirement is properly recorded. For example, if a current problem is that an enquiry takes 30 seconds, we need to record the required response time, rather than stating that 30 seconds is too long.

In areas where satisfactory support is already provided by an existing system we will still need to document requirements. This is partly because we should not take it for granted that requirements will stay exactly as they are, but mainly that without defining all requirements there is no guarantee that the final system will be complete.

It is important not to view Requirements Definition as a way of just minuting our fact-finding discussions. In the early investigation stages of a project the sheer number of requirements that emerge can be difficult to manage. While we must not alienate users by dismissing requirements too early (especially in brainstorming sessions), we must always attempt to explore the validity of a requirement with users before recording it formally in the catalogue. We should also attempt to identify duplicate requirements before documenting them. Despite the danger of wasted effort, our policy should always be one of 'if in doubt, record it'.

Our aim should be to capture a picture of requirements that is as accurate and complete as possible. The precise number of catalogue entries generated is largely a matter of individual judgement and style (and an organisation's internal standards). For example, take a requirement for three reports for ZigZag's management showing yesterday's sales by customer, by supplier and by artist. We could view this as a single requirement covering all three reports or three related requirements. Both views would be equally valid. The important thing is that the need for all three reports is documented.

▶ SUMMARY

1. Requirements Definition is the technique of identifying and documenting the requirements of the new system. It is an iterative technique, carried out throughout the life of the project.

2. The Requirements Catalogue is the product of Requirements Definition. It is largely text-based, and provides a complete picture of the requirements for the new system. Each catalogue entry is cross-referenced with other, more precise SSADM products, which collectively specify the solution.

3. Requirements are classified as either functional or non-functional. Non-functional requirements can apply to specific functional requirements, to subsets of the system or to the entire system or project.

4. While many requirements will arise as a result of problems with existing systems, the Requirements Catalogue must be focused on what the business requires of the new system.

▶ EXERCISES

4.1. During interviews with ZigZag's warehouse manager the following notes were taken:

- ◆ The whereabouts of stock is not pinpointed precisely enough and the pickers (the people who collect the items for customer order from stock in the warehouse and assemble the package ready for despatch) have to spend considerable time locating stock for despatch. It takes an average of 20 minutes to pick all items in an order and take them to despatch, when it should take around 15. The problem is that stocks are put away within large zones or aisles, on shelves in racks labelled by hand. While most people in the warehouse are experienced and know where to look for most things, new pickers have problems, and if the order is for an uncommon item even the best pickers have trouble finding the stock. With the picking of many more small orders for e-commerce customers the problem is likely to be exacerbated.

- ◆ The new system should be able to record exactly which locations each product is stocked in (and it may be held in more than one shelf or rack if the stock is large). Then when the customer order is printed for the picker it could list exactly where to find the stock by shelf and rack number. To do this we would need to record where the stocks have been put when deliveries are received from suppliers, but that would only add 10 minutes or so to each delivery receipt, and we only get 20 a day, whereas we make around 1000 despatches per day. Pickers would also need to update the system if they found the stocks were not where the system said they were. That should not happen too often – possibly only 10 times a day.

◆ Another problem for pickers is that they have to key all items despatched against the customer order, line by line. It would save another 2 minutes or so if pickers only had to confirm that they had completed the picking of the entire order. They would then only need to change individual lines where they had a problem (such as the stock being damaged) and were unable to pick the requested items.

   a. Use the notes above to identify requirements for the new ZigZag warehouse system.

   b. Making assumptions as to relative priorities, create a full Requirements Catalogue entry for each requirement you identified.

4.2. Figure 4.3 shows an extract from the ZigZag requirements summary. Classify each entry as either a functional or a non-functional requirement.

Requirements Summary			
Id	Name	Priority	Owner
104	System must support up to 100,000 live products	E	AS
105	Record proposed purchase order	E	FB
106	Confirm purchase order	E	FB
107	Record customer order	E	JK
108	Arrange despatch of customer orders	E	FB
109	Customer orders to be kept 3 months then archived	E	JK
110	Provide delivery to despatch audit trail	D	MP
111	Provide supplier performance monitoring facility	D	MP
112	Record delivery data including rejections	E	MP
113	Schedule deliveries (to nearest half hour)	E	MP
114	Overdue delivery report	D	MP
115	Re-schedule delivery	E	MP
116	Make use of existing PCs in stock office	D	HS
117	Delivery due dates to be converted to DD/MM/YYYY	E	HS
118	Facilitate alternative product selection	E	JK
119	Report on stocks nearing withdrawn from sale date	D	FB
120	Provide possible stock-out warnings	N	FB
121	Monitor supplier invoices	D	AS
122	Produce stock report	D	FB

Fig 4.3 Extract from the ZigZag requirements summary

The definition of requirements does not happen in isolation. Usually Requirements Definition is performed in parallel with Logical Data Modelling and Data Flow Modelling so that the findings from the application of each technique should inform the others. For this reason the reader will find more Requirements Definition exercises interspersed with the exercises of the next two chapters. The exercises below should be considered as starting points for identifying requirements; during Investigation the Requirements Catalogue is a 'live' document on which new requirements are recorded as and when they are identified.

4.3. Produce a list of requirements for the Fresco ticket agency of Exercise 3.8.

4.4. Produce a full Requirements Catalogue entry for each Fresco ticket agency requirement identified. (As we develop other systems analysis products for Fresco you may return to this catalogue to update it.)

4.5. The Natlib scenario of Exercise 3.13 leads to the identification of the system's main 'doing' activities. Study the Business Activity Model produced for Exercise 3.13 and try and imagine some 'planning', 'enabling', 'monitoring' and 'controlling' activities for Natlib. As each activity is identified place it on your list of requirements and produce a Requirements Catalogue entry for it.

4.6. What would be the best way to gather information for Natlib? Write a report explaining the pros and cons of using each fact-finding technique in the case of Natlib. Your report should end with a conclusion of which fact-finding technique(s) you consider most suited for Natlib and a Gantt chart of your fact-finding project plan.

5

Modelling data

INTRODUCTION

The sole purpose of an information system is to support or automate business activities by storing and processing relevant business information or data. It is therefore critical to the success of any IS development that the meaning, structure and business rules of the required data are fully analysed, understood and modelled.

The analysis of data takes two main forms. First, we can analyse the data used by existing systems. Most of the data required for the future system will be the same in content or meaning as that used currently. The actual format or structure of the data will rarely if ever remain unchanged, however. For example, the existing system may store an order date in the format DDMMYYYY, while we may ultimately decide to record the same information in a new database that uses the format YYYYMMDD. While the meaning of the data remains the same, its physical format has changed. During the Investigation phase we are concerned with understanding the underlying (i.e. logical) data requirement rather than making decisions about its physical implementation (a process we undertake during the Physical Design phase of the project, prior to building the system). This frees us to explore the business requirements for data without being constrained by the physical limitations of any particular technical or physical environment.

The second form of data analysis concerns requirements for new business data. These requirements may arise from the need to store and process entirely new data in order to support new or modified business activities and rules, or from the removal of physical limitations within the existing systems.

We may decide to concentrate on building a model of required data straight away, without any formal analysis of current system data. Indeed, many experienced analysts will take this approach, particularly on smaller systems. For less experienced analysts, or for complex systems developments, or where data requirements are expected to remain largely unchanged, the approach of starting with an analysis of existing data (or even re-using existing data models) will provide the most rigorous approach and will also prove more efficient than starting with a blank sheet of paper. This approach will also help in driving out restrictions in data support arising from existing technical constraints.

> **Tip**
>
> Where we choose the current data analysis approach, it is important to concentrate solely on modelling the existing situation first, while recording required changes in the Requirements Catalogue. We will then apply *approved* new requirements to the current data model to transform it into the required data model for the new system. Any attempt to produce a hybrid current/required data model that transforms over time into the required data model will almost always end in confusion.

In SSADM the vehicle for analysing the logical structure of an organisation's information is the Logical Data Model (LDM). A Logical Data Model is a way of graphically representing what that information is really all about, how it relates to other information and business concepts, and how business rules are applied to its use in the system. The LDM is possibly the most important and ultimately the most rigorous product of an entire SSADM project.

> The Logical Data Model is sometimes referred to as an Entity Relationship Diagram.

> ▶ **Learning objectives**
>
> After reading this chapter, readers will be able to:
>
> ◆ identify system entities and attributes;
>
> ◆ create a Logical Data Model;
>
> ◆ appreciate the central role of data modelling to systems development;
>
> ◆ understand the documentation that supports and validates the data model.

> ▶ **Links to other chapters**
>
> While Logical Data Modelling is central to all techniques, and therefore chapters, within the book, there are particularly strong links with the following:
>
> ◆ Chapter 2   Logical Data Models can be used in defining Feasibility Options, and in defining the scope of the Feasibility Study.
>
> ◆ Chapter 6   The Logical Data Model must support, and map to, the Data Flow Models defined during process modelling.
>
> ◆ Chapter 7   A Logical Data Model for the required system will be a key component of the selected Business System Option.
>
> ◆ Chapter 10   Relational Data Analysis will be used to validate and enhance the Logical Data Model.
>
> ◆ Chapters 11 and 12   The behaviour of entities will be modelled as a basis for process design.
>
> ◆ Chapter 14   The Logical Data Model is transformed into a Physical Data Model, as part of Physical Design.

▶ LOGICAL DATA MODELLING

An organisation's data will be physically stored in many different places, e.g. paper files, computer files. This data will almost inevitably contain duplications and compromises as a result of the physical restrictions of storage, processing or practicality. For example, an actual purchase order will hold information about products (*product name*, *product number*, *product price*), suppliers (*supplier name*, *supplier address*), the order's heading (*purchase order number*, *purchase order date*), as well as the quantity of each product ordered (*quantity ordered*). While we may have a single physical grouping of data on one purchase order form, what we actually have is information about several different things – products, suppliers and purchase orders. In other words, the underlying logical view is of a number of separate data groupings, each describing a different business concept or object. We will also find that information on, for example, products is physically held in many other places, such as on customer orders, invoices and despatch notes. This all leads to a confusing mess of duplication and interconnecting information, which in turn leads to problems in maintaining data consistency and integrity.

Logical Data Modelling aims to unravel this mess by getting at the underlying picture of just what it is that the system actually holds data about, and exactly how this data truly interrelates.

Logical Data Models consist of two parts: a diagram called the Logical Data Structure (LDS), and a set of associated textual descriptions that explain each part of the diagram.

▶ LDM CONCEPTS AND NOTATION

Logical Data Models use four main concepts and symbols:

▶ Entities

Any object or concept about which a system needs to hold information is known as an entity type (or entity for short).

To be a valid entity we must wish to hold information on more than one *occurrence* of it. Entity occurrences are real-world instances of an entity type. For example, the entity type Supplier will have occurrences such as:

Supplier No	3621	2327
Supplier Name	Off Beat Recordings	Bella Sonic
Supplier Address	12 High Street etc.	Lake Industrial Estate etc.

In other words an entity type is a generic description or definition of an object, of which there will be a number of real-world examples.

The symbol for an entity in an LDS is a round-cornered rectangle (Figure 5.1) containing the entity's name (which must be unique).

An entity must have a number of properties to qualify as such:

1. There must be more than one occurrence of the entity.

2. Each occurrence should be uniquely identifiable.

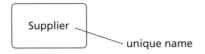

Fig 5.1 Entity

3. There must be data that we want to hold about the entity.

4. It should be of direct interest to the system.

Each item of information (or data) that we hold about an entity is known as an *attribute* or *data item*. Examples of attributes for Supplier might be Supplier Number, Supplier Name, Supplier Address and Supplier Telephone Number.

The detail of an entity's attributes is not formally included on the LDS itself. This is held in separate textual descriptions, which will be discussed later. To start with we are concerned with identifying the major entities of the system. As the analysis progresses we will find more entities until a sound data model is achieved.

▶ Relationships

Entities do not exist in isolation, but are related to other entities; in physical data structures these relationships are signified by physical links such as pointers or placement in the same file or document; in logical models relationships represent business associations or rules and *not* physical links.

Any entities that are related are linked by a line on the LDS. The line is labelled with the name of the relationship, and is named in both directions. Figure 5.2 tells us that purchase orders are *placed with* suppliers and that suppliers are *suppliers for* purchase orders. It does not tell us – yet – how many suppliers a purchase order can be placed with, or whether a purchase order can be created without a supplier being allocated. These are business rules, which we add to an LDS by annotating the line with the relationship's *degree* and *optionality*.

Fig 5.2 Relationship

Degree

The number of occurrences of each entity type participating in a given relationship is denoted by the degree or cardinality of that relationship, and illustrated on the LDS by adding 'crow's feet' to the relationship's line.

There are three types of degree, as shown in Figure 5.3:

◆ **Many-to-many (*m:n*).** This tells us that each occurrence of A is related to *one or more* occurrences of B, and each occurrence of B is related to *one or more* occurrences of A.

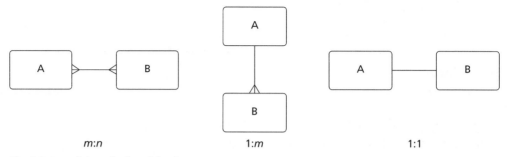

Fig 5.3 **Possible relationship degrees**

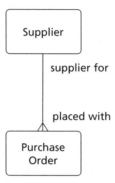

Fig 5.4 **Relationship with degree**

- **One-to-many (1:*m*).** This tells us that each occurrence of A is related to *one or more* occurrences of B, but each occurrence of B is related to *only one* occurrence of A.

- **One-to-one (1:1).** This tells us that each occurrence of A is related to *only one* occurrence of B, and each occurrence of B is related to *only one* occurrence of A.

For example, in Figure 5.4, each supplier can be the supplier for one or more purchase orders (most suppliers would supply products on a regular basis), but each purchase order can be placed with only one supplier (i.e. the relationship is 1:*m*). This reflects one of the business rules of ZigZag. If the rule were that each purchase order could be placed with more than one supplier, the relationship would be *m:n*.

As we shall see in later sections, relationships that are 1:1 or *m:n* are usually converted to 1:*m* on closer analysis; and indeed for many of the later design techniques of SSADM they must be. For now we are concerned with high-level data usage only, so all types of relationship are acceptable.

Optionality

Each relationship is further annotated to show if it must exist for *all* occurrences of the participating entity types. If there can be occurrences of one entity that are not related to at least one occurrence of the other, then the relationship is said to be *optional* for that entity. The relationship line is then converted to a dashed line at its optional end (which could mean both ends if both entities are optional participants).

Suppliers do not necessarily have to be the supplier for any currently recorded purchase orders (they may have been in the past, but records of those orders have now been deleted). So Supplier is an optional participant in the relationship.

However, every purchase order must be placed with a supplier, so its participation is mandatory.

The diagram can now be redrawn as in Figure 5.5.

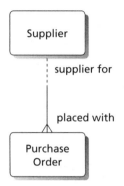

Fig 5.5 Relationship with optionality

The relationship now reads:

Each Supplier *may be* the supplier for one or more Purchase Orders.

And in the opposite direction:

Each Purchase Order *must be* placed with one and only one Supplier.

▶ DEVELOPING THE LDS

To start with, we are only interested in producing a high-level model of the current system's underlying data structure. The level of detail should be sufficient to confirm the project's scope.

Due to its largely conceptual nature Logical Data Modelling can be one of the most intense activities of an SSADM project. In many projects development of the LDM is started by holding brainstorming sessions with small groups of analysts and users. It is quite possible that several overview data models will be created, all of which may support the information needs of the organisation. Analysts may have put a lot of effort into their creation and feel quite threatened when other team members propose alternatives. This presents quite a challenge to the management of modelling sessions, and procedures should be set up beforehand to deal with disputes.

As the analysis progresses the technique becomes more rigorous to ensure that the LDM represents all of the data used by or required for the system, and that it is logically correct. By that time the data model will conform to stringent rules and there will be no ambiguity as to its structure and contents.

▶ Identifying entities

Logical Data Modelling is a more rigorous technique than the relatively informal Data Flow Modelling technique and so provides one of the best ways of gaining a thorough understanding of current or required systems support.

With a little practice analysts often find that the best method of data modelling is to draw up possible LDSs almost instinctively, either directly from system documentation or interview results, or even during interviews themselves. Relationships are added as each entity is identified and then checked with users on the spot. This approach has a lot to recommend it, particularly at this level of detail or for small systems, as diagrams are produced and verified quickly.

Where a system is complex or unfamiliar to the analyst, a more systematic approach is probably more suitable, especially for the novice. One such approach, which at least has the advantage of safeguarding against missing the obvious, is described below.

To identify entities in the current environment we can begin by looking at our physical data stores to find out exactly what it is that they hold information about. If we take the customer order file and discuss it with users, we find that it not only contains details of each individual order, but of the customers themselves, i.e. *customer address*, *customer telephone number*, etc., and so encompasses at least two entities, namely Customer and Customer Order. Continuing this for each data store gives us a list of candidate entities which includes:

Supplier	Purchase Order	Customer	Delivery (from a supplier)
Product	Stock	Customer Order	Despatch (to customers)

There are no hard and fast rules for the spotting of candidate entities within each data store. Knowledge of the business area is the most useful aid to understanding the information usage and needs of a system.

However, it does not really matter if the list is not correct first time – indeed it would be very surprising if it were. The whole process of Logical Data Modelling is, like so many other SSADM techniques, an iterative one, and there will be ample opportunity along the way to change and adjust the model.

Once the list has been drawn up we should verify it with key users during preliminary scoping interviews.

The key questions to ask of each entity are:

◆ Are any of the candidates merely attributes of another entity?

◆ Do any of the candidates represent a subset of occurrences of another entity?

◆ Do all of the entities have a unique identifier?

During this process we may discover new entities, merge existing entities or discard candidates as being outside the area of investigation.

▶ Identifying relationships

We now examine each entity to see if it is directly related, *in a way that is of interest to the system*, to any of the other entities. The best way to do this is in discussion with users, either taking each entity in turn, or starting with a key entity and moving around the LDS 'network' as the relationships are identified.

There will often be relationships between entities that exist in the real world, but which are not of relevance to the system under discussion; e.g. a customer of ZigZag may well be employed by one of its suppliers. This is NOT something that ZigZag will be interested in recording!

Entity Matrix

One way to check that all possible pairings of entities are considered is to use an Entity Matrix as in Figure 5.6. Remember that we are looking for relationships or associations between entities as they currently exist, and not as we, or users, would wish them to. If we come across a relationship that is not supported in the way that we feel is best, that is just too bad. It must be documented on the LDS as it is, but we will make a note in the Requirements Catalogue of the 'better' relationship for use in the new system. An example of this will come a little later (between Purchase Order and Delivery).

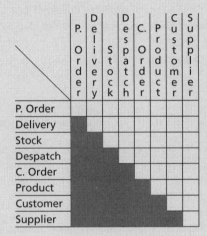

Fig 5.6 Entity Matrix

By working through all of the cells in the matrix, putting a cross where a direct relationship exists between entities and leaving the cell blank where it does not, every possible pairing will be looked at. Each pairing need only be checked once, so the lower left-hand part of the matrix can be blocked out.

Obviously this will not give us the full description of a relationship (as it provides no names, optionality or degree), but it will tell us where a potential relationship exists.

For the ZigZag system, as an example, we will look at the top line in detail (the Purchase Order line):

Purchase Order–Purchase Order

There is no direct relationship between one Purchase Order occurrence and any other, so the cell is left blank.

Purchase Order–Delivery

Currently ZigZag has a rule, based on the limitations of their systems, that each Purchase Order must be delivered as a whole. If there are any items that are unavailable for delivery they will be cancelled (and possibly reordered). Each Delivery is checked and recorded against a single Purchase Order, so there is a clear direct relationship between the two entities.

Purchase Order–Stock

A Purchase Order will usually result in new stock being delivered. However no *direct* links are recorded by ZigZag between specific stocks of a product and the order responsible for their delivery, so no direct relationship exists between Purchase Order and Stock.

(As each purchase order is directly related to a delivery, and each delivery will in turn be related to the specific stocks it delivered, there is in fact an indirect relationship between Purchase Order and Stock via Delivery.)

Purchase Order–Despatch

A Purchase Order results in a Delivery, Customer Orders are put together in Despatches: so Despatch is related to Customer Order, not Purchase Order, and so no relationship exists.

Purchase Order–Customer Order

No direct relationship.

Purchase Order–Product

Each Purchase Order requests the delivery of quantities of products, so there is clearly a relationship. (A little later we will discover that this is really an indirect relationship, but for the time being, *given the eight entities we have identified so far*, a relationship between Purchase Order and Product should be recognised.)

Purchase Order–Customer

No direct relationship.

Purchase Order–Supplier

Each Purchase Order is placed with a single specified Supplier, so there is a direct relationship between the entities.

Putting the results of these decisions on the top line of the Entity Matrix gives us Figure 5.7.

	P. Order	Delivery	Stock	Despatch	C. Order	Product	Customer	Supplier
P. Order		X				X		X

Fig 5.7 Placing relationships on the Entity Matrix

The rest of the grid is completed in a similar fashion (see Figure 5.8), by following the same procedure of examining each pairing.

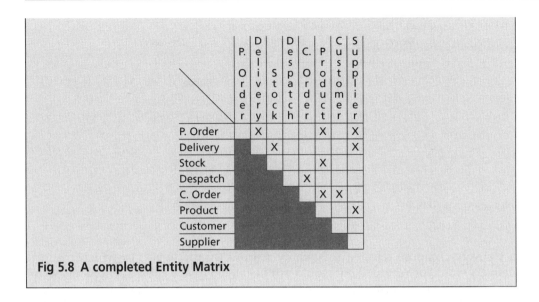

Fig 5.8 A completed Entity Matrix

▶ Describing relationships

Having identified where we think relationships exist, we now consider their degree, optionality and names. We do this by identifying the business rules that apply to each entity pairing. The basic process is the same for all pairings, so we will look at just one example.

Stock–Delivery

We first consider the relationship from the Stock perspective: each Stock occurrence will consist of a quantity of a single product, all of which was delivered on the same delivery. If within the depot we have a quantity of a given product, some of which was delivered in one delivery and some in another, then we will have more than one Stock. This is an example of one of ZigZag's business rules, and one that will continue in the new system. For example, say we have in stock 100 copies of a recording of Puccini's *Tosca*. It is important to ZigZag to know if they belong to the same batch (hence delivery) or not. If 30 of the above CDs were received last week and 70 this week they should be placed in different 'stocks' so that the older ones are despatched first and, more importantly, if defects are found then the batch concerned can be isolated and returned to the supplier. So we would have one stock of 30 copies of *Tosca* delivered say on 10 December 2001 and another stock of 70 delivered on 17 December. Thus each Stock occurrence is related to just one Delivery.

If we now look at the relationship from the Delivery perspective we note the following: each delivery may contain a number of different products, each of which will be stored as a separate stock (remember that each Stock occurrence is a quantity of a single product). Thus each Delivery is related to one or more Stock occurrences (see Figure 5.9).

We now consider the optionality of the relationship: each Stock must have been delivered by a Delivery. So the relationship at the Stock end is mandatory. However,

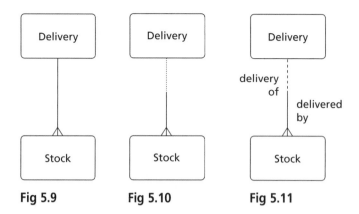

Fig 5.9   **Fig 5.10**   **Fig 5.11**

a Delivery could be rejected for quality reasons by the depot, in which case the delivery would be recorded but would not be related to any subsequent Stock occurrences. So the relationship is optional at the Delivery end (Figure 5.10).

Choosing a name is often the hardest part of the procedure. It is important to name a relationship in both directions as it forces us to examine the true nature of the relationship, sometimes leading to the discovery of additional relationships or even entities. We should always try to choose phrases that accurately reflect the users' view of the relationship. In our example it is not too difficult to find reasonable names: 'delivery of' and 'delivered by' (see Figure 5.11).

The relationship now reads:

Each Stock must be delivered by one Delivery.

And from the other direction:

Each Delivery may be a delivery of one or more Stocks.

Continuing this process for all of the relationships identified on the matrix gives us a first-cut overview LDS for the current system (see Figure 5.12).

Presentation tip

There are a few guidelines on LDS presentation that are worth mentioning. It makes the diagram easier to follow if we avoid crossing lines as much as possible. For 1:*m* relationships we should also aim to keep the entity at the '1' end above the entity at the '*m*' end. Clearly there will be occasions when these two aims clash, so do not spend too much time on this.

The overview LDS provides us with a good basis for building a more complete model of current data. We begin the process of creating a detailed model by looking at this model and discussing it with users to check our understanding of the scope of current data and to see if there are any lower-level entities which can be added immediately.

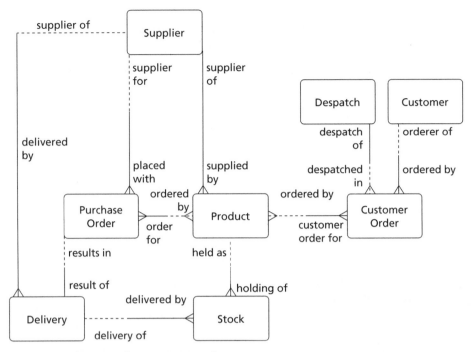

Fig 5.12 Overview LDS (current system)

In the case of ZigZag this results in the addition of Product Type, Depot Zone and Supplier Invoice. We 'discover' these four entities by various means, as described below.

The first two, Product Type and Depot Zone, are discovered through the following train of thought: ZigZag holds different types of product such as CDs, cassettes, videos and DVDs. These product types are stored in different distinct areas or zones to help staff in organising the depot. Each one of these zones has shelving capable of storing particular sized products. Each zone can hold many product types, e.g. a zone that can hold CDs will also be able to hold DVDs as they have similar dimensions; and each Product Type may be assigned to more than one zone, if they are of a type that can fit into shelving in more than one zone.

Figure 5.13 shows these two newly 'discovered' entities and their relationship with Product.

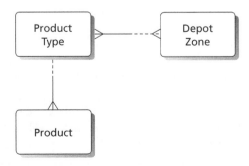

Fig 5.13 The representation of Depot Zone and Product Type in the ZigZag LDS

Fig 5.14 The representation of Supplier Invoice and Delivery in the ZigZag LDS

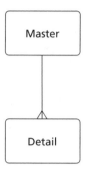

Fig 5.15 Masters and details

This identification of the Supplier Invoice entity results from discussions with the Purchase Order Clerks who keep files of matched (against a delivery) and unmatched supplier invoices. Since ZigZag only accepts invoices of goods that have already been delivered and since it insists on one invoice per delivery, the Supplier Invoice entity has a 1:1 relationship with the Delivery entity (Figure 5.14). We will be returning to these two entities later on.

Before we go any further we must now introduce a few new concepts.

▶ Master and detail entities

As mentioned earlier, most relationships are 1:*m*. The entity at the '1' end is known as the *master* and the entity at the '*m*' end as the *detail* (Figure 5.15).

The terms 'master' and 'detail' refer only to an entity's role in a particular relationship; it is quite possible for an entity to be the master in one relationship and the detail in another, e.g. Product in the ZigZag current system LDS (Figure 5.16).

Keys

Logical Data Models are based on the principles of the relational data model. For a definitive discussion of relational databases see Date (2000). Although we do not usually employ formal relational data analysis techniques until later in the project, it will be helpful to look at a few relational concepts informally now.

Fig 5.16 Product is a master of Stock but a detail of Supplier

At this stage in our analysis we should be able to select at least one identifier for each entity type, i.e. an attribute that enables each occurrence of an entity to be uniquely identified, e.g. for Customer we could use *customer number*.

Any attribute or set of attributes which together uniquely identify an entity is known as a *candidate key*. One of these candidates (there will often only be one) should be selected as the *primary key*. Whenever we require direct access to an entity, the primary key is used to identify which occurrence we are interested in. For example, if we needed to access the Supplier entity to find out a supplier's address, we would use the primary key of *supplier number* to identify the correct occurrence.

If we have a relationship between two entities we need to be able to associate the occurrences at one end with the related occurrences at the other. In a relational model (such as the LDM) we do this by including the primary key of the master in the set of attributes of the detail. The copy of the master's primary key in the detail entity is known as a *foreign key*. To illustrate this we will examine the relationship between Purchase Order and Supplier (Figure 5.17).

To access all purchase orders placed with supplier number 271, we look for all occurrences of Purchase Order with a supplier number attribute value of 271. Coming in the opposite direction, to access the supplier for purchase order 5001, we look for the single occurrence of the Supplier entity whose primary key is equal to the supplier number given in the foreign key of purchase order number 5001, i.e. supplier number 271.

Primary keys belong to one of three types:

1. a **simple key**, consisting of a single attribute;

2. a **compound key**, consisting of two or more foreign keys;

3. a **hierarchic or composite key**, consisting of one or more foreign keys and a qualifying non-foreign key attribute.

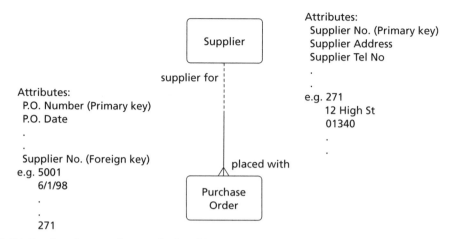

Fig 5.17 Foreign keys enforce relationships

The entities Purchase Order and Supplier, as shown in Figure 5.17, both contain simple keys.

> We will sometimes use the following convention, where the primary key is underlined and the foreign key preceded by an asterisk to show the contents of each entity:
>
> Supplier (<u>supplier number</u>, supplier address, supplier tel. no.)
> Purchase Order (<u>P.O. number</u>, P.O. date, *supplier number)

We will find examples of compound and composite keys when we start resolving many-to-many relationships.

▶ Resolving many-to-many relationships

In our overview LDS for the current system we have several *m:n* relationships. These are fine for the purposes of presenting a high-level summary of data usage, but must be resolved during more detailed analysis by replacing them with, at least, two 1:*m* relationships. The main reasons for this are:

◆ Many design techniques can only be carried out on hierarchical (i.e. master–detail) relationships which are hidden by *m:n* relationships.

◆ *m:n* relationships make navigation around the model very difficult or even impossible (and, although we are not really concerned with technical issues at this point, *m:n* relationships cannot be implemented in a relational database).

◆ *m:n* relationships very often hide information about the participating entities or the relationships themselves.

To illustrate how to resolve them we will look at Product and Purchase Order, which are related as follows:

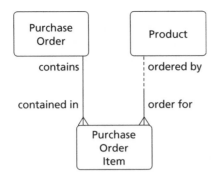

Fig 5.18 Purchase Order Item links Purchase Order to Product

Each Product may be ordered by one or more Purchase Orders.
Each Purchase Order must be an order for one or more Products.

This relationship as it stands is confusing. Product and Purchase Order are related because each purchase order will typically contain a number of products that ZigZag is ordering from a particular supplier. Furthermore, the quantity being ordered is also recorded; otherwise the order would be meaningless.

If we now consider where to place the attribute that shows the amount of product ordered we will find that it does not fit naturally in either of these two entities.

But, if we look more closely at a sample purchase order of ZigZag (see Appendix A), we will discover that details of quantities and products are held in individual purchase order lines. So in this case we can choose a more natural link entity, which we will call Purchase Order Item (rather than Purchase Order Line which sounds a bit too similar to the physical printed line on the order form), as in Figure 5.18.

The key for Purchase Order Item will be *Purchase Order Number* plus *Product Number* – a compound key. We can then establish the quantity of each product being ordered on a particular Purchase Order by looking at all Purchase Order Item occurrences with the appropriate Purchase Order Number as part of its key.

A similar argument can be applied to the relationship between Customer Order and Product, giving us the link entity of Customer Order Item, which will record the quantities of each individual product within the customer order.

Whenever we introduce a link entity we need to ensure that the relationships we recorded previously with its master entities are still valid. For example, in our overview LDS we recorded a many-to-many relationship between Despatch and Customer Order. This may at a high level appear reasonable as customer orders are bunched together in van-loads (represented by a Despatch occurrence which records the date and time of despatch), and it is common for some items in an order to go into one van-load and some into another. However, the contents of each purchase order item within the order is always despatched in its entirety in the same van-load (e.g. if 3 copies of Puccini's *Tosca* are ordered within a single customer order, they will all be delivered together). Therefore, each Despatch is actually related to many Customer Order Items, rather than to whole Customer Orders.

Table 5.1 Table of Depot Zone and Product Type pairings.

Depot Zone	Product Type
101	DVD
101	CD
105	DVD
105	CD
102	VHS
102	BV
102	SPB

Depot Zone and Product Type provide another more complex illustration of many-to-many relationships:

Each Depot Zone may store one or more Product Types.
Each Product Type must be storable in one or more Depot Zones.

The attributes that make up Depot Zone are *Depot Zone Number, Shelf Height, and Depot Zone Description etc. Depot Zone Number* is a unique identifier that is assigned to each Depot Zone, and is the label attached to the end of each row of shelving in the zone. *Depot Zone Description* would include values such as CDs and DVDs, Videos and Books, and Tapes etc., which describe the sorts of products that the shelving in each zone can accommodate.

The attributes of Product Type include *Product Type Code* and *Product Type Name*, where the *Product Type Code* is an abbreviation of the *Product Type Name*, e.g. 'BV' for 'Blank video', 'DVD' for 'DVD' etc.

So, for example, we might have the following cases:

Depot Zones 101 and 105 store 'DVD' and 'CD' product types;
Depot Zone 102 stores 'VHS', 'BV' and 'SPB' (small paperback book) product types.

To make these associations we would have to set up lists of foreign keys in both entities, of arbitrary length. As well as being against the rules of Relational Data Modelling, this would cause a significant maintenance overhead, make navigation around the model very difficult and in any case would not be acceptable for many of the design techniques which follow.

So what we do is to create a *link entity*, each occurrence of which will store a valid association or pairing of a Depot Zone occurrence with a Product Type occurrence, such as shown in Table 5.1. We will call this new entity Depot Zone Allocation as in Figure 5.19. The primary key of Depot Zone Allocation is a combination of the primary keys of Depot Zone and Product Type.

In effect the link entity acts as a list of associated entity occurrences. So if we want to know which Depot Zones can hold CDs we just read through the list picking out all those that have 'CD' as part of their primary key.

This whole process is known as *resolving many-to-many relationships*.

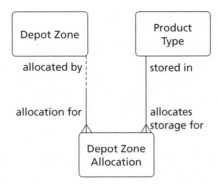

Fig 5.19 Depot Zone Allocation links Product Type to Depot Zone

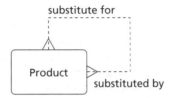

Fig 5.20 Pig's ear

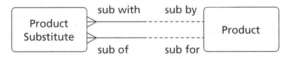

Fig 5.21 Pig's ear resolution

Note: Two or more different relationships between the same entities are also known as *parallel relationships*.

▶ Pig's ears

When ordering a product (e.g. *Tosca* recorded by the London Symphony Orchestra) it is possible that a supplier will not have sufficient stocks to satisfy the order. In this case ZigZag Purchase Order Clerks need to identify alternative products (e.g. *Tosca* by the Berlin Symphony Orchestra) that can be substituted for the unavailable one in the order. This means that each product could be related to a number of alternative substitute products, and could itself act as a substitute for a number of other products. This is an example of a recursive relationship, where an entity is related to itself. It is also sometimes called a 'pig's ear', due to its appearance (see Figure 5.20).

However, this is an *m:n* relationship which must now be resolved. The method for doing so is precisely the same as we would use if the entities at each end were different, i.e. by using a link entity (Figure 5.21).

The key for the Product Substitute entity will be a compound key consisting of appropriate pairs of Product keys.

▶ Resolving one-to-one relationships

As with *m:n* relationships, 1:1 relationships are useful in overview LDSs to present a high-level picture of data usage. However, these too must now be resolved, either by merging the entities involved or by replacing the relationship with a 1:*m* (of which a 1:1 relationship could be viewed as a special case).

The problems associated with 1:1 relationships are less clear-cut than those with *m:n* relationships:

◆ 1:1 relationships often obscure an underlying single entity.

◆ There may be a missing link entity.

◆ Later design techniques may require all relationships to be master–detail.

In the ZigZag overview LDS there are two 1:1 relationships; between Delivery and Purchase Order (Figure 5.12), and between Supplier Invoice and Delivery (Figure 5.14).

Discussions with users reveal that deliveries are identified by the purchase order they are satisfying, and that the only information currently held about them details which parts of the purchase order they have successfully delivered. It is quite easy in this case to view Delivery as a logical extension (or conclusion) of a Purchase Order, so we will merge the two entities and transfer all of Delivery's relationships to Purchase Order. To do this successfully, Purchase Order will contain attributes *delivery date* and *supplier's delivery reference* while Purchase Order Item will contain *quantity delivered*.

Closer inspection of the second 1:1 relationship of the overview LDS reveals that invoices actually contain invoiced items, each of which details the payment required for the contents of a single Purchase Order Item. This means that we actually have two 1:1 relationships involving purchase orders and invoices (Figure 5.22).

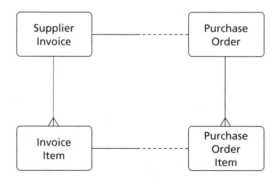

Fig 5.22 Possible 1:1 relationships from the ZigZag LDS

The information carried on Invoice Items and Purchase Order Items describe very similar things, namely quantities of products: in the case of Purchase Order Items, of products ordered and delivered; in the case of Invoice Items, of products invoiced. So the two entities can be merged, with the *invoiced quantity* attribute made optional within Purchase Order Item, as it will have no value until after a delivery takes place (see Figure 5.23).

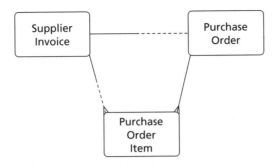

Fig 5.23 Purchase Order Item 'absorbs' Invoice Item

It is sometimes impractical or illogical to merge entities, either because they describe very different (if related) objects, or because users are unhappy that important information is being obscured, as in the case of trying to merge Supplier Invoice with Purchase Order. In these situations the 1:1 relationship cannot be replaced with a 1:*m*. Instead, in order to provide information needed for later design techniques, we need to declare a master from which we can determine the detail. This master will be the entity that is usually created first. In our Supplier Invoice/Purchase Order example, Purchase Order will *always* be created first, and so is declared the master. The Supplier Invoice will then contain a foreign key of Purchase Order Number. We can denote the master–detail relationship by retaining the 1:1 relationship on the LDM and placing a comment in the entity description, or by showing a 1:*m* relationship on the LDM (again with an appropriate comment in the entity description) as in Figure 5.24.

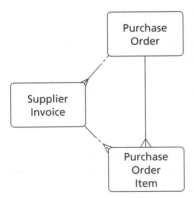

Fig 5.24 Supplier Invoice becomes a detail of Purchase Order

▶ Removing redundant relationships

In some ways an LDS is a little like a route map: by following relationships around the LDS we can navigate between any pair of entities, and usually by a variety of routes. The shortest route is clearly via a direct relationship, but it is usually possible to travel indirectly via other entities.

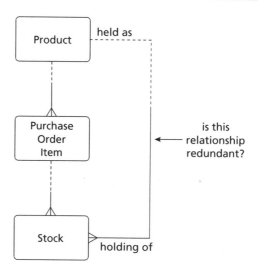

Fig 5.25 Possible redundancy

One of our aims when drawing up an LDS should be to include only the minimum number of relationships needed to apply all of the business rules relating to data. Any unnecessary relationships are termed 'redundant', and will involve us in a maintenance overhead if implemented.

If we find an indirect relationship that enables us to navigate between exactly the same occurrences (and no others) of the two entities linked by a direct relationship, then that direct relationship is redundant. The fact that the route is longer is not important; the only thing that matters is that the business rule is preserved in our model.

For example, if we take the entities Product, Purchase Order Item and Stock, we may wish to consider whether the relationship between Product and Stock shown in Figure 5.25, is redundant.

Clearly this relationship is also represented by the indirect relationship via Purchase Order Item, i.e. it can be deduced by linking Stock and Product via Purchase Order Item:

◆ Each Stock is present because of a Purchase Order Item, which in turn must be for a specific Product. Hence we can establish which product constitutes any given stock.

◆ Each product may be mentioned as part of one or more purchase order items, each of which gives rise to one or more occurrences of Stock. Hence we can identify any stocks of a given product.

The direct relationship in Figure 5.25 would only duplicate this information and so is redundant. (The same cannot be said for, for example, the direct relationship between Purchase Order and Purchase Order Item in Figure 5.24. Removal of this relationship will lose vital business information.)

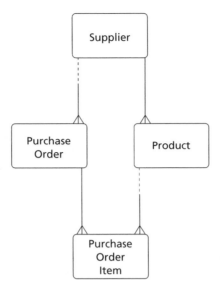

Fig 5.26 Are any of the above relationships redundant?

The major difference between relationships and a route map is that each relationship carries with it a meaning, and so different 'routes' between entities will often have different meanings, thereby connecting different occurrences. When removing a redundant relationship we must always check that the entity occurrences it links are the same as those linked by the indirect relationship, i.e. that the deduced relationship is the same as the 'redundant' one, and that we are not losing important information.

Consider the subset of the ZigZag LDS in Figure 5.26. Is the relationship between Supplier and Purchase Order redundant?

◆ For each supplier we can access all the products they supply, and from here we can identify all the purchase order items that contain these products. Hence we can find all of the purchase orders for a given supplier.

◆ For each purchase order we can find all of the products on that order, and for each product we can identify the supplier.

So at first sight the Supplier/Purchase Order relationship may appear redundant. However, if we were to remove the relationship we would lose the information that each Purchase Order can and must be placed with only *one* supplier. Using the indirect relationship it would be possible for a purchase order to contain many items, each item for a product of a different supplier. This would contravene an important business rule. Therefore, the relationship is NOT redundant.

Applying all of the above refinements to the Overview LDS of Figure 5.12 leads to the Current System LDS in Figure 5.27 (where relationship descriptions have been removed just to make it easier to read).

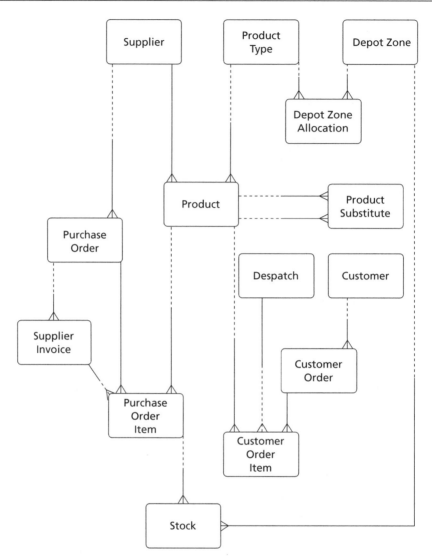

Fig 5.27 Current Environment LDS

Further modelling concepts

Mutually exclusive relationships

It is possible that an entity has two relationships that cannot both exist for the same occurrence. In other words an entity occurrence will either have one relationship to another entity or the other, but not both. For example, imagine that ZigZag allowed customers to collect a customer order item *instead* of having it despatched to them (this is not in fact a requirement within the case study). An entity would then exist called Collection. The relationships between Customer Order Item, Despatch and Collection would then read:

Each Customer Order Item may be collected by one Collection OR despatched in one Despatch.

Relationships of this type are known as *mutually exclusive*. The notation used to indicate that two or more relationships are mutually exclusive is an arc placed across each of the affected relationships, as in Figure 5.28.

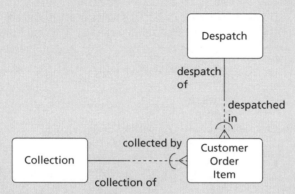

Fig 5.28 Mutual exclusivity

If one or other of the relationships *must* exist, but not both, we make the relationship lines mandatory (i.e. solid). If one or other of the relationships *may* exist, but not both, we make the relationship lines optional (i.e. dashed). If we have a number of groups of mutually exclusive relationships on an LDS it can get difficult to identify which relationships are covered by the same set of arcs. In this situation we would label each member of a set with an identifier (e.g. a lower-case letter). Each group (usually but not always a pair) of mutually exclusive relationships must apply to the same entity (as in Customer Order Item above), and can apply to both master and detail entities.

Entity sub-types

Entities may be found that have occurrences that split into distinct sub-types that share the majority of attributes (including key attributes) and relationships, but have some attributes and relationships that are distinct to those sub-types. For example, if we extended the scope of the ZigZag case study to include purchasing of consumables (items used by but not stocked or sold by Zigzag, e.g. stationery, cleaning products) the purchase order items generated would be of two types. First, we would have Stock Purchase Order Items, and, secondly, Internal Purchase Order Items. We would represent this idea on the LDS by extending the entity notation to include sub-types as demonstrated in Figure 5.29.

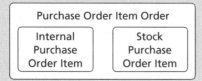

Fig 5.29 Entity sub-types

In Figure 5.29, 'Purchase Order Item' is called the super-type while 'External Purchase Order' and 'Internal Purchase Order' are sub-types.

Sub-types share the key of the super-type as well as any common attributes. In other words, sub-types do not have a separate key that distinguishes them from each other.

In ZigZag, a Stock Purchase Order Item may lead to the delivery of Stock, but an Internal Purchase Order Item will never lead to a stock record (since the Stock entity is there to show the stock that is available for sale, not the stock available for internal use). In data modelling terms, this means that a Stock Purchase Order Item alone contains the key of Stock as a foreign key, and so has a relationship with Stock. Both sub-types (and hence the super-type) have relationships with Product and Purchase Order. This is illustrated graphically in Figure 5.30.

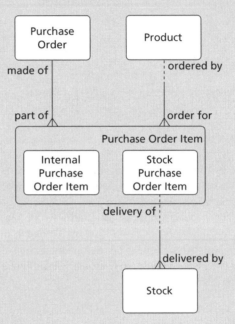

Fig 5.30 Entity sub-types

Note that mutually exclusive relationships can also be modelled as mutually exclusive sub-types. For example, we could model Customer Order Item as two sub-types, *Collected Customer Order Item* and *Despatched Customer Order Item*.

Entity aspects

Sometimes an entity needs to be modelled in more than one sub-system. When such an entity is identified, the term 'aspect' is used for the entity in each sub-system. For example, the ZigZag sub-system we are dealing with contains the entity Supplier Invoice. This entity is recorded in our system because it is the people working in the warehouse who receive an invoice for reconciliation. We only record the invoice in our system and never do anything about it. As far as we are concerned, it is the Accounts department that will retrieve the invoice we have reconciled and arrange for its payment. This means that for us in the ware-house a Supplier Invoice is something to reconcile while for Accounts it is something to arrange payment for. If we were to unify the two systems we should be aware of the different behaviours of the entity in each relevant sub-system.

Aspects in different sub-systems need to be co-ordinated. For example, Accounts should not pay for an invoice if we in the warehouse have not reconciled it against delivered and accepted goods.

Aspects are denoted by a dash after the entity name, as shown in Figure 5.31.

Supplier Invoice –Stock Control	Supplier Invoice –Accounts

Fig 5.31 Entity aspects

It is possible to show different aspects of an entity as linked to one 'basic' aspect via one-to-one relationships.

Strictly speaking, every entity is the aspect the system we are building has of the real-life entity being modelled.

▶ LDM documentation

Each entity on the LDS is defined through its contents and its relationships. It is therefore vital that the contents of each entity be explicitly stated. If a CASE tool is used the contents of each entity is stored in a database linked to the LDS. In the absence of a decent CASE tool an Entity Description form such as that depicted in Figure 5.32 (and generated as a report by CASE tools) should be raised.

The kinds of thing we might include in Entity Descriptions are:

◆ A textual description of the entity if needed.

◆ Attribute names. In the early stages we will note only the most significant attributes. As the analysis progresses we will document all known attributes.

◆ Primary and foreign keys.

◆ Relationship details (optional as these details are already shown on the LDM).

◆ Volumes and growth rate. Approximate values for the average, maximum and minimum number of occurrences are critically important for use in the selection and physical design of the database later in the project.

◆ User access. Details of which users have access to the entity, and of what type (i.e. create, read, etc.).

◆ Archiving Instructions. Details of when occurrences should be archived or deleted.

Many of these details will already exist as non-functional requirements in the Requirements Catalogue.

The contents of an entity description are normally the subject of installation standards as prescribed by the prevailing Policies and Procedures. In some organisations only brief descriptions will be required, while in others they may run to several pages for each entity and include descriptions of relationships and details of each individual attribute.

Entity Description				
Entity Name Purchase Order				
Description A request for purchase and delivery of goods from a single supplier.				
Attribute		**Primary Key**	**Foreign Key**	**Mandatory/ Optional**
Purchase Order Number		Yes		M
Purchase Order Date				M
Supplier Number			Yes	M
Purchase Order Status				M
Delivery Date				O
Delivery Start Time				O
Delivery End Time				O
must/may be	**either/or**	**Link Phrase**	**one & only one/ one or more**	**Entity Name**
must be		placed with	one & only one	Supplier
may		result in	one or more	Supplier Invoice
must		contain	one or more	P.O. Item
Entity Volumes: Max. 15000 Min. 6000 Average 10000				
User			**Access**	
P.O. Clerk			Read, Create, Delete, Modify	
Despatch Scheduler			Read	
Purchaser			Read, Create	
Growth Rate: 15% per year				
Archiving Purchase Orders should be archived to tape six months after the last related line has been delivered or cancelled.				

Fig 5.32 Entity Description

Alternatively, attributes can be defined in detail in a Data Catalogue made up of Attribute Descriptions. Each Attribute Description will usually contain:

◆ A description of the attribute.

◆ Cross-references. Each attribute should be cross-referenced with other forms that refer to it, and in particular with a list of the entities that contain it.

◆ Grouped Domain. A domain is a set of values that may validly be taken by an attribute, e.g. Product Quantity has the domain Integer, Delivery Time has the domain Time. If a domain is shared by a number of attributes it is called a *Grouped Domain*. Purchase Order Status has a domain of allowable values such as provisional, placed, confirmed, delivered, etc.

◆ Length: the maximum allowable number of alphanumeric characters for each value of the attribute.

◆ Derivation. An attribute may result directly from input to the system or be derived from input or from other attributes.

◆ Validation. If appropriate we can add details of validation checks for the attribute.

◆ User access. The users who will be allowed access to values of this attribute.

The data model is complete when the contents of each entity are clearly understood.

Some of the main attributes of a *selection* of entities from ZigZag's current stock control system are shown in the box. In this list, primary keys are underlined and foreign keys are preceded by a star '*'.

Selected entities from ZigZag's current stock control system

PRODUCT
Product Number
*Product Type Code
Product Name
Product Description
Release Date
Sell-by Date
(special promotional products)
Sell-from Date
Standard Purchase Price
Standard Selling Price

PURCHASE ORDER
Purchase Order Number
*Supplier Number
Supplier's Delivery Reference
Purchase Order Date
Purchase Order Status
Delivery Date
Delivery Start Time
Delivery End Time

SUPPLIER
Supplier Number
Supplier Name
Supplier Address
Supplier Tel. No.
Supplier Contact Name

SUPPLIER INVOICE
*Purchase Order Number
Supplier's Invoice Number
Invoice Date

PRODUCT SUBSTITUTE
*Product Number [substitute]
*Product Number [substituted]

PRODUCT TYPE
Product Type Code
Product Type Name
Product Type Description

PURCHASE ORDER ITEM
*Purchase Order Number
*Product Number
*Invoice Number
Quantity Required
Quantity Confirmed
Quantity Delivered
Quantity Accepted
Invoiced Quantity
Agreed Unit Price
Required-By Date
Required-By Time-Period

STOCK
Stock Id
*Purchase Order Number
*Product Number
*Zone Code
Quantity Stocked
Quantity Stocked
Quantity Reserved

DEPOT ZONE
Depot Zone Number
Depot Zone Description

DEPOT ZONE ALLOCATION
*Depot Zone Number
*Product Type Code

Small project documentation

Where projects are small in size, or where data requirements are relatively straight-forward and well understood (as is typical with student projects), a simplified form of documentation often helps with communication and clarity.

In addition to the LDM and an attribute listing similar to that in the box above, two documents can be used to summarise entity and attribute descriptions, as in Figure 5.33.

▶ Validating the LDM

Once we are fairly confident that our LDM is complete (that Entity Descriptions are sufficiently detailed and include the main attributes of each entity) it should be informally validated to ensure that it supports key processes or enquiries with the current systems (if the LDM is a current environment model), or key requirements recorded in the Requirements Catalogue (for required system LDMs).

Entity Name	Short Description or Comments (optional)	Min Volume	Max Volume	Ave Volume	Growth Rate
Supplier	Numbers rise at Xmas	400	750	500	5%
Product		10000	100000	25000	25%
Depot Zone	Storage Area or Aisle within Depot	26	40	32	20%
Purchase Order	Numbers rise at Xmas	1500	16000	10000	15%
Purchase Order Item		90000	420000	200000	25%

Attribute Name	Short Description or Comments (optional)	Domain	Length
Purchase Order Number	Automatically generated by system	Integer	9
Purchase Order Date	Date order placed	DDMMYYYY	8
Purchase Order Status		P(Provisional) V(Placed) C(Confirmed) D(Delivered)	1

Fig 5.33 Entity Description table and Data Catalogue table for small projects

▶ Access paths

The Data Flow Model will be developed in parallel with the LDM, as described in Chapter 6. As elementary processes from the DFM are developed, they should be checked informally against the LDM, in a similar way to the enquiry described here.

Finally, we need to check that the LDM can provide access to all of the data items required by each update or enquiry process. Most processes will need to access a number of data items, which will be specified by some selection criteria. These items will often be represented by the attributes of more than one entity. We must now ensure that we can navigate around the relationships of the LDS, applying the selection criteria to filter out the entity occurrences we need to provide all of the necessary data. These navigations are called 'Access Paths'.

For example, when allocating a zone in which to store the stock of a particular product received in a delivery (a process called 'Allocate Stock Zone'), we will need to find out which depot zones have been designated for the storage of that type of product.

The entry point to the LDS is via the *product number* in the entity Product. We can then access its product type, and then the possible zones in which this product can be stored by reading through all the occurrences of Depot Zone Allocation for that product type. We can show this informally by annotating the LDS as in Figure 5.34.

▶ SUMMARY

1. Any object or concept about which a system needs to hold information is known as an entity type (or entity for short). To be a valid entity we must wish to hold information on more than one *occurrence* of it. Entity occurrences are real-world instances of an entity type.

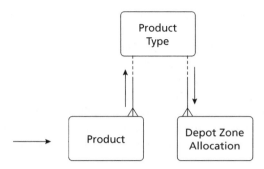

Fig 5.34 Informal access path

2. Each item of information (or data) that we hold about an entity is known as an *attribute* or *data item*.

3. Entities do not exist in isolation, but are related to other entities. Relationships are recorded in the Logical Data Model if they represent associations between entities that are of interest to the system.

4. The Logical Data Structure is a diagrammatic representation of entities of the system under study and their relationships.

5. The Logical Data Model consists of the Logical Data Structure together with textual descriptions of each entity and attribute.

▶ EXERCISES

5.1. Produce a Logical Data Model for the following scenario.

Markalot is a new university. You have been employed by Markalot as a systems analyst to design a system for the efficient recording of student assessments. The main objective of the system is to be able to identify the lecturer who has allocated the highest marks during an academic year. Typically, a Markalot student enlists for taught modules. At the start of an academic year each taught module is allocated to a lecturer who becomes responsible for teaching and assessing the module. For each module 100 multiple-choice questions are set. Students are asked to sit one exam per module. Each exam for a module lasts one hour and consists of 25 of these questions randomly chosen by the system. Students are expected to input their answers directly on the system, which should be able to output an 'instant' result as soon as the exam is finished. After each exam, the 5 questions which students find the most demanding are marked as 'not to be used again' and replaced by 5 fresh questions. When students score over 50% they pass the module. If they score between 30% and 49% they are allowed a reassessment next time the exam is run. To enter this reassessed exam they have to pay an examination fee. The date they make the payment is recorded. When students score less than 30% in an exam they are forced to retake, and pay for, the whole module next time it is offered. At the end of each exam students get a transcript of their attempt which shows their answers as well as the correct answers to the questions.

5.2. Produce an overview LDS for the following scenario.

Treebanks is an exclusive racquet sports club. Members join the club to play particular sports, e.g. tennis, squash, badminton. They may also (in addition) become members of teams, each of which is dedicated to a single sport.

A file is kept on each member detailing which sports they have signed up for, and which teams they belong to.

Each playing court at Treebanks is designed for a specific sport, and playing sessions can be booked by either a team or by an individual member.

Bookings are recorded on a booking sheet, which is divided into one-hour sessions. A booking may also be made for Treebanks equipment to be used on court, e.g. racquets, balls. Each piece of equipment is labelled with a unique number and some, such as umpiring chairs, are tied to particular courts. An index card file is used to keep track of equipment, and when a piece of equipment is booked its number is added to the booking sheet and its index card updated.

Each session belongs to a specific price band (at the moment there are six of them), according to its time, day of the week, etc. These bands are regularly changed.

5.3. Suggest keys and other attributes for each of your entities from Exercise 5.2.

5.4. Produce an overview LDS for the Natlib system of Exercise 3.13 and suggest keys and other attributes for each of your entities.

5.5. Produce a Logical Data Model for the Hergest scenario of Exercise 3.14.

5.6. Suppose that the Hergest educational institute of Exercise 3.14 wishes to tighten its control over its distribution of its Certificates of Attendance by actually recording the students who are present in each session. How would the data model you created in Exercise 5.5 be affected by this new requirement given that all the institute's courses consist of exactly 10 two-hour sessions and that a student does not receive a certificate of attendance if he or she has not attended at least 7 sessions of a course?

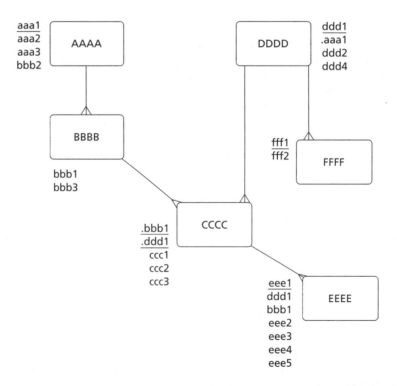

Fig 5.35

5.7. Further suppose that the Hergest educational institute of Exercises 3.14 and 5.6 offers assessment at the end of each course. This assessment takes the form of a written test. If students score more that 50% on this test they are awarded a certificate. This certificate states the course taken, its level, the mark attained by the student and the day of the award. Adjust your data model of Exercise 5.6 to accommodate these new requirements.

5.8. How would your data model of Exercise 5.7 change if the courses of the educational institute are not of the same size and duration?

5.9. Provide a list of requirements that are supported by your data model of Exercise 5.8. Clearly state which requirements are simply maintaining the functionality of the description of Exercise 3.14 and which are additional to it.

5.10. Produce a Logical Data Structure for the Bodgett & Son current environment depicted in Exercise 3.15.

5.11. Assuming the underlined keys are correct, identify and correct the errors in the LDS shown in Figure 5.35.

6

Data Flow Modelling

INTRODUCTION

Data Flow Modelling is a widely used and mature analysis technique, and is recommended by most structured methods. Data Flow Models (DFMs) are easy to understand and, with a little practice, reasonably quick and straightforward to develop. They consist of two parts: a set of Data Flow Diagrams (DFDs) and a set of associated textual descriptions. It is the DFDs that provide us with the truly effective analysis tool.

Within a given environment the Business Activity Model indicates the human activities that concern us, but it does not contain the detail we need to build a computerised information system. The technique of Data Flow Modelling is used to progress the analysis of the system's processes by providing a more detailed model of all the system's data processes.

DFDs illustrate the way in which data (or information) is passed around the system, and how it is transformed and stored within the system. They have a number of points to recommend them, not least of which is their ability to represent both logical and physical views of a system's existing or required functionality.

When used to show the information content of data-oriented business activities, DFDs will provide a model that closely matches the users' perception of the current system. Their lack of ambiguity and ease of understanding makes them a powerful communication tool.

Another great strength of DFDs is their hierarchical nature. The top level of the hierarchy (level 1) is used to show how the system as a whole operates, at a summary level. It will show the major flows of data or documents, the functional areas or high-level processing of the system, and the principal data stores. The next level down will consist of a number of DFDs, each of which will break down or decompose one of the higher-level processes to reveal its inner workings, i.e. its internal data flows, sub-processes and data stores. The hierarchy can extend to an unlimited number of levels or decompositions, but in practice two or three levels are usually enough.

This hierarchical structure means that we can adopt a top-down approach to analysing processing that enables us to develop an understanding of the overall system, before delving deeper into lower levels of detail. We will stop the process of decomposition when we have reached a level where further decompositions are unlikely to improve our understanding. The other advantage of producing a hierarchy of diagrams is that we can present and review different parts of the model at appropriate levels of detail to different users in a way that reflects their area or level of knowledge and responsibility.

To start with, our interest is really confined to the top-level DFD. We will first concentrate on the major functions of the system, i.e. those that our Business Activity Model suggests directly support the primary functions of the business area. A description of how we decompose DFDs will come later in this chapter. The key thing at this point is to understand that we do not need to capture every minute detail on our top-level DFD. The mechanisms for doing this will be introduced and applied later in the project.

Finally, before we look at DFD development in detail, always bear in mind that DFDs are an aid to understanding a system, not an end in themselves. It is very easy to become obsessed with the idea of producing the perfect DFD, and to regard it as a work of art. It is not; it is an analysis tool and will rarely be right first time. We should expect to amend it many times during the life of a project, as our understanding increases, and our willingness to alter DFDs is a sign of our flexibility and responsiveness.

▶ Learning objectives

After reading this chapter, readers will be able to:

◆ identify system entities and attributes;

◆ produce a Data Flow Diagram (DFD);

◆ convert Business Activity Models, Resource Flow Diagrams and Document Flow Diagrams to DFDs;

◆ document a Data Flow Model;

◆ produce Context Diagrams;

◆ scope a project;

◆ 'logicalise' a Data Flow Model.

▶ Link to other chapters

◆ Chapter 3   Aspects of the Business Activity Model are converted to a Data Flow Diagram.

◆ Chapter 7   The current Data Flow Model is transformed into its final (required system) form.

◆ Chapter 8   The required system Data Flow Model is used to identify system functions.

▶ DFD CONCEPTS AND NOTATION

Data Flow Diagrams use four main symbols.

1. External entities

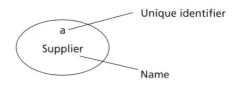

Fig 6.1 External entity

External entities are people, organisations or other computer systems that act as sources of data to or recipients of data from the system or area under investigation.

The same notation will be used for external entities in physical and logical DFDs. In physical DFDs an external entity will be any source or recipient of data that lies outside the physical system, whether that system is computerised or not. Consequently many of the people who actually input data into an existing computer system (or use output from it) will be considered internal to the physical system as a whole, i.e. actually part of the system itself (albeit a manual part) – as the analysis progresses and a clear differentiation between the system and its users takes shape, many of the people who have been perceived as part of the 'physical' system will end up as *external* to the 'logical' system.

For example, ZigZag purchasers decide on and provide details of orders to be placed with suppliers, but purchase order clerks actually assemble and store the details for passing to the suppliers. In this instance the purchaser is the external entity, i.e. the data source, while purchase order clerks are considered to be, at least initially, within the boundary of the system.

The name given to an external entity refers to a *type* not an *occurrence* of the external entity. For example, in the ZigZag system 'Supplier' is an external entity that acts as a recipient of purchase orders. It is a generic type; an occurrence of Supplier would be Sony or EMI.

Data entities and external entities

The external entities of a Data Flow Diagram should not be confused with the entities that form the building blocks of the Logical Data Model. It is rather unfortunate that the word *entity* is used in both concepts but the observant reader should manage very quickly not to confuse 'entities' with 'external entities'.

---

2. Processes

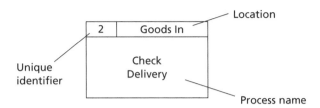

Fig 6.2 Process

Processes represent business activities carried out on and triggered by data. They should not be confused with computer programs. A process may sometimes equate directly with a program, but even then will be defined in user terms rather than in computer jargon. In other words, it should reflect the business activity it supports.

DFDs normally only show processes that transform or change data in some way, rather than merely formatting it for report purposes. The exception to this is where the production of reports or queries forms a significant part of a system's functionality in the eyes of the users.

The name given to a process should be as brief as possible, but still convey its true purpose. Names such as 'process delivery' are too vague; after all the symbol tells us that we are 'processing' something; 'check delivery' is much more meaningful.

The unique identifier should not be confused with a sequence number; it is not the purpose of DFDs to show the *order* of processing.

In physical DFDs it is important to give the location of the business activity that triggers the process, even if this location refers to a large imprecise area such as 'depot'. One of the properties of a physical process is that it takes place in a definable location or is carried out by a defined job holder (in which case the location will be a job title). If a process is always split into two distinct locations, then it is likely that we have two separate but dependent processes. Clearly, physical locations will become more precise as we move down into lower levels of detail.

3. Data Stores

Data Stores are, as their name suggests, stores or holdings of data within the system. In physical DFDs we will use four types of data store:

◆ **D**: computerised data store, i.e. a computer-held reference data file (D stands for 'digitised');

◆ **M**: manual data store, e.g. filing cabinet, record book;

◆ **T(M)**: manual transient data store. This represents a temporary store where data is held until read *once*, and then removed or deleted, e.g. an in-tray or a mailbox;

◆ **T**: Computerised Transient Data Store – a computerised temporary version of the above, e.g. a temporary sort file.

The name given to a data store should reflect its contents, and not just its storage mechanism. So we would never call a data store 'Filing Cabinet', as this gives no clue

Fig 6.3 Data store

as to its content. Figure 6.3 illustrates the ZigZag data store 'Purchase Orders'. This is a manual store consisting of copies of the paper-based purchase orders as sent to suppliers.

4. Data flows

Fig 6.4 Data flow

Data flow arrows show the flows of data to, from and within the system. They represent the inputs to and outputs from processes and data stores inside the system, and the information passing into and out of the system as a whole. Data flows therefore act as links between other objects in a DFD.

Only certain objects can be directly connected by data flows:

◆ two processes
◆ a data store and a process
◆ a process and an external entity.

We cannot connect a data store with another data store, as this would imply that data magically floats around the system of its own accord. The truth is that some activity or process must take place to actually move the data. External entities cannot pass data directly to a data store as this would mean that they were inside the system boundary, and so not external entities at all. In some circumstances we might want to clarify a system by showing data flows between external entities, but these flows are strictly speaking outside the business area and so we will show them as dashed lines.

Flows may be either one-way or, rarely, two-way. Care must be taken that data flows only contain real data that is required to support the business, i.e. they do not contain system control messages or read requests. This is particularly important with two-way flows. When reading from a data store it is tempting to include a flow to the store containing the key of the required data; in fact only the flow from the store should be shown on the DFD as this represents the data that is of interest to the business activity.

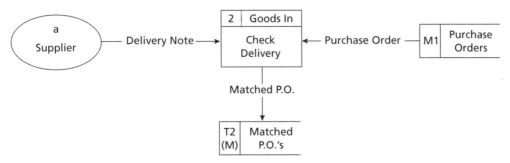

Fig 6.5 A sample DFD section

All data flows should be labelled, and in documenting physical flows the label will often be the name of an actual document, e.g. Purchase Order Copy 1, Purchase Order Copy 2. Whatever name is chosen it should give some clue as to its data contents.

By combining the different DFD objects we can build up a picture of how data is passed around and used in the system. For example, the DFD extract in Figure 6.5 illustrates the following business activities:

1. Suppliers send delivery notes to the Goods In section.

2. These are checked against the relevant purchase order.

3. Matched purchase orders are placed in a temporary file to await processing.

▶ DEVELOPING DFDs

In totally manual environments we begin with modelling *physical* flows of data within the current system, and only at an overview level.

With experience this can often be done quickly and with a great deal of accuracy straight away from the results of information gathering. As you read through the textual summary the DFD is built up as each data flow, process and data store is identified.

For less experienced practitioners it can be helpful to use one or more of the intermediate techniques given below.

Regardless of how we produce them, it is essential that DFDs be developed and reviewed with users as we go along. In fact, if possible, we should have users present in the project team taking an active role in their initial production.

▶ Context Diagrams

A Context Diagram is really a DFD that shows the entire system as a single process, with data flowing between it and the outside world as represented by external entities. Its main purpose is to help in fixing the boundary of the information system (and there-fore the area under investigation) and to show its interaction with external entities.

The Business Activity Model will provide the first feel for the system boundary. We can now proceed with more care by identifying more clearly the people, institu-tions and even other systems that will send or receive information from the system under investigation.

External Entity	Source or Recipient (S or R)	Data Flow
Supplier	S	Delivery Note
	R	Purchase Order
	S	Delivery Details
	S	Invoice
Purchaser	S	Purchase Order Quantities
	R	Stock Report
	R	Rejected P.O. Copy #2
	R	Matched P.O. Copy #1
Customer	R	Despatch Note
Sales and Marketing	S	Customer Order
	R	Matched C.O. Copy #1
Accounts	R	Matched Invoice

Fig 6.6 Identification of flows crossing the system boundary

To develop a Context Diagram we carry out the following tasks:

1. Identify all sources and recipients of data from the system, i.e. external entities.
2. Identify the *major* data flows to and from the external entities.
3. Convert each source or recipient into an external entity symbol.
4. Add the data flows between each external entity and a single box representing the entire system.

One way of documenting the results of tasks (1) and (2) is by using a table as in Figure 6.6.

Remember that, to start with, we are only interested in a level of detail sufficient to confirm the system boundary and project scope, and only with respect to *current* system operations.

Using this table of sources and recipients we can draw the Context Diagram shown in Figure 6.7.

Presentation tip

To aid communication with users, we sometimes show a flow of information between two external entities. These flows are shown on the diagram with a dashed line. In Figure 6.7 we have used this convention to show two such external flows.

The first is the payment, which goes from the Accounts department to the supplier. If we are presenting our analysis findings to a group that includes members of the Accounts department or financial backers of the project it helps to include the 'payment' flow even though it

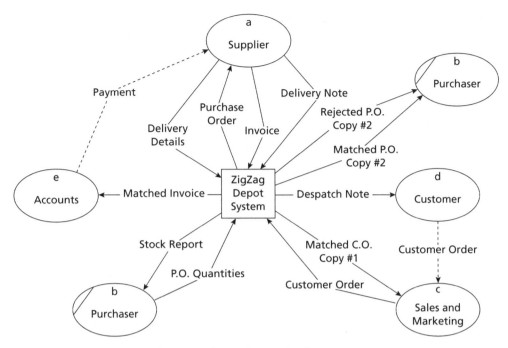

Fig 6.7 Context Diagram (current physical overview)

is outside the current project's scope. The inclusion of this flow helps close the payments loop when presenting the Context Diagram and helps members on the periphery of the project to understand the project without feeling alienated.

The second external flow is the customer order that customers send to the Sales and Marketing department which then forwards it to the depot. This flow will prove important when the new system is developed to encompass e-commerce retail customers. We will have occasion to meet this flow again later – e.g. see Figure 6.9. In a presentation the inclusion of this external flow allows the champion of e-commerce in the group to expand on the opportunity to have the customer's orders flow directly to the depot.

The inclusion of external flows helps give a more complete picture of the position of the current project within the business. Their inclusion indicates that the project team is aware of how the current project fits within the business – a fact that helps the project team gain the confidence of the project's decision makers.

Notice that Purchaser appears twice in Figure 6.7 with a diagonal line through it. This line is used to reduce clutter whenever an external entity is drawn more than once in any diagram.

Of the external entities shown on the diagram, some are more obvious than others. Customer and Supplier are clearly outside our system, since they lie totally outside the business; Accounts and Sales and Marketing lie inside the business but represent a different business area from the one we are investigating. They are therefore represented as external systems with which our system somehow communicates;

Purchasers are on the brink and we considered placing them inside our system but they finally ended outside the system for reasons stated when we developed the Business Activity Model. (We have to be conscious that we are looking at the stock movements of goods within the ZigZag depot. The system we are developing becomes aware of possible new stock at the time a purchase order is set up. This purchase order is created within the system by the purchase order clerk, who works on the products recommended by Purchasers. The recommendation itself is not stored. This leads us to place Purchasers as external entities responsible for sending in a recommendation which the PO clerk will turn into specific, recorded purchase orders.)

As we go along with the analysis we will have occasion to identify and add more external entities.

▶ Document Flow Diagrams

Having defined the system boundary we can now move on to examine the flow of documents within the existing system using a Document Flow Diagram.

Document Flow Diagrams illustrate the flow of physical documents associated with the area under investigation. In this context documents may take the form of pieces of paper, conversations (usually over the telephone) or even data passed between computer systems.

To create a Document Flow Diagram we carry out the following tasks:

1. Identify all recipients and sources of documents, whether inside or outside the system boundary.
2. Identify the documents that connect them.
3. Convert each source and recipient into an external entity symbol.
4. Add data flow arrows to represent each connecting document.
5. Add the system boundary to exclude the external entities identified in the context diagram.

As with context diagrams, tasks (1) and (2) can be documented in a table as shown in Figure 6.8. Using this table we can draw the Document Flow Diagram shown in Figure 6.9 (with a system boundary transferred from the context diagram in Figure 6.7).

▶ Converting Document Flow Diagrams to DFDs

To transform the Document Flow Diagram into a DFD we follow each document flow in turn, asking the following questions:

- ◆ What process generates this document flow?
- ◆ What process receives this document flow?
- ◆ Is the document stored by a process?
- ◆ Where is the document stored?
- ◆ Is the document created from stored data?
- ◆ What business activity triggers the process?

Source	Document	Recipient
Supplier	Invoice	Purchase Order Clerk
Supplier	Delivery Times	Stock Clerk
Supplier	Delivery Note	Goods In Clerk
Supplier	Delivery Details	Purchase Order Clerk
Stock Clerk	Stock Storage Report	Stock Keeper
Stock Clerk	Stock Report	Purchaser
Stock Clerk	Stock Report	Despatch Supervisor
Stock Clerk	Matched Purchase Order Copy #2	Purchase Order Clerk
Sales and Marketing	Customer Order Estimate	Purchaser
Sales and Marketing	Customer Order	Despatch Clerk
Purchaser	Purchase Order Quantities	Purchase Order Clerk
Purchase Order Clerk	Rejected Purchase Order Copy #2	Purchaser
Purchase Order Clerk	Purchase Order Copies ×2	Goods In Clerk
Purchase Order Clerk	Purchase Order Amendment	Supplier
Purchase Order Clerk	Proposed Purchase Order	Supplier
Purchase Order Clerk	Invoice and Matched Purchase Order Copy	Accounts
Goods In Clerk	Rejected Purchase Order ×2	Purchase Order Clerk
Goods In Clerk	Matched Purchase Order Copy #2	Stock Clerk
Goods In Clerk	Matched Purchase Order Copy #1	Purchase Order Clerk
Despatch Supervisor	Matched Despatch Report	Despatch Clerk
Despatch Clerk	Matched Customer Order Copy #2	Accounts
Despatch Clerk	Matched Customer Order Copy #1	Sales and Marketing
Despatch Clerk	Despatch Report	Despatch Supervisor
Despatch Clerk	Despatch Note	Customer
Customer	Customer Order	Sales and Marketing

Fig 6.8 ZigZag sources and recipients of documents

Each time we identify the need for a process or a data store we will check to see if it has already been identified elsewhere in the diagram. If not, we will add it to the emerging DFD. An understanding of the activities that give rise to each process helps keep the whole system development in context.

The best way to explain this is by using an example. So we will look at the flows between Sales and Marketing, Despatch Clerk and Despatch Supervisor (Figure 6.10). We start with the flow from Sales and Marketing towards the Despatch Clerk. Through the Business Activity Model and the Work Practice Model we notice that the Despatch Clerk, upon receiving the customer order, consults the stock files,

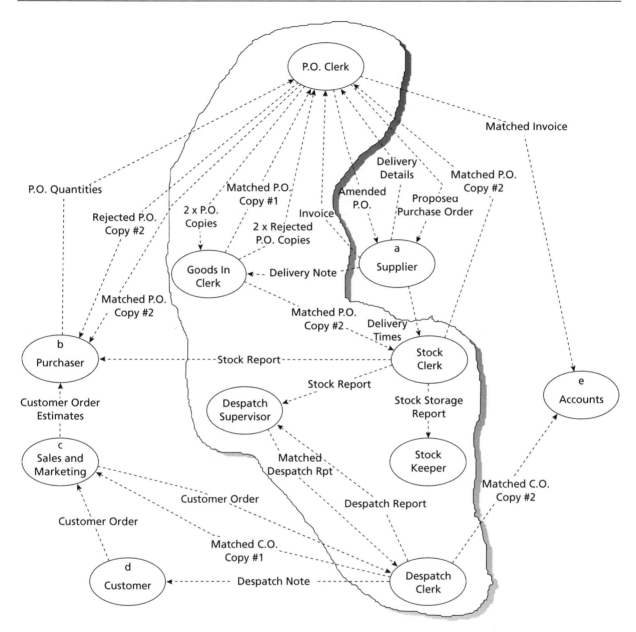

Fig 6.9 Document Flow Diagram with system boundary

finds the stock to be used to fulfil the customer order, makes copies of the customer order and places those copies in the customer order files (Figure 6.11).

Moving to the document flow from the Despatch Clerk towards the Despatch Supervisor, we notice that the Despatch Supervisor *receives* despatch details in the form of a report and *assembles* the physical despatch. At first sight it appears that we have the situation depicted in Figure 6.12.

But, closer scrutiny reveals that the Despatch Supervisor simply assembles the despatch, annotates the despatch report and sends it back to the Despatch Clerk.

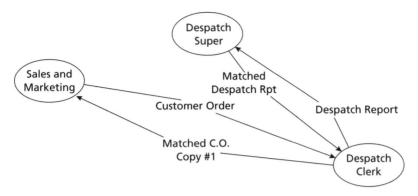

Fig 6.10 Document Flow Diagram extract

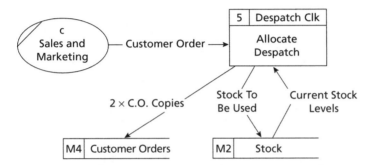

Fig 6.11 Conversion of flow from Sales and Marketing towards the Despatch Clerk

Fig 6.12 Apparent conversion of flow from Despatch Clerk towards the Despatch Supervisor

This means that the Despatch Supervisor does not use any data store to record the assembly and despatch of the customer orders. All the recording is done by the Despatch Clerk a little later on. We therefore see that the activity of assembling the despatch is not a proper *data* process, since all data processes *have* to use a data store. This leads us to the realisation that the 'Assemble Despatch' activity is *outside* the information system, as indeed we may have already observed when producing the Business Activity Model. The relevant data flow section then becomes that of Figure 6.13. (It may take some time for novice analysts to appreciate what has

Fig 6.13 Actual conversion of flow from Despatch Clerk towards Despatch Supervisor

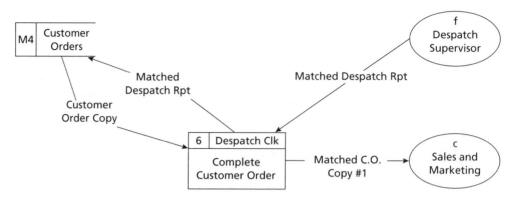

Fig 6.14 Conversion of flow from Despatch Supervisor back to Despatch Clerk

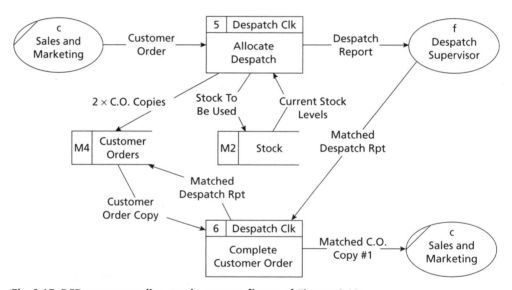

Fig 6.15 DFD corresponding to document flows of Figure 6.10

happened here. When we produced the context diagram of Figure 6.7 we identified some fairly clear-cut external entities, but a closer look at each and every document flying about in the vicinity of the business area we are studying, coupled with the understanding we have gained from building the Business Activity Model, leads to the identification of an as-yet-undiscovered external entity, namely that of Despatch Supervisor.)

Once the despatch is assembled details are passed from the despatch supervisor back to the despatch clerk, who matches them with the filed customer order copy and sends this back to Sales and Marketing. Details of the matched order are placed in the Customer Orders file (see Figure 6.14).

This now accounts for all of the document flows in Figure 6.10, and putting together the results of these transformations we get Figure 6.15.

By transforming all of the flows in the Document Flow Diagram in this way, and combining the results, we can draw up a top-level DFD, as in Figure 6.16.

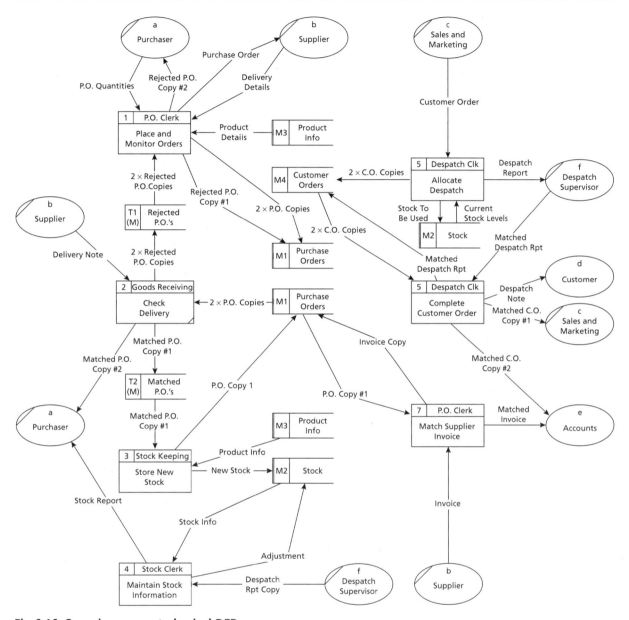

Fig 6.16 Overview current physical DFD

Notice that when a data store appears more than once on the diagram, we place a double bar at the left-hand end of its box. This serves a similar purpose to the diagonal line in an external entity box in helping to reduce clutter.

Although the main purpose of a DFD is to aid understanding of a system, it is important to make it as clear as possible by reducing the number of crossing lines. There is nothing more confusing than a DFD with data flow arrows intertwined with each other. Do not become obsessed with eliminating crossed lines, as it can be a time-consuming process when the DFD is likely to be amended frequently; just bear in mind that other people will need to understand the diagram easily. Another presentational guideline is to place data stores in the centre (as much as is possible), then surround them with processes, and finally end up with the external entities on the edge of the diagram.

When we produce a first-cut DFD the chances of achieving the optimum grouping and breakdown of processes and flows straight away are pretty slim. So be prepared to go through several iterations, merging or splitting processes, data stores and data flows, until they are all at the same general level of detail. Any objects that are combined at this stage will be expanded upon later when we decompose the top-level DFD into lower levels in the hierarchy. So as a guide try to keep the number of processes down to below ten or twelve. If there are many more than this, it will be quite difficult to get an overall picture immediately.

For a given system, it is extremely unlikely that any two analysts will produce the same DFD, even if they both model the business with equal accuracy. DFDs are relatively unambiguous, but the breakdown and grouping of processes and data flows is very flexible, especially at the top level, and is largely a matter of subjective judgement. The crucial test of the accuracy of a DFD is whether it supports the functionality of the system. As long as users and analysts are happy that it does, it is 'correct', regardless of its artistic merits.

Overview DFDs such as that in Figure 6.16 are in some ways intentionally incomplete. For example, in the ZigZag system there are processes carried out to maintain information about the depots themselves. We will be adding those as and when we identify them, usually with the help of the data model, which is developed concurrently.

The current DFD helps in identifying areas that need further analysis. For example, a perusal of the DFD of Figure 6.16 indicates that process 3 (Store New Stock) seems to bounce data from one data store to another with no obvious external prompt. This means that something is amiss. We will be looking at this process a bit later on.

Also, data store M3 (Product Info) only has data coming out of it. Where does this data come from? Who is responsible for it? Is there a process that feeds this data store?

A requirement of the new system stipulates that we are to continue using product and supplier data as created by the system in Sales and Marketing. This gives a hint as to where the information in this data store comes from, but still leaves our other questions unanswered. We will therefore be keeping an eye on this data store as we go along.

To an experienced analyst Document Flow Diagrams may appear a rather long-winded procedure, but for beginners, especially where the current system is heavily document-based, they can be a very useful way of getting started. In some cases it may also be easier to produce the Document Flow Diagram first, and then use it to create the Context Diagram once the system boundary has been agreed.

▶ Converting Resource Flow Diagrams to DFDs

Document Flow Diagrams are not the only possible tool we can use to help us produce Data Flow Diagrams. In Chapter 3 we saw how to model the results of initial analysis by producing Resource Flow Diagrams (see Figure 3.7).

In an environment where a number of different physical resources move around frequently, it may be a good idea to start by modelling the flow of resources instead of the flow of documents. With a resource flow in hand we can ask questions similar

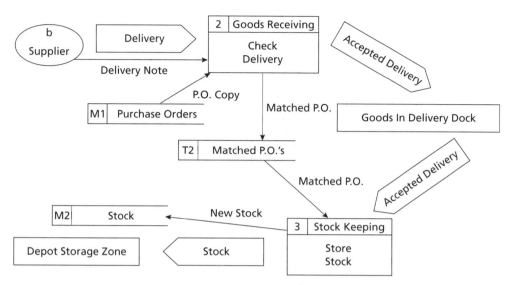

Fig 6.17 Converting the Resource Flow of Figure 3.7 into a DFD

to those we asked when we were converting a Document Flow Diagram into a Data Flow Diagram, namely:

◆ What process records the receipt of this resource?

◆ What process records the placement of the resource in a resource store?

◆ What process records the removal of the resource from a resource store?

◆ What new or old data accompanies the resource?

◆ What previously stored data is used in each movement of this resource?

Such questions in the area of ZigZag's receipt of goods transform the Resource Flow Diagram of Figure 3.7 into the Data Flow Diagram section shown in Figure 6.17.

The names and numbers used in Figure 6.17 may differ slightly from those used in the overview Current Physical DFD, but if we look closely at our level-1 DFD in Figure 6.16, we can see that the Resource Flow Diagram does actually confirm its accuracy in the delivery receipt area. If it would add to the understanding or acceptance of users we could annotate the level-1 physical DFD with resource flows and stores for presentations.

▶ Converting Business Activity Models to DFDs

If a BAM has been produced as part of modelling a system's processing, and if the project team has also decided to produce a DFD, then this DFD should be based on the analysis that led to the BAM. Indeed, it would be folly to ignore the BAM and to try and produce the DFD 'from scratch'.

A BAM is transformed into a DFD by asking of it questions such as:

◆ Does the activity use data?

◆ Is the activity responsible for the storage of new data?

◆ Does the activity require already stored data?

It is important to appreciate that each product of SSADM has to be used by at least one subsequent product. If a product is generated and then never used again, we should question the prudence of producing that product in the first place.

Fig 6.18 The Goods Receiving activities of ZigZag (see Figure 3.4)

Activities that are rich in data lead to the identification of the kind of data-oriented processes that form a well-defined DFD. Activities that do not involve data (because, for example, they are purely physical activities) do not result in DFD processes but may end up giving us a clue about external entities.

If, as an example, we focus again on the goods receiving section of ZigZag, we can extract from the BAM of Figure 3.4 the section shown in Figure 6.18. Of the five activities in Figure 6.18 three, namely 'Place Goods in Delivery Dock', 'Remove Goods from Delivery Dock' and 'Store Goods in Depot', are purely physical, involving the carrying of goods around the depot. The other two, 'Check Delivery' and 'Allocate Stock Location', are rich in data. The former requires access to the original purchase orders, which are now annotated with information about the goods being delivered. The latter requires the allocation and recording of the final destination where each newly arrived stock is to be placed in the vast depot.

The Work Practice Model (see Figure 3.12) further informs us that these two activities are part of the job descriptions of the stock keeper and the stock clerk respectively.

In addition, ZigZag's organisation chart (see Figure 3.10) indicates that the checking of deliveries is part of the 'goods receiving' section of the warehouse, while 'allocating stock locations' is part of 'stock keeping'.

Putting all these observations together gives us a DFD section like the one in Figure 6.17.

Beginning an analysis of a system's processing

The reader will have observed that Chapters 3 and 6, which are dedicated to modelling the current processes of a system, have led to the recommendation of four models: Resource Flow Diagrams, Document Flow Diagrams, Business Activity Models and Data Flow Models.

We have demonstrated how to use any of these diagrams as a starting point and we have also shown how to use some of these diagrams to assist the production of others.

As with most of systems analysis there are no fixed rules as to what to do first or second or even to do them at all. For each given project, the project team should be able to make a judgement of what models to use and in what sequence to create them.

It is quite possible for one project to produce only a Business Activity Model and then proceed, with a solid data model in hand, into specification.

Alternatively, a project may gain from the attention to detail afforded by Data Flow Modelling, and a project team may dispense with all other models and concentrate on Data Flow Modelling.

Yet another choice would be to use either of the processing models to help in the production of any combination of the others.

Strictly speaking, only the BAM, the context diagram and the current physical DFD are required for the purposes of documenting the processing side of things; the other diagrams we have discussed are essentially working documents, and are unlikely to be carried forward into subsequent steps.

Sometimes, the policies and procedures of the organisation for which the system is designed may dictate the type of models it prescribes to all its projects.

Note that each of the processing models provides a different point of view of the system, and it is often the case that some of these products are produced in parallel.

Figure 6.19 illustrates how the different process modelling tools feed from each other. The Data Flow Diagram and Business Activity Model are more important than the other two.

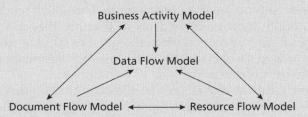

Fig 6.19 Relationship between various process models

▶ DECOMPOSING DFDS

The top-level (level 1) DFD provides us with an overview of required processing in the business area, but there is still a lot of information on the detailed workings of complex processes, which is not shown at the top level.

Let us look, for example, at process number 1 of our overview DFD (see Figure 6.16). The name of this process alone indicates that at least two things are represented

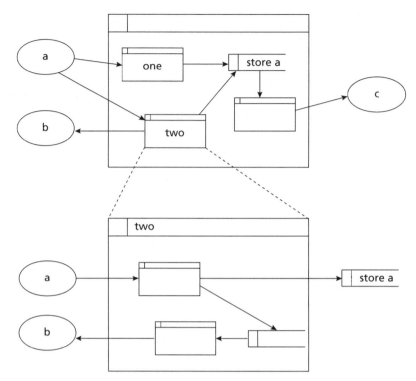

Fig 6.20 DFD decomposition

here: the 'placement' and the 'monitoring' of purchase orders. Further analysis suggests that, after receiving from a purchaser a list of products that have to be ordered, the purchase order clerk searches the product files to find the suppliers of the requested products; sets up purchase orders for each supplier; sends those purchase orders; receives notification from suppliers of whether they can fulfil the order; finds alternative substitute products when the supplier is out of stock; confirms the purchase order; arranges a delivery date; and forwards information about rejected deliveries back to the purchaser.

We could now create level-2 DFDs that illustrate the internal data flows, data stores and lower-level processes of each individual process on the top-level DFD (as in Figure 6.20). These lower-level processes could then be further decomposed into yet lower levels of DFD and so on, until we arrive at a full description of the required processing.

To start with we will decompose only the most complex processes to a level that helps us to understand the fundamental requirements of the new system. As the analysis progresses, all processes will be decomposed until they reveal clearly the workings of the system.

▶ Drawing lower-level DFDs

We take the process box from the higher-level DFD to form the boundary for the new DFD. The name and identifier of the parent process become the name and identifier of the lower-level DFD (Figure 6.21).

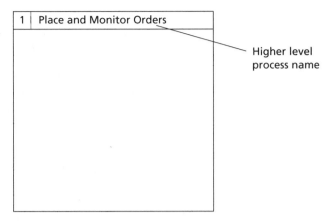

Fig 6.21 Starting a lower-level DFD

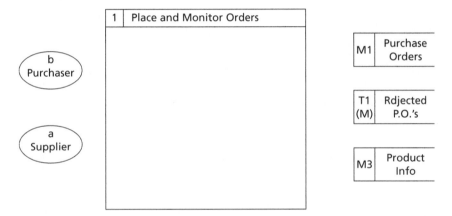

Fig 6.22 Adding external entities and data stores to retain consistency with the DFD of Figure 6.16

The sources and recipients of data for the DFD are those objects (external entities, data stores and other processes) that provided data for or received data from the process at the higher level (see Figure 6.22).

External entities and data stores may be decomposed at the lower level to provide a more detailed description of their role or contents. To maintain links between levels, the identifier of decomposed elements consists of the higher-level identifier plus an alphabetic suffix (see Figure 6.23).

Summary data flows may also be decomposed into several detail flows, but there must be *no* flows between objects at lower levels that are not represented at higher levels by at least a summary flow.

If any entirely new data flows, data sources or data recipients are uncovered while drawing low-level DFDs, higher-level DFDs must be amended to include them, as the objects communicating with the process must be consistent on all levels.

Processes *within* lower-level DFDs are numbered by adding a numeric suffix to the identifier of the parent process (see Figure 6.24).

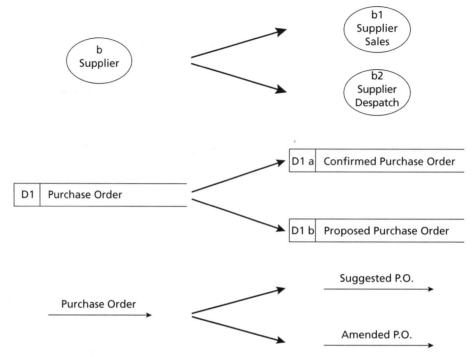

Fig 6.23 The decomposition of DFD elements

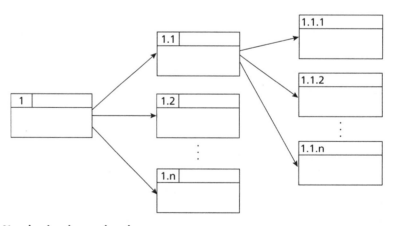

Fig 6.24 Numbering lower-level processes

Data stores that are only used entirely *within* a lower-level DFD are numbered by adding a numeric identifier to the identifier of the DFD (see Figure 6.25).

Lower-level DFDs are drawn in much the same way as the top-level DFD: by following data flows into the diagram and identifying receiving and generating processes, plus any associated data stores. In fact, the only difference is that for the top level we follow flows into the system as a whole, while for other levels we follow flows into an individual process.

```
┌──────┬──────────────────────┐
│ D2/3 │                      │
└──────┴──────────────────────┘
```

Fig 6.25 Internal data store – the third digital data store that is solely used by process number 2

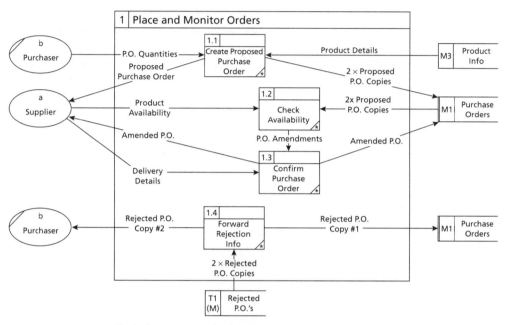

Fig 6.26 Place Purchase Order DFD (level 2)

Inevitably, the detailed study of a process will lead to the identification of yet more data flows. For example, the 'Place and Monitor Orders' level-2 diagram (Figure 6.26) contains flows, such as 'Product Availability', which are absent from our original level-1 overview DFD. To maintain consistency, the original level-1 diagram has to be adjusted by taking in the new level-2 insights. In fact, any useful CASE tool should do this, or at least warn of any inconsistencies between DFD levels.

▶ Elementary Process Descriptions

If a process is at a level where further decomposition would not reveal any additional requirements or system understanding we classify it as an Elementary Process, and mark its process box with an asterisk (Figure 6.27).

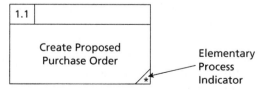

Fig 6.27 Elementary process

Elementary Process Description
Process ID: 1.1
Process Name: Create Proposed Purchase Order
Description: Purchase order details are received from purchasers for individual products. The Product files, which contain information about products and their suppliers, are read to determine the supplier for each product. All the product orders for each supplier are batched together, and a single Proposed Purchase Order placed with them. At this stage a Purchase Order is termed 'proposed', and will remain so until the supplier confirms that stock is available. Note: In future, the system should allow for more than one supplier per product (see requirement 118). When this is so, the primary supplier of a product should be identified first.

Fig 6.28 An Elementary Process Description

For each elementary process we will complete an Elementary Process Description (EPD), summarising its operations and activities as in Figure 6.28.

No textual description is necessary for higher-level processes as they can always be described by the sum of the elementary processes that constitute them.

▶ THE CURRENT PHYSICAL DATA FLOW MODEL

The overview DFM will usually provide a good starting point for developing the full Current Physical DFM. We will begin by identifying any high-level processes missing from the overview. For ZigZag this only involves adding processes to maintain the Supplier and Product files.

Once this is done we need to look for any additional data flows associated with the extra processes or which were left out of the overview model for reasons of clarity (such as routine amendment flows). We may also discover new data stores for inclusion on the top-level DFD, but it is more likely that lower-level data flows will be found within lower-level DFDs. Figure 6.29 illustrates the enlarged level-1 DFD for ZigZag.

We now examine each process on the level-1 DFD to check if it requires decomposition. Some level-1 processes will already be sufficiently well understood or straightforward for decomposition to be unnecessary, e.g. Allocate Despatch. For these processes we should complete their definition by filling out an Elementary Process Description (EPD).

Remember that for physical DFDs we are interested in documenting exactly how processes are actually carried out in the real world. It can be very tempting to omit or skim over details that seem to be due to 'trivial' physical constraints (such as batching up documents in an in-tray). This temptation should be strongly resisted. Missing out 'trivial' details will almost certainly confuse users, who will feel that the resulting diagrams are unrealistic, and could lead to important constraints being missed which should be documented in the Requirements Catalogue for inclusion in the new *physical* system much later in the project. When a good Business Activity

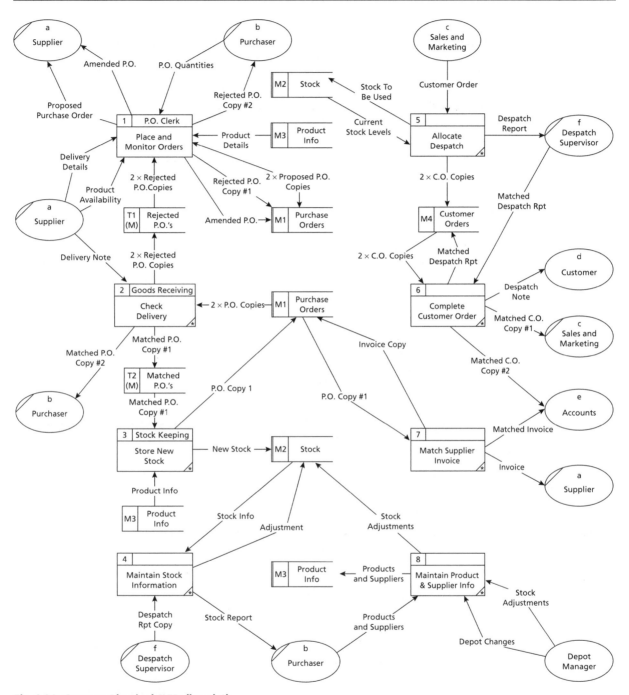

Fig 6.29 Current Physical DFD (level 1)

Model is developed, it helps users see the transition between physical activities and data processes.

If any of the level-2 DFDs are still unclear or their sub-processes complex, we may need to produce level-3 DFDs. Further levels (4 and so on) are rarely needed. As a rule all processes at the same level of the hierarchy should be at equivalent levels of detail or complexity, and so require the same number of further decompositions. If we find that some of the high-level processes need decomposing to more levels than the others, then we could consider splitting them up at the top level. However, before we get caught up in excessive 'over-modelling', we should once again bear in mind that the overriding aim of Data Flow Modelling is to reflect the user's perception of the system, and not to produce a pure model.

A few things to look out for when checking a DFM are:

◆ If any processes have more than eight flows in or out there is likely to be a fair amount of activity going on inside it to handle these flows. So think about whether they would benefit from decomposition.

◆ If any processes have less than four flows in or out then they are probably at a reasonably low level already, so think twice about decomposing them.

◆ Processes that act as a 'dead-end' for data flows (i.e. there are no flows out of them) are certainly incorrectly defined. Processes transform data, they do not store it, so there must always be a flow out to a data store, process or external entity.

◆ Processes that appear to generate data (i.e. there are no flows into them) are likewise incorrect.

◆ Data stores that have flows going into them (updates) but no flows out (reads) should be checked to see if they are actually used anywhere. It is possible that they are only used for reporting purposes, in which case the relevant processes may not be shown on the DFM.

◆ Data stores with no flows going into them should also be looked at closely. The only circumstances where these are valid are where the data store is maintained by another system (i.e. it is a reference-only data store for the system under investigation).

◆ All data flows on the lowest-level DFDs should be single-direction flows.

▶ Data Flow Modelling documentation

Once we have decomposed the DFD hierarchy to its lowest level, the DFM is completed by adding the following textual descriptions:

◆ Elementary Process Descriptions.

◆ External Entity Descriptions detailing the role and responsibilities of external entities (Figure 6.30).

◆ Input/Output Descriptions (I/O Descriptions) for bottom-level data flows that *cross the system boundary*. These detail the data items contained in each flow and will be used later in the project to define how the system communicates or interfaces with the outside world (i.e. to define dialogues with the system). The structure, with regard to optionality or repetition of data items, is recorded in comment form at this stage. Two of the I/O Descriptions for process 1 from Figure 6.26 are shown in Figure 6.31.

Once the DFM is complete we should check that any problems identified by users regarding existing system support are fully recorded in the Requirements Catalogue. We can then proceed to translate the physical DFM into a logical DFM of current processing in order to identify its underlying functionality.

External Entity Description		
ID	**Name**	**Description**
b	Purchaser	Originator of Purchase Order requirements following a review of stock levels and customer demand. (Also person responsible within ZigZag for the setting up and negotiating of products, product prices and suppliers, but this information is dealt with via Sales and Marketing who then inform the depot of products and their suppliers)

Fig 6.30 Sample External Entity Description

I/O Description				
From	**To**	**Data Flow Name**	**Data Content**	**Comments**
b	1.1	P.O. Quantities	Product No. Qty Required Req-By Date Req-By Time Period	The purchaser may state a date by when the product should be delivered or a time period, e.g. 3 weeks from now.
1.1	a	Proposed Purchase Order	P.O. Number Supplier Name Supplier Address Depot Name Depot Address Req-By Date Req-By Time Period Product No. Product Name Qty Required Product Price	Each Purchase Order will contain several lines (usually up to about 12).

Fig 6.31 Sample I/O Descriptions

▶ LOGICALISING DATA FLOW MODELS

The Current Physical Data Flow Model depicts a detailed snapshot picture of the current system's processing, including all of its physical constraints and peculiarities. Many of these physical elements represent historical or administrative decisions which no longer apply, or which should not be carried forward into the new system. But at the heart of the DFM is a large amount of functionality that provides real active support for the business. It is very likely that our new system will be required to carry forward this support.

A physical DFM shows the current system in all its glory – good and bad. As the project moves on it is our task to convert the physical picture into a logical one that reflects the underlying business functions of the existing system. We can then decide which elements will be retained in the new system, and add additional functionality (as specified in the Requirements Catalogue) to it to form a complete model of the required processing. We will carry out these latter activities later on when we under-take the Business Systems Option, and the more detailed Function Definition; for now we are purely interested in deriving a logical DFM of current processing.

If we have produced an effective Current Physical DFM, all of the detail we require on current processing will be available from the lowest-level DFDs. It now 'merely' requires 'logicalising'. We could start all over again with a top-down an-alysis of processing to produce a logical set of DFDs, but this would clearly make no sense at all as it would mean repeating a large amount of analysis. Instead, we will work from the bottom up, by removing the physical elements of elementary processes and bottom-level DFDs, and then reconstructing the hierarchy by logic-ally regrouping these transformed processes.

It is quite tempting to carry out a sort of intuitive logicalisation exercise, and this may work for experienced analysts. But for most people it is a much better idea to follow a methodical transformation, by considering each DFD object carefully.

▶ Rationalising data stores

Data stores in a physical DFM reflect the way in which data is actually stored. In a logical DFM we replace these with logical data stores based on the organisation of data in the LDS.

To begin with it appears tempting to devote a single data store to each entity on the LDS. Unfortunately this could lead to an unmanageably large number of data stores, some of which would just represent link entities. It would also mean duplica-ting details of the data structure that are more properly documented in the LDS. To avoid these problems we will create logical data stores that represent *groups* of closely related entities (although some entities will stand out on their own and so remain the only entity represented by a data store).

As usual with most systems analysis there are no strict guidelines for grouping entities, but we can start by asking:

1. Are there any entities that can easily or naturally be described using a single phrase? (For example, Product Type, Product and Product Substitute could be called 'Product Information'.)

2. Are there any groups of related entities (e.g. Purchase Order and Purchase Order Line) that are created at the same time?

3. Are any groups of entities (e.g. Customer, Customer Order, Customer Order Line and Despatch) associated with the same major inputs or outputs?

At the end of the day we will probably create groupings by identifying entities that just 'feel right' together. Obviously what feels right to one person may not feel right to another, so there will be many different possibilities for grouping the LDS. One such way for the ZigZag LDS is shown in Figure 6.32.

To avoid data duplication the LDS must be divided into discrete groups such that:

◆ Each data store contains one or more related entities.

◆ Each entity is held entirely within a single data store (i.e. must not be duplicated or fragmented).

The only exceptions to this are decomposed data stores. Data stores on a phys-ical DFD may be decomposed to clarify which parts of a data store a process is accessing. If the decomposition is based on different subsets of data items then they may in fact be equivalent to different entities. Rationalising these data stores may mean that each decomposed data store is now equivalent to a single logical data store, so the decomposition is no longer necessary. However, in cases where the decomposition was based on data item value or status it may still be valid, so there will be two logical data stores containing the same LDS extract.

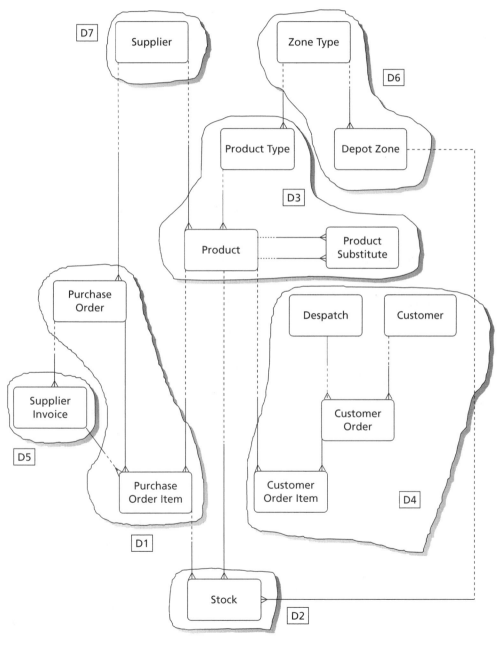

Fig 6.32 A grouping of the ZigZag LDS (first seen in Figure 5.27)

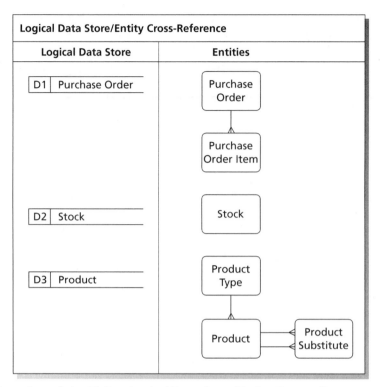

Fig 6.33 A section of the ZigZag Logical Data Store/Entity Cross-reference

We document the correspondence between logical data stores and entities using a Logical Data Store/Entity Cross-reference, as in Figure 6.33. The left-hand column lists the data stores, while the right-hand column shows the corresponding LDS subset. Each entity should appear only once. We do not show decomposed data stores on the cross-reference, but just the parent data store, e.g. if we were to retain the value-based decomposition of Purchase Order into Firm and Provisional Purchase Orders, we would only show the parent Purchase Order data store.

The Logical Data Store/Entity Cross-reference is one of the most important documents of analysis since it is the only means through which the consistency of our data and process models is enforced.

▶ Removing Transient Data Stores

Transient data stores in physical DFDs almost always exist to satisfy some physical constraint. They either represent temporary halts in a data flow or a postponement in storing information. All transient data stores that represent a temporary halt should be removed from the logical DFD and replaced with a simple data flow. Transient data stores that represent a postponement in recording information have no meaning in an automated system, but have to be handled with a bit of care to ensure that the new system will preserve all healthy current working practices. To illustrate how to handle transient data stores we will consider the 'Store New Stock' process in Figure 6.34. (Remember that this process sticks out like a sore thumb in the DFDs of Figures 6.16 and 6.29.) We will now demonstrate how the process of

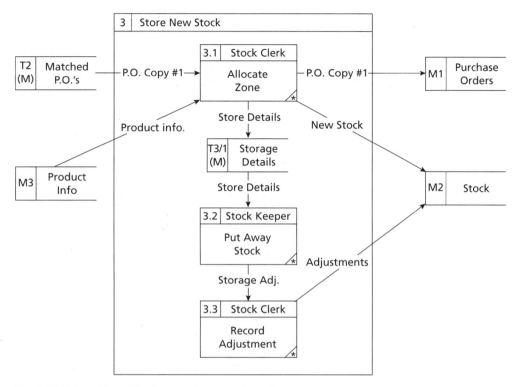

Fig 6.34 Store New Stock contains transient data stores

logicalisation can clarify certain aspects of the current DFD that may have escaped our notice until now.

The internal transient data store represents the activity whereby the stock clerk writes the storing instructions on a piece of paper, which is then placed in a tray to be collected by the stock keeper. It therefore only represents a delay in action and is, logically, superfluous. Its removal gives us Figure 6.35. (Note that the removal of

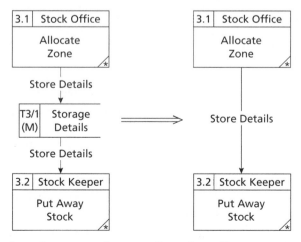

Fig 6.35 Transient Data Stores may become direct Data Flows

this data store does not lead to any loss of information; the 'New Stock' data flow contains the relevant storage details.)

The removal of the second transient data store T2(M) which lies outside the process (see Figures 6.16 and 6.34) is slightly trickier to handle. This is because it contains information about products that have been successfully delivered. We need to make sure that this information has been recorded before we dismiss the transient data store. When goods are delivered and ZigZag accepts them, the acceptance has to be recorded in the appropriate place. If we now recollect, purchase orders and deliveries are linked in a 1:1 relationship which led us to merge delivery information in the Purchase Order and Purchase Order Line entities of the LDS. According now to the Logical Data Store/Entity Cross-reference, these two entities have ended up in the 'Purchase Orders' data store. We therefore have to make sure that the 'Check Delivery' process, which is responsible for accepting deliveries, does show a data flow into that data store. If it does, the 'Matched P.O.s' transient data store is superfluous and can be replaced by a data flow. If it doesn't, and indeed it doesn't in our DFD of Figure 6.16, we have to adjust the DFD before replacing the transient data store with a data flow. We will return to this point a bit further on.

▶ Rationalising Elementary Processes

Processes in a physical DFM will accurately describe what is being done with the information in a system, but will also reflect the way in which activities are physically organised. In particular, they will provide details of the location of processes and reflect organisational structures and job responsibilities.

The purpose of a logical DFM is to describe the underlying logical processing, in effect to represent the ideal organisation of data processing activities that would occur if no physical constraints existed.

Transforming physical processes into logical processes is an iterative activity, involving discipline in identifying purely physical elements of processing. There are no strict rules or tasks that we can apply to rationalising processes, but SSADM does offer a fairly comprehensive set of guidelines:

◆ Remove details of the location of a process since location only indicates a physical constraint. While removing details of locations from processes we will often find that the organisational structure or job holder concerned is actually altering the data that is input to the physical process. In these cases the location will become an external entity. In this way the activity of logicalisation firms up on the system boundary – see Figure 6.36.

◆ Merge any process that merely re-formats or re-arranges data (e.g. it indicates a sort or a batching process) with the process that triggers it.

◆ Check that each process actually transforms data, and does not just report on it. Reporting requirements should be documented in the Requirements Catalogue, unless they form a major part of the system's functionality, in which case they will remain as elementary processes.

◆ Try to combine processes that *always* occur together in sequence. In particular look for processes linked by a single data flow, which only occur separately because of their physical locations (Figure 6.37).

◆ Combine processes that are duplicated in the physical DFM. Again, this will usually occur for reasons of physical location, or overlapping job responsibilities.

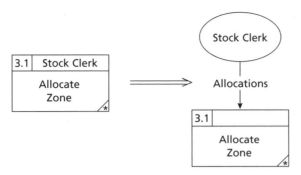

Fig 6.36 Locations may become external entities

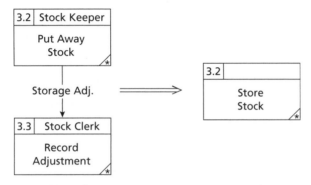

Fig 6.37 Processes that always occur together are combined

◆ Where more than one process contains a common activity, create an Elementary Process Description for the duplicate processing and cross-reference with all processes that use it, e.g. complex calculations are frequently carried out as part of several processes.

◆ If a process contains an element of human judgement or subjective decision-making, this element is represented by data flows to and from an external entity. The same applies to processes that require a user to check the results of a system instruction (Figure 6.38).

In Figure 6.38 we have also taken the opportunity of renaming process 3.2. We chose to do so because the original name had a physical connotation that may obscure the information content in which we are interested; the new name is more precise in information terms.

◆ Data flows into the system that clearly act as a trigger for a set of processing will ideally be received by a dedicated process.

◆ Data flows are totally stripped of any physical connotations. For example, data flow 'P.O.Copy #1' which is used by process 3.1 – Allocate Zone – contains information of goods that have been received. We therefore strip this flow of its 'paper copy' connotations by renaming the flow 'Accepted Delivery Details'.

◆ All data stores are shown as digitised to reflect that they represent groupings of entities that are, after all, 'logical' and therefore totally stripped of any physical connotations.

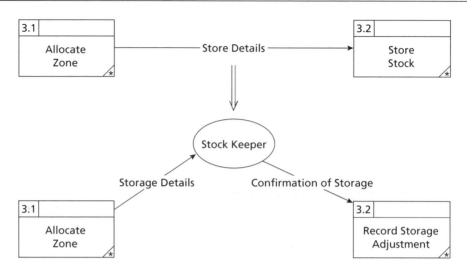

Fig 6.38 Human decisions are reflected through external entities

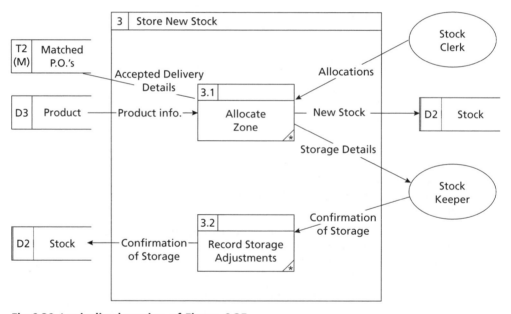

Fig 6.39 Logicalised version of Figure 6.35

Putting together the above transformations we arrive at the logicalised version of Store New Stock in Figure 6.39. Notice that we have concentrated solely on this process and left the transient data store T1 which links processes 2 and 3 unaffected for the time being.

In rationalising bottom-level processes we are only concerned with processing that is to be carried forward into the new computer system. So we should remove any purely human activities and replace them with external entities.

The logicalisation of the 'Store New Stock' process has led to a change in work practices which should not escape the astute system analyst: whereas in the Business Activity Model and in Figure 6.34 the stock keeper informs the stock clerk of any stock allocation adjustments and the stock clerk updates the files, in the proposed new system it is the stock keeper who records, directly, any storage adjustments.

▶ Rebuilding the DFM hierarchy

It is likely that some or all of the original boundaries of the lowest-level DFDs (and thus of high-level processes) were determined by physical factors, such as location. These factors are no longer relevant, so we will now regroup our logicalised elementary processes to form new low-level DFDs based on logical considerations. We will then use these to define higher-level processes, which will in turn reflect a more logical view of functionality.

Possibly the most effective way of starting this is to combine all of the lowest-level DFDs into one very large DFD, in which every process is elementary. This should be fairly straightforward as each DFD will show which others it interfaces with around its border. We can then draw new boundaries around logically related processes to form new bottom-level DFDs. The problem with this is that the 'big' DFD will often be too large or cumbersome to draw. So we will usually end up by examining the bottom-level processes in their separate, but interlinked, diagrams.

However, these are purely matters of presentational convenience, and whichever way we choose to approach it our main task is to establish new logical groupings of elementary processes. Groupings can be difficult to define, and are usually based on subjective views of functionality. Once again, strict rules are difficult to apply: it is far more a matter of experience and judgement. However, to help us there are three basic classes of grouping to look for.

1. Functional groupings

The most important thing when grouping processes is to reflect the users' perception of functional areas. Current systems may already be organised along functional lines, at least in part, so we should be able to carry some groupings forward from the physical DFM. We may also have identified some functional areas during Business Activity Modelling.

Obvious functional areas for ZigZag include: Purchase Order Placing; Stock Receiving; Customer Order Despatch. In some cases high-level physical processes will have been linked by flows running through a transient data store, e.g. process 2 (Check Delivery), which is linked to process 1 (Place and Monitor Orders) and process 3 (Store New Stock) in the ZigZag DFD of Figure 6.29. With the removal of the interceding data stores these processes will end up being linked directly by data flows, which implies a close functional relationship. More often than not these processes will now be merged. In the case of ZigZag, the following picture emerges when we remove the transient data stores and strip down to elementary process level the vicinity concerned with the arrival of new stock (see Figure 6.40).

Following our earlier discussions, we will dismiss the transient data stores after we make sure that we don't lose any of the delivery information contained in them. The diagram of Figure 6.41 shows two direct links between processes. We first deal with the link between processes 2 and 1.4. What task is process 1.4 performing? It forwards information about rejected goods back to the purchaser who originated the order while keeping a record of the rejection. But that task can be handled by a data flow from process 2 directly into data store D1. Process 1.4 therefore only adds

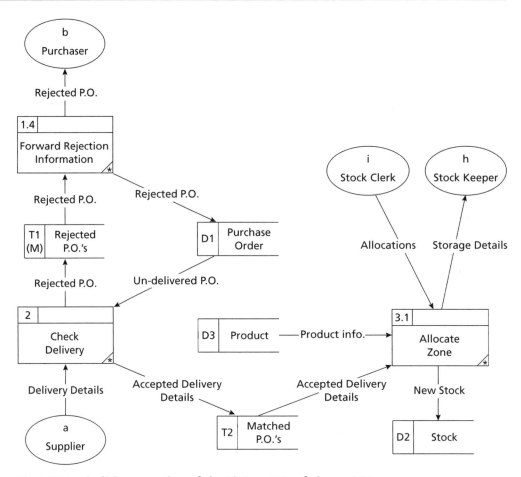

Fig 6.40 Logicalising a section of the ZigZag DFD of Figure 6.29

to the bureaucracy of the organisation and we can, after consultation with the users, remove it totally (this will of course have repercussions on the working practices of ZigZag because now the people checking a delivery are responsible for forwarding relevant rejection information to the purchasers, not the people who set up the purchase orders; but note that, because of the versatility of the new system, everybody concerned can have access to this information because we will *have* recorded it in the system).

Turning now to the data flow between processes 2 and 3.1, we find that things are slightly more involved. We detect this from the fact that process 3.1 – Allocate Zone – involves a second trigger from the Stock Clerk external entity which results in an update of the Stock data store. Here we ask ourselves whether there is a time lapse between accepting a delivery in the delivery bay and actually deciding where to store it. Consultation with the users reveals that such a lapse does indeed exist; zone allocation takes place independently and what the stock clerks really want is information about recently arrived goods. We solve this 'problem' by allowing the Stock Clerk external entity to access the Purchase Orders data store (see Figure 6.42).

(The final remark above actually means that a new requirement to 'provide information of unallocated deliveries' has to be added to our Requirements Catalogue.)

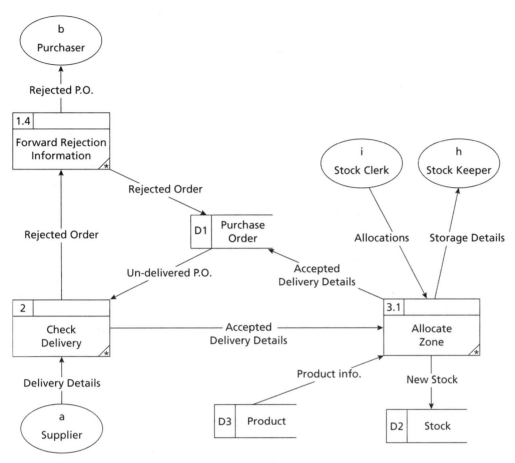

Fig 6.41 Removing the transient data stores of Figure 6.40

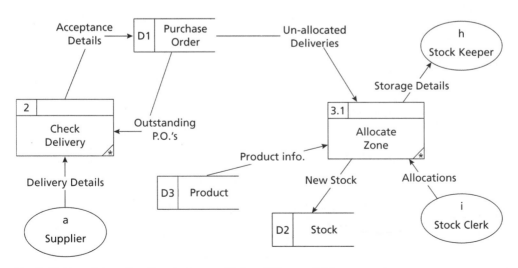

Fig 6.42 Resolving the direct process links of Figure 6.41

The above discussion indicates that there is a close functional proximity between the 'Store New Stock' and 'Check Delivery' processes on the level-1 DFD in Figure 6.29. We can therefore consider grouping them to form a new level-1 process called, say, 'Receive New Stock' (see Figure 6.43). Note also how the Goods In people become an external entity to indicate that it is they who decide what to accept or reject.

(Figure 6.43 involves the processes of Figure 6.40, which due to merging appear with new identifiers. Readers should take some time to follow the manifestations of the processes involved to satisfy themselves that all the information of the original diagrams has been captured by Figure 6.43.)

Close functional relationships between processes are also sometimes indicated by interaction with the same groups of external entities.

2. Process type groupings

Processes often fall into distinct types:

◆ those that support the business functions of the system;

◆ those that maintain system information or control. In the ZigZag system there are processes within each functional area concerned only with archiving (i.e. the removal of data from the system to some sort of back-up, such as a tape library). So we could group all of these together in a low-level DFD called 'Housekeeping'.

3. Data access groupings

If functional or process type groupings prove difficult to identify, we may find it useful to group together processes that access the same or similar data. Indeed, for systems

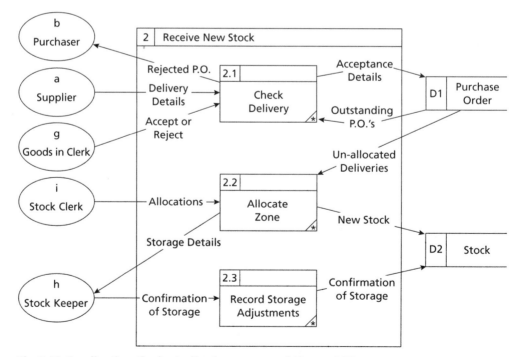

Fig 6.43 Readjusting the logicalised processes of Figure 6.29

whose main purpose is the maintenance of reference data this may be equivalent to grouping processes along functional lines. If we find that a process could belong to more than one grouping, we should place it in the one that contains the most important data accesses (creates and deletes rather than reads) for that process.

Perhaps the easiest way to identify which processes access which entities is to draw up a Process/Entity cross-reference, such as the one in Figure 6.45.

Rebuilding the hierarchy for ZigZag leads to the level-1 DFD shown in Figure 6.44.

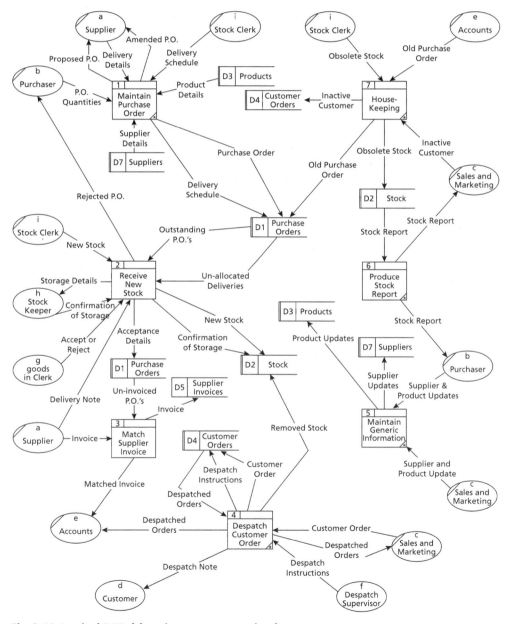

Fig 6.44 Logical DFD (showing current services)

In general, process groupings should be based on functionality, and only for those processes that do not clearly fall into neat functional areas should we look for process type and then data access groupings.

Process/Entity Cross-reference

DFDs should support the processing of entity occurrences in the following ways:

◆ There must be at least one DFD process that records the creation of each entity.

◆ There must be at least one DFD process that records a value for each attribute.

◆ There must be at least one DFD process that records the deletion of each entity.

The only entities not subject to the above rules are those that are used for reference only, and which have been created by other systems (e.g. Product is created and maintained by the Sales and Marketing system).

The correspondence of processes and entities can be checked by drawing a process/entity matrix (Figure 6.45), which lists entities along the top and elementary processes down the side. We then take each process in turn and examine any data flows going to data stores. These will represent updates to data (and therefore to entities). We should decide which entities are created, read, amended (updated), or deleted by the flow, and mark the relevant cell in the process/entity matrix with a C, an R, an A (U), or a D. (The process/entity matrix is also known as a CRAD (CRUD) matrix.)

Process \ Entity	Product	P. Order	Depot
1.1 Create Proposed PO	R	C	R
1.2 Check Availability		R	
1.3 Confirm PO		A	
1.4 Forward Reject Info.		A	

Fig 6.45 Process/entity matrix (extract)

Each entity (that is not a reference-only entity) should have at least one C and one D in its column. If there are any gaps the DFM should be checked for missing processes or data flows, and updated if necessary. For example, the process/entity portion in Figure 6.45 suggests that we have missed the process that is responsible for deleting a purchase order. We therefore proceed to enquire what mechanism does ZigZag use to archive old or obsolete purchase orders. This discussion with the users leads to the identification of process 1.5: 'Archive Old Order'. (Because deleting information from the system is quite a drastic measure we will be returning to it during Specification.)

▶ Completing the Logical DFM

Once we have drawn all of the lower-level DFDs and verified them with users and against the LDM, we should update the EPDs, External Entity Descriptions and I/O Descriptions to reflect the new logical model.

Many of the problems associated with physical constraints in the current system will in effect 'disappear' with logicalisation. However, some will still be applicable to the new system, e.g. legal requirements to use particular forms or carry out duplicate manual processing. We should record any physical constraints that are still valid in the Requirements Catalogue.

▶ SUMMARY

1. Data Flow Diagrams illustrate the way in which data (or information) is passed around the system, and how it is transformed and stored within the system.

2. The Data Flow Model consists of a hierarchy of Data Flow Diagrams and associated textual descriptions.

3. Data Flow Diagrams are most often developed from scratch or from Business Activity Models. They may also be developed from intermediate diagrams, such as Document Flow Diagrams and Resource Flow Diagrams.

4. Data Flow Models are extremely flexible, and can be used to represent both logical and physical views of a system's existing or required functionality.

▶ EXERCISES

6.1. Produce a Document Flow Diagram for the Fresco system of Exercise 3.8.

6.2. Produce an overview physical DFD for the Fresco system of Exercise 3.8.

6.3. Use the Business Activity Model from Exercise 3.13 to produce a Data Flow Diagram from the Natlib library.

6.4. Decompose all processes of your Data Flow Diagram from Exercise 6.3. How has it affected your original answer to Exercise 6.3?

6.5. Produce a list of requirements from your models of Exercises 3.13 and 6.4.

6.6. Suggest situations where you would consider not using the Data Flow Modelling technique.

6.7. Produce a *logical* Data Flow Model for the Hergest educational institute of Exercise 3.14 as supported by the data model produced for Exercise 5.7.

6.8. Produce a Data Flow Model for the system described by the requirements of Exercise 5.9.

6.9. Use the data model produced for the scenario of Exercise 5.1 to produce a Logical Data Flow Diagram for Markalot.

6.10. Produce a list of requirements that conforms to the Markalot scenario and your answers to Exercises 5.1 and 6.9.

6.11. The level-1 DFD depicted in Figure 6.29 contains seven elementary processes. This suggests that the diagram could be levelled otherwise. By bunching together processes 2 and 3, 5 and

6, and 4, 7 and 8 set up a new level-1 DFD made up of three processes. Progress your new level-1 processes to elementary level. Which depiction of the system do you prefer?

6.12. Study the text concerning Figures 6.40 and 6.41. Refer to ZigZag's Work Practice Model to see whose jobs will be affected. Adjust the WPM to reflect the changes introduced by logicalisation.

6.13. Trying to follow what happens to each current physical process, data flow and data store during logicalisation is a project management nightmare (which probably suggests that a situation has to be quite complicated before we decide to produce a full Current Physical followed by a full Logical DFM). Produce a list of all current physical elementary processes implicit in Figure 6.29 and correlate them with the elementary processes of an equivalent list for the logical DFM of Figure 6.44.

6.14. Produce a Current Physical level-1 DFD for Bodgett & Son (Exercise 3.15).

6.15. Use the suggested solution to Exercise 6.14 to decompose process 1 (Estimate Job).

6.16. Logicalise the DFD produced in Exercise 6.14.

Business System Options

INTRODUCTION

The purpose of the systems analysis carried out so far has been to agree the functionality of the new system. During this analysis we produced a comprehensive statement of user requirements in the Requirements Catalogue and a detailed description of the business problem in the form of current system models.

Our next task is to develop potential system solutions to this business problem (called *Business System Options*, or *BSOs*), and to evaluate their impact and benefits. Final selection will be made by the Project Board assessing the relative merits of the alternative Business System Options, and it will adopt the one that best balances the matching of business requirements with return on investment. This may involve dropping some of the less important requirements if they cannot be cost-justified, or they place the project at risk of late delivery. The result will be to narrow down the area of study to just those requirements that can be justified and decide on the shape of the system that we will be specifying in detail.

In reality, however constrained the project brief may appear, there will always be a number of different options for satisfying any set of user requirements. The thing that will vary from project to project is the degree of difference between the options.

For example, where a packaged solution has already been identified during a Feasibility Study different options will concentrate on how the package is to be enhanced, customised and implemented. One option may suggest a 'vanilla' implementation (i.e. one with no changes whatsoever), installed by the software supplier. Another might offer a closer match with requirements through the bespoke development of an additional feature, again implemented by the software supplier. Finally, we could suggest that the package be substantially modified in some areas, and the implementation carried out by an in-house development team.

This scenario is far removed from one where the project brief has not dictated a particular type of solution, and where one option might be a package implementation, while another could be a fully bespoke development.

Selection of a Business System Option represents one of the most important decision points for the project. The costs incurred in bringing us to this stage are still relatively small, being made up of largely internal effort from a small project team, together with representatives of the business and suppliers of software or hardware. While the costs of any external consultancy can be significant, they remain very low relative to the costs incurred in delivering a detailed design and implementation (which will usually involve the purchase of hardware and/or software). It is therefore important now, before passing the 'point of no return' on the investment, to reassess the original business case for the project in the light of our increased understanding of requirements, benefits and likely development costs. It is also important that sufficient time and effort be devoted to BSO definition and evaluation, as changes to the option selected will become increasingly expensive to make as the project progresses.

▶ **Learning objectives**

After reading this chapter, readers will be able to:

- ◆ understand how to define Business System Options;
- ◆ present Business System Options;
- ◆ document Business System Options, including the transformation of the LDM and DFM into required system models;
- ◆ conduct a Function Point Analysis.

▶ **Links to other chapters**

- ◆ Chapter 2   The BSO element of the Feasibility Study will act to constrain and guide the development of the full BSO.
- ◆ Chapter 4   The Requirements Catalogue is the key input to the development of BSOs.
- ◆ Chapters 5 and 6   The selected BSO will include Required System Logical Data and Data Flow Models.
- ◆ Chapter 13   The outline TSO (Technical System Option) included as part of the selected BSO will be carried forward for formal definition once the system has been more fully specified.

▶ DEVELOPING BUSINESS SYSTEM OPTIONS

A prescriptive definition of what a BSO should contain is impossible to give, as BSOs will vary greatly with different types and sizes of project, and more particularly with the differing needs of Project Boards, organisational standards and project team make-up. However, they all attempt to do the same thing, which is to describe a potential computerised information system and its costs and benefits. This description will be both textual and diagrammatic. The final version will usually contain diagrams such as BAMs, DFDs and LDSs, but will usually be supplemented with less formal diagrams whose main purpose will be to illustrate the BSO in order to aid the understanding of the Project Board and its business representatives.

One aim of the BSO evaluation process is to fix the requirements of the new system. After the Project Board has made its decision, no new requirements will be accepted or dropped without an Exception Report requesting an executive decision in accordance to the policies and procedures of the organisation (often a costly exercise, particularly if changes are made late in the project). In other words, the requirements that are chosen during BSO constitute a contract against which the future system will be judged.

Figure 7.1 shows the process of choosing a business option with all the likely iterations included.

Often the Project Board will find none of the proposed options entirely acceptable, and will request the development of new options or the refinement of a presented option, before final selection is made. In this way we may get iterations of the process in Figure 7.1.

> **Note**
>
> Many of the activities involved in defining possible system solutions (e.g. cost–benefit analysis, risk management, supplier negotiation) are those of general systems analysis and will be highly dependent on circumstance and organisational standards as defined in the organisation's policies and procedures. These activities will not be discussed in any great detail here (for further information see References and Further Reading).

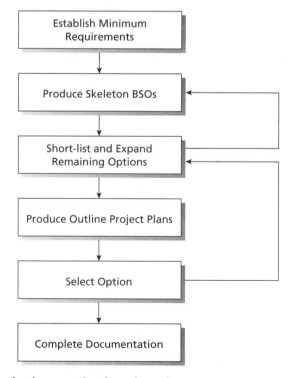

Fig 7.1 Choosing a business option is an iterative process

▶ Establish minimum requirements

Each BSO must satisfy at least the minimum requirements of the new system. At this stage the Requirements Catalogue will contain a large number of very detailed entries, each of which will have been individually prioritised. To grade and sort all of these by priority and to list every requirement satisfied by each BSO would be extremely cumbersome and rather confusing as a basis for discussions with users. It is more useful to group requirements into prioritised functional areas, each with an overall textual description and a list of its Requirements Catalogue entries. The set of minimum requirements for the new system can then be expressed as a list of functional areas that need to be covered by all BSOs.

The minimum requirements are those without which the system will be infeasible, either because it will not have any overall business benefit, or because it will not be able to support some of the functions of the business. There will almost always be some critical shortfall in the current system (due to capacity constraints, new business functions, new regulations, etc.). So the minimum will tend to cover the functions of the existing system plus solutions to those shortfalls.

It is important to note that the minimum requirements of a system are not the same as those classified as Essential in the Requirement Catalogue. Some of these Essential requirements may only be essential in the longer term, e.g. in the ZigZag catalogue the requirement to handle 100,000 products will certainly be essential in the medium term, but the existing limit could be lived with for a short while. So, strictly speaking, such requirements could be omitted from the minimum requirements list, albeit with a warning about the consequences. It may also be the case that certain requirements are only essential should the Project Board choose to implement an overall area of functionality, e.g. customer ordering over the Internet for ZigZag could be rejected as financially infeasible once our analysis has revealed its true requirements and costs.

Where a Feasibility Study or the PID has identified that a package solution is to be implemented, the minimum requirements will cover the core functions of the package (all packages have a set of core functions without which they cannot operate).

▶ Develop skeleton BSOs

We will need to develop a number of skeleton BSOs (six is about as many as any project team can handle at any point in time, and three would be about average). These skeleton BSOs will largely be in textual form, and may vary in functionality from a BSO that satisfies just the minimum requirements to one that satisfies virtually all justifiable (non-conflicting) requirements.

The most common way of developing options is to again use brainstorming sessions or workshops, to which we would invite (in addition to the project team) business representatives, systems developers and technicians, and for larger projects possibly consultants. Where a package has been pre-selected we would also include representatives of the software suppliers or their agents.

The process of defining options is one of the most creative of the entire project. It is extremely important that all ideas be given consideration, however 'off-the-wall' they may at first appear. The amount of effort required to produce a reasonably well-thought-out set of skeleton options can be considerable, so we must be **careful** to allow sufficient time within the project plan.

In reality the Feasibility Study or PID will have imposed some constraints on just how creative we can be. So options must at least adhere to these, *unless* Requirements Definition has highlighted requirements that were not revealed by the previous studies. Where new business functions are found to require additional support over and above that envisaged previously, constraints may need challenging. For example, we may find that the development software specified in a PID cannot on closer examination deliver the required user interface.

We will usually find that many of the requirements identified by users as high-priority cannot be cost-justified or delivered within the time constraints imposed by the PID. We will need to handle these carefully, as many users (and the Project Board) will feel some emotional attachments to these requirements and dropping them from options will be politically sensitive. Nonetheless it is important to apply cost–benefit analysis to all requirements, and even if we feel it is politically necessary to include requirements that appear unjustifiable we should highlight their cost or time impacts within the options.

Types of Business System Option

Package

Packages may offer a solution that can be less costly and faster to implement than bespoke solutions. They may also offer the only realistic solution where the organisation has little or no internal systems development capacity. However, they are rarely (if ever) an ideal fit with user requirements. There will be two basic ways to address this functional shortfall, and both may be worth exploring during the definition of BSOs.

First, the package can be enhanced with additional bespoke functions or the core code modified. This will certainly add to cost and development time, and may lead to maintenance overheads in the future. It will, however, deliver a closer fit with requirements, and may be essential to deliver mandatory requirements. In any event, packages will usually require some enhancement to handle interfaces with any existing systems that will remain after implementation. In order to comply with the conditions of ongoing technical support and maintenance it is often the case that any modifications to the package must be carried out by the software supplier themself. This will add significantly to the costs of the project. So any requirements met in this way will need to be carefully cost-justified.

Secondly, requirements can be modified to fit the package. This is invariably unpopular with users, but may be unavoidable if core areas of the package are involved. For requirements that are non-essential or where a reasonable alternative already exists within the package, this approach will almost certainly involve lower cost and risk than modification.

In reality most projects that result in package implementation will involve a blend of these two approaches.

Where a package is being considered, but has not been pre-selected, the Requirements Catalogue should be used to short-list potential software (with only packages capable of delivering the minimum requirements being considered). Each of the short-listed packages will then form the basis of a skeleton BSO (along with any other viable options).

Bespoke software

Where no packages exist that match requirements (e.g. for new or unique business functions), or where an organisation has substantial internal capability, a bespoke solution may be considered. This will almost certainly provide a better match for requirements, but will typically incur higher costs and involve longer time to deliver.

▶

If the bespoke development is carried out in-house, then costs for bespoke software can often be favourable compared with packages that require any significant modification. However, when the development is outsourced or when packages with close fits with requirements can be identified (typically the case for Accounting or Human Resources systems) costs can outstrip those of a package solution. The closer functional fit of the bespoke development will then need to provide additional benefits sufficient to justify the bespoke development.

Careful cost estimation and planning of the delivery will be necessary, usually leading to longer BSO definition timescales.

Current system with modifications

It is reasonable to assume that for the vast majority of projects the existing solution is unsatisfactory, and therefore unlikely to provide a viable BSO as it stands.

However, in many cases there will be an option to leave the current system largely in place and either modify it directly, or add to it with a minor package or a bespoke software implementation covering its shortfalls.

In many projects this will be an option that the Project Board, whose members are not well acquainted with the day-to-day problems of the existing system, will want to be presented, particularly if they cannot see one or two overwhelming problems that rule it out entirely.

Although this will tend to be unpopular with the project team and direct system users, who will want to deliver a more nearly perfect solution, it is often the lowest cost and risk option, at least in the short term.

In the longer term, the lack of real support for important business requirements, capacity issues, and ongoing maintenance costs of old systems will often mean that this is a temporary solution only. The decision to modify the current system will also be influenced by the level of documentation that supports this system.

Phased delivery

During the exploration of options it may be found that a greater number of requirements can be cost-justified that can be delivered within an acceptable timescale. In this case it may be preferable to present options that deliver a series of phased solutions. The first will deliver at least the minimum requirements, with subsequent phases delivering logical units that support new functional areas.

It is important that each of the phases stands up in its own right (i.e. can be justified); otherwise they will invariably be rejected by the Project Board (or academic supervisor). It is also important to develop plans that show with some degree of certainty when all phases will be delivered. Otherwise, business representatives will be suspicious that the later phases will never be delivered, and that the project team is merely presenting them to keep users on board with a cut-down solution.

End-user development

Where the solution appears to be the implementation of an end-user tool, BSOs will concentrate heavily on the selection of the tool and methods of training and support.

In such projects there will also commonly be a requirement to populate a database for the tool to operate on. So the BSOs we develop will also need to take account of the systems development effort needed to build an appropriate data model, and, most significantly, to build data population interfaces. The costs of these activities are often underestimated, as is the ongoing maintenance of end-user systems.

Each BSO description, even at the skeleton stage, should include the following non-functional details:

♦ **Approximate cost of the option.** This must include internal and external resource costs, as well as any software and hardware costs. At this stage costs will be very rough. As a rule of thumb it is wise to add as much as 50% contingency to costs within skeleton BSOs. This figure will reduce as more detail is added, and our knowledge of costs improves. However, even the final BSO should include as much as 25% contingency (remember that projects never stay within budget).

♦ **Development timescale and outline plan.** Again, at this stage include large amounts of contingency, particularly for any bespoke elements of the option. Where elements are package-based or outsourced, firm dates and estimates from suppliers, with penalties for late delivery, will reduce the amount of necessary contingency, but not below 20% at this stage (later on 10% would be acceptable).

♦ **High-level technical description.** Although BSOs will be largely logical in nature, as they refer primarily to the functional requirements delivered by each option, they will almost certainly require differing technical infrastructures to support them. There may also be restrictions imposed by technical strategies or limitations that are impossible to dismiss, as they will limit the *functional* possibilities of any new system. No attempt should be made at this stage to fully specify the technical environment (as BSOs do not contain sufficient design detail to support this), but enough detail must be provided in order to produce cost estimates. The level of detail included in the final BSO will also need to be sufficient to select hardware platforms and suppliers if the project has the remit to do so (the PID may specify which technical environment should be used), particularly in the case of a package option where a platform change (for example, from NT to UNIX) may be necessary. Any non-functional requirements that will definitely need to be considered later in evaluating Technical System Options should be noted in the Requirements Catalogue.

♦ **Human–computer interface.** The human–computer interface (HCI) will often be the most understandable and sensitive area of the system to users. If a package option is proposed, then samples of the HCI (screen shots, a demo, or report samples) will help illustrate the option. For a bespoke system there may be different options available (e.g. browser, Windows) which will need to be considered by users and the Project Board, and which may have different development costs and resource requirements.

♦ **Organisation of the system.** This will include information on the required type of access to the system (e.g. on-line or off-line), interfaces with other systems and the distribution of the system.

♦ **Approximate data and transaction volumes** (essential for hardware costing).

♦ **Major benefits** to the business of each business area supported.

♦ **Impact on the organisation and other existing systems.** This must include training needs for the Project Team, and, more importantly, the users. It should also cover data conversion requirements and any transition arrangements necessary to wind down the old system, while bringing the new system up to speed (e.g. the completion of orders in the old system, possibly on paper, or transfer of half-finished orders to the new system for completion).

Id	Name	BSO 1	BSO 2	BSO 3	BSO 4
104	System must support up to 100,000 live products		X	X	X
105	Record proposed purchase order	X	X	X	X
106	Confirm purchase order	X	X	X	X
107	Record customer order			X	X
108	Arrange despatch of customer orders			X	X
109	Cust. Orders to be kept 3 months then archived			X	X
110	Provide delivery to despatch audit trail				X
111	Provide supplier performance monitoring facility		X	X	X
112	Record delivery data including rejections	X	X	X	X
113	Schedule deliveries (to nearest half hour)	X	X	X	X
114	List overdue deliveries		X		X
115	Facilitate rescheduling of overdue deliveries		X		X
116	Make use of existing PCs in stock office	X	X	X	X
117	Delivery due dates to be converted to DD/MM/YYYY	X	X	X	X
118	Facilitate alternative product selection	X	X	X	X
119	Report on stocks near withdrawn from sale date				X
120	Provide possible stock-out warnings				
121	Monitor supplier invoices				X
122	Produce stock report		X		X

Fig 7.2 Extract of mapping of ZIgZag's Business Systems Options to Requirements Catalogue

It is useful to relate these details to individual functional areas within the BSOs, so that the relative merits of each of those areas can be discussed.

For ZigZag, business options might include: the minimum of a few enhancements and stand-alone spreadsheets in order to overcome critical operational problems (we will call this BSO1); an option of fully automating current stock systems functionality with added capacity (BSO2); the minimum plus the handling of customer orders over the Internet (BSO3); and an option encompassing all requirements with a priority of Essential and Desirable (BSO4).

It is convenient to summarise options in a grid that shows the constituents of each option (see Figure 7.2).

▶ Produce short-list of options

None of the skeleton options will be entirely out of line with the wishes of users, partly because our investigation of requirements should mean we have a good understanding of what will be acceptable. For the same reason there are unlikely to be huge variations in the options at this point, so to continue with as many as six will mean a lot of effort for little return.

It is important to resist any temptation to reduce the list to a single 'obvious' option, as this will frequently be unduly influenced by the current system, or by one particularly vocal personality.

Instead we should discuss the relative benefits and drawbacks of all of the options with users and senior business representatives, including members of the Project Board, with the aim of reducing the list to just two or three. By including Project Board members in discussions it will help us to avoid developing any options that the Project Board is likely to dismiss outright, and by including their input early on should increase their commitment to the selected option. It will also save time when we come to present the options formally, as the Project Board will have some familiarity already with the key issues.

In the case of ZigZag, we will assume that the business is committed to the development of a new system in order to provide for future growth, so we will drop the first BSO (current system with 'patches') and the second as it does not support the strategically important e-commerce initiative.

▶ Expand remaining options

Once we have narrowed the list down, we need to flesh out the remaining options. Remember that our aim is not only to describe the functionality and organisation of the proposed system, but to provide sufficient information to enable the options to be assessed by the Project Board.

One option will inevitably be the preferred option of the project team, either because it delivers the optimum cost–benefit outcome or because it represents the best functional or strategic fit. It is quite acceptable to identify this preferred or 'recommended' option (indeed many Project Boards will insist on it, as it acts as a focus for their decision making), but this should not distract us from fully exploring and defining the alternatives(s). The process of adding further detail to each BSO can alter the project team's views on its relative benefits. So if we concentrate too much on one preferred option we will miss out on this valuable process.

The resulting BSO descriptions must be largely textual in nature in order to be widely understood, but any differences in functional support and system boundary can be emphasised by using BAMs (or even DFDs and LDSs), as well as informal diagrams and illustrations.

During BSO we may also wish to provide an extra chunk of DFM or LDM to help in the understanding of new requirements. If this is done, we have the extra bonus of being able to provide a better estimate of development costs.

Work Practice Modelling can also be an invaluable tool for exploring and documenting options. For example, through WPM we may produce models for the picking of customer orders that reflect three differing approaches:

1. A picker takes (or is allocated by the system) a number of orders to pick and then passes them on to a specialist despatcher to package up and address for despatch.

2. Each order is picked and packaged ready for despatch by a picker, one order at a time.

3. Pickers specialise in picking the smaller e-commerce orders, in which case they adopt approach 1 above, or wholesale orders, in which case they adopt approach 2 above.

By modelling these scenarios and discussing the results with users, three very different options for satisfying business requirements will emerge.

The short-listed BSOs will be much more detailed than the skeleton BSOs, and in particular will require greater attention to the following:

Common ways of analysing tangible costs and benefits are outlined in Chapter 2 and detailed in Robson (1997).

◆ **Cost–benefit analysis.** Costs should include hardware and software purchase costs, systems development costs, installation, migration, operating and training costs. Benefits are usually more difficult to assess, especially in the area of cost savings if measures have not been made of the costs of operating the existing systems. Most benefits are intangible, and come in the guise of service improvements, increased accuracy, availability, timeliness, usability and functionality.

◆ **Impact analysis.** The proposed system is certain to have an impact on the working practices and business organisation (e.g. it may create or eliminate jobs, or change existing job descriptions or practices, all of which adds to risk and training costs). Work Practice Models used to explore different options can be used to develop an informal descriptive impact analysis, which will be essential for users in evaluating the BSOs.

◆ **System development and integration plans.** Each option will require a different development strategy, and may possibly lead to large-scale re-planning of the project. Any problems with integrating the new system with existing systems should be highlighted.

Once again, it is worth emphasising that when developing BSOs we must always take care that the finished product is easily understood by users (who may not be familiar with SSADM's formal notation) as it is they who will make the final decision on the system solution.

Function Point Analysis

Function Point Analysis provides a neat way of using the models we have already created to calculate a verifiable estimate of the future system's development costs. The main premise behind Mk II Function Point Analysis (Drummond, 1992) is that the time and effort needed to design, construct, test and implement a process, any process, depends on the number of inputs to that process, the number of outputs from that process, and the number of entities that need to be visited to conclude that process.

When the analysis has been done in the manner suggested by the default structure of SSADM, most user requirements map directly to an elementary process; this elementary process has data flows going in and/or out of it, each containing a documented number of data items which we can count; it also uses a number of data stores which contain a number of entities as shown on the Data Store/Entity Cross-reference. We can therefore use a spreadsheet to add all the inputs, all the outputs and all the entity accesses of *all* the requirements in the Requirements Catalogue. This gives us N_i, N_o and N_e, the system's total number of inputs, outputs and entity visits, respectively. We use these values to ascertain the system's worth in 'unadjusted function points' using the formula

$$UFP = 0.58 \times N_i + 0.26 \times N_o + 1.66 \times N_e$$

We then calculate the 'size' of the system by multiplying UFP by a 'Technical Complexity Factor' TCF:

Table 7.1 Industry average productivity values

System size (S)	Productivity for a development using 3GL (third-generation language) (p)	Productivity in a 4GL (fourth-generation language) environment (p)
50	0.099	0.158
100	0.106	0.169
150	0.111	0.178
200	0.116	0.185
250	0.118	0.189
300	0.119	0.191
350	0.119	0.190
400	0.117	0.187
450	0.113	0.181
500	0.109	0.174
600	0.098	0.156
700	0.085	0.137
800	0.074	0.118
900	0.065	0.104
1000	0.058	0.093
1100	0.055	0.088
1200	0.054	0.087

$S = \text{TCF} \times \text{UFP}$

The TCF factor ranges from 0.65 for systems which are built by experienced teams using tested technology, to just over 1 when new methods, platforms or technology is to be used.

With S in hand, the time in weeks needed for the whole development is given by

$\text{Weeks} = 2.22 \times \sqrt{S}$

while the effort required to develop an on-line system is

$\text{Effort} = S/p$

where the productivity p is found from Table 7.1.

For off-line systems the effort is increased by 50%.

The number of people required to complete the job is given by

$H = (0.044 \times \text{Effort})/\text{Weeks}$

This last formula can also be used in conjunction with Table 7.2 to estimate the number of people who have to be involved in every SSADM module and beyond.

All the weightings of Function Point Analysis given here correspond to industry averages. Each development team can use these as an estimate or as a comparison of their performance when compared against the industry's averages.

Table 7.2 Industry average percentages per development module

	Effort	Elapsed time
Requirements Analysis	11%	20%
Requirements Specification	11%	15%
Logical System Specification	5%	5%
Physical Design	10%	10%
Code and Unit Testing	46%	25%
System Test	12%	15%
Implementation	5%	10%

▶ Select Business System Option

We now present the short-listed BSOs to the Project Board or user representative body. The activities involved in preparing and carrying out the presentation will be dictated by the internal standards of the organisation and the circumstances of the project. For example, if the project team represents an external software house, the presentation may well take the form of a sales session, whereas if the team is an internal one the format of the presentation will probably follow organisation guidelines.

Tips for presenting BSOs

The manner in which the BSOs are presented can greatly influence the outcome of the selection process. Not only can a poorly presented set of options lead to ill-informed decisions, it can also undermine the commitment of the Project Board to the project as a whole. Some ways of assisting the Project Board in assessing options include:

Slides

Most organisations will require a formal presentation of the options. As a general rule aim for 20 slides or less. The content of the slides should not be a script for the presenter to read from. Instead, try to use tables and illustrations that support the spoken content of the presentation.

Concentrate on highlighting the key *differences* in functionality, cost–benefit and time-scales, as it is these that will determine the selected option. Consider how individuals choose a new car. They will draw up a short-list based on cars that share a set of common features that broadly address their needs (e.g. numbers of doors and seats, rough cost). They will then select the precise model by concentrating on the key differentiators (e.g. fuel consumption, performance, actual cost, image). Project Boards will make their decisions in a similar fashion.

Ensure that the presentation and slides relate each BSO to the objectives outlined in the PID. Be clear as to why the recommended option is preferred by the project team.

Reports and handouts

Always ensure that a copy of at least the executive summary of the BSO report is circulated prior to the presentation. This will enable the presenter to reduce the amount of time they need to spend on introductory or background material. The aim of the BSO presentation is to enable the Project Board to examine and explore the BSOs, so any advance understanding will maximise the time available for decision making.

Prototypes and demonstrations

Where a package solution is proposed, or where the human–computer interface is a major differentiator, demonstrations of certain functions can be useful. They tend to enliven the presentation and make the project seem more 'real'. However, they should be used with caution, as demonstrations can be time-consuming and all too often encounter technical problems, which can lower confidence in the project. Particularly where prototypes are used, care should be taken that the Project Board is not left with the impression that the system is virtually complete.

Walkthroughs and prior briefings

If possible, arrange briefings with key Project Board members or other senior users just prior to the presentation. This will provide an opportunity to iron out any weaknesses in the presentation of the options and highlight problems that may arise with one or more of the options. Where there are key influencers outside of the Project Board, obtaining their input and support in advance of the presentation can be invaluable in adding strength to the case for an option.

Whatever the circumstances, our overall aim will be to explain each option clearly, placing emphasis on its relative strengths and weaknesses, to a level that will enable management to select a single BSO as the basis for the rest of the project.

In practice the Project Board will often adopt a hybrid BSO, combining features of two or more of the original BSOs. The Project Board will often also generate new ideas leading to minor changes to the selected option; it may even request the definition of a completely new BSO, although if our analysis and discussions prior to formal selection have been thorough this should be unlikely.

BSO presentation also provides the Project Board with an opportunity to reassess the viability of the project as a whole, and possibly to cancel it. This will be especially true if a formal Feasibility Study was not undertaken.

In many projects the BSO presentation will be the first time that project costs are understood with any degree of confidence. In these cases we will also be seeking to secure finance for the rest of the project. Even where finance is already in place, the process of BSO definition may have revealed changes to the estimated cost of the project, so again we will need to seek approval to continue with an increased project cost.

We should record the results of the presentation in detail, paying special attention to the reasons for selection of the chosen option.

We must ensure that the selected BSO is fully documented, as it will provide the basis for the rest of the project and for the final system. We begin by annotating the Requirements Catalogue to show which requirements are to be implemented by the new system. BSO selection will usually result in some of the requirements' being dropped. We should make sure that the reasoning behind their withdrawal is added to the resolution section of the Requirements Catalogue entries.

ZigZag's Business System Options

During the presentation of the two remaining options (BSO3 and BSO4), discussion focused primarily on some of the more 'minor' requirements, as there was widespread support for the major new components, such as e-commerce, which are common to both options.

For example, it was felt that requirement 111, which deals with the monitoring of supplier performance, was a bit loose. Management was not yet clear how to measure this performance and it was felt that an ill-thought-out report might lead to wrong conclusions. The requirement was therefore dropped.

The following features were considered and formally endorsed:

◆ on-line access to all information, including facilities for *ad hoc* queries.

◆ automation of routine operations, such as zone allocation.

It was clear that the major objectives associated with cost savings, extra services and improved efficiency are satisfied by both options. There will be savings in staff as many operations will be automated, and it is expected that most paper filing will be eliminated. It was felt that redundancies would be avoided, as expected retirements will reduce staff levels naturally, and growth in business will mean that staff are needed in other areas.

Following discussion of concerns that opening up the system to the Internet would make ZigZag vulnerable to outside attack, strong backing was given to greatly increased security measures and access controls.

Stock-keeping activities will not be greatly impacted except that, with the introduction of automatic zone allocation, smaller zones or 'locations' can be introduced leading to more precise stock storing and picking instructions (a major labour-saving measure, particularly with the increased picking and assembly of small orders for e-commerce).

Most of the data will be held centrally, with workstations in all of the depot offices and in the Purchasing and Sales and Marketing divisions. Any reports or queries still based on paper will be printed locally, with the exception of stock listings for the Purchasing and Sales divisions, which will be produced centrally.

Purchase order details will be entered into the system on-line by the Purchasing division, and wholesale customer orders by the Sales and Marketing division. All subsequent processing and inputs will be carried out in the individual depots.

The management at ZigZag decides to back a major redevelopment based on BSO4. They feel that the features of the new system can be cost-justified and that the development timescales are acceptable. The organisational impact is in line with their wishes, and staff whose jobs are automated or eliminated will be redeployed.

▶ REQUIRED SYSTEM LOGICAL DATA MODELLING

Once we have selected a BSO, we need to update the Logical Data Model to support all of the new information and processing requirements of the new system. The Logical Data Model produced prior to BSO selection will reflect the current system, and while we may have reworked elements of it while defining our BSOs, we will probably not have completed the process for the entire model. In any event, the final BSO may be a hybrid and so any reworking done during BSO definition will have to be reviewed.

The resulting Required System LDM forms the backbone of the Requirements Specification and of the subsequent Logical Design. It will be constantly referenced and adjusted as specification progresses, using the results of detailed low-level analysis and design techniques.

▶ Transforming the Logical Data Model

We conclude the top-down analysis of data by transforming the Current Environment LDM into the Required System LDM. We do this by examining the Requirements Catalogue to identify what elements of the data model should be removed and what new elements have to be added to support new functionality.

Removing redundant elements

The selected BSO may apply to only part of the current system, in which case we would need to remove those parts of the LDM that are not required for the new system.

In the case of ZigZag, if we had chosen to reject e-commerce and retain the existing wholesale customer ordering system, we would now remove the entities Customer, Customer Order, Customer Order Line and Despatch, as they would no longer be within the scope of the project.

Where the business activities to be covered by the required system include those of the existing system, it is unlikely that any elements of the Current System LDM would be redundant in the Required System LDM (with the exception of relationships which may become redundant as a result of adding new entities, and hence new indirect relationships).

Adding new elements

It is, of course, more likely that the transformation of the Current into the Required System LDM will result in additional entities, attributes and relationships.

There are a few additions to the ZigZag model worth looking at. Some can be handled with the concepts we know already and some will force the introduction of a few new concepts.

The ZigZag Requirements Catalogue specifies that a Purchase Order will be allowed to result in several deliveries, and that a Purchase Order Item may itself be split over more than one delivery. We will model this requirement by creating an entity Delivery, distinct from Purchase Order.

As each Delivery can deliver more than one Purchase Order Item, there is a many-to-many relationship between the two entities. We resolve this by viewing each Delivery as containing a number of Delivery Items, each of which will be the result of one Purchase Order Item (Figure 7.3).

The reappearance of the Delivery and Delivery Item entities resurrects the arguments that led to Figures 5.12–5.27 in Chapter 5. The relationship, therefore, that allows the system to provide a delivery-to-stock audit trail, as requested by requirement 110, is that shown in Figure 7.4.

Three other requirements resulting from the introduction of e-commerce orders in ZigZag's selected BSO are:

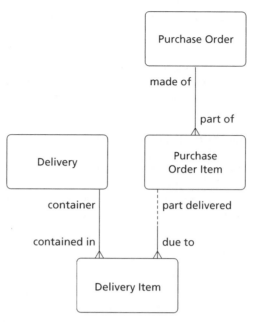

Fig 7.3 Data model supporting several Deliveries per Purchase Order Item

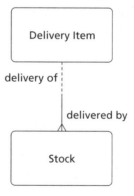

Fig 7.4 Each Stock is the result of a Delivery Item

◆ The ability to split a Customer Order or Customer Order Item over a number of Despatches. In the current system if an order cannot be completed because of lack of stock of a product, the order is despatched without the missing items. The customer (usually a shop) will then re-order later if the item is still required. For e-commerce any missing items will be despatched later, when stocks are available again.

See exercise 1 in Chapter 4 for a full description of this requirement.

◆ More precise locations for stocks of products within depot. Tighter control and efficiency in the picking of large numbers of small orders will be made possible by directing pickers to precise shelves within the larger depot zones (i.e. aisles or areas).

◆ Potentially different prices for e-commerce customers and wholesale customers, requiring a new E-Commerce Price attribute within the Product entity and a new Customer Type attribute in the Customer entity.

As Customer Order Items can now be despatched in more than one Despatch, we have a many-to-many relationship between the two entities. We resolve this by the introduction of an entity called 'Pick' which will record the amount of a Customer Order Item sent in each Despatch. In order to provide an audit trail against each Customer Order and to monitor productivity, we will also record which picker carried out each Pick.

Each Depot Zone will be broken down into a number of precise Locations, detailing the aisle number, shelf number and rack (i.e. container) of each Stock. Each Location is capable of holding more than one Stock of the same or different Products. If a large number of items of a single product are delivered to the depot, the system may need to allocate more than one Location for its storage. This will result in the creation of more than one Stock, each of which will be recorded against a single Location.

These two requirements of the last two paragraphs are therefore modelled as in Figure 7.5.

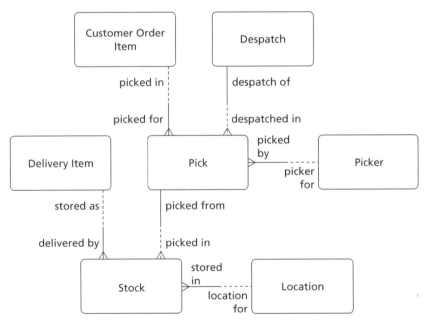

Fig 7.5 Data model supporting Pick and Location

One further significant change to the LDM is the introduction of more than one supplier for a given product. This will allow ZigZag to purchase products from different suppliers if the regular (or preferred) supplier is out of stock. The result is a many-to-many relationship between Product and Supplier, resolved by the introduction of the resolution entity Supplier Product (see Figure 7.6).

As we add each data-related entry in the Requirements Catalogue to the LDS, we should update the entry with a cross-reference to relevant entities or relationships.

The completed LDS for ZigZag is shown in Figure 7.6.

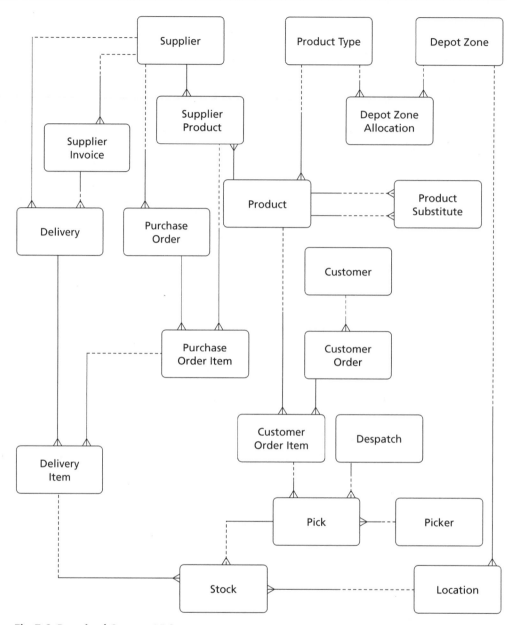

Fig 7.6 Required System LDS

▶ REQUIRED SYSTEM DATA FLOW MODELLING

The Logical DFM prior to BSO definition provides us with a picture of the underlying functionality of the current system. We now amend or enhance this model to produce a Required System DFM. We do this by carrying forward those elements of the Logical DFM that are to be retained in the new system, and by adding new processing as detailed in the Requirements Catalogue and included in the selected BSO.

▶ Transforming the DFM

We now take the updated Requirements Catalogue and amend the DFM to create the Required System DFM. Although the DFM will not be used in specifying the new system, it will be maintained as a means of placing the system design in context, and of describing how functions fit together at a high level. To transform the Logical DFM we take the following steps:

1. Remove any processing which will not be required in the new system.
2. Add level-1 processes and data stores (based on the LDS extensions which are being applied in parallel) to support new business activities.
3. Amend or develop lower-level DFDs to reflect changes in processing, such as new level-1 processes, new or amended support for existing activities, processes to maintain new entities, and changes to process groupings.
4. Amend the level-1 DFD to reflect lower-level changes.
5. Complete I/O Descriptions, External Entity Descriptions and EPDs for any new elements of the system, and update existing descriptions to reflect any changes.
6. Annotate the Requirements Catalogue with details of how the Required System DFM satisfies requirements.

Figure 7.7 shows the required level-1 DFD for ZigZag. The contents of the required DFM's data stores are dictated by the Data Store/Entity Cross-reference of Figure 7.8.

This is an important point worth repeating: the flows in and out of data stores contain data items which should be contained in those data stores, but the contents of each data store is defined through the content of the entities it encompasses, as shown in the Data Store/Entity Cross-reference; this means that we cannot finalise the DFM before we finalise the LDM, and we cannot finalise the LDM before finalising the DFM. We therefore see once again that Logical Data Modelling and Data Flow Modelling go hand in hand, cross-checking and validating each other.

Once the Required System DFM is complete, we should validate it against the Required System LDM once more using the Logical Data Store/Entity Cross-reference, by checking that Elementary Processes exist to create and delete every entity.

Identifying system events

While developing the Required System DFM it is worth thinking ahead to Function Definition and event identification. The reader may recall that we can have two types of event:

◆ a business event that triggers a series of business activities
◆ a system event that leads to new data entering the computerised system.

The DFM is a good technique for identifying the latter, system events.

Functions consist of units of processing carried out in response to real-world events, such as requesting stock transfers, or receiving deliveries. This processing may encompass more than one elementary process on the DFM. We should attempt to develop the DFM in such a way that it is possible to identify which elementary processes, when taken together, will handle the effects of an event, from input to output.

Events in DFMs (other than time-based events, such as the end of a tax year) are represented by data flows into the system. So when creating the Required System DFM,

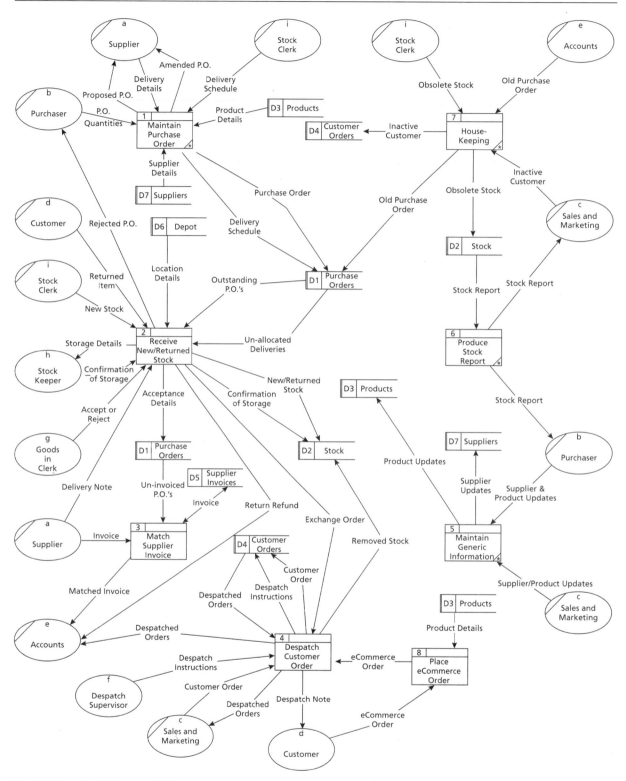

Fig 7.7 Required System DFD

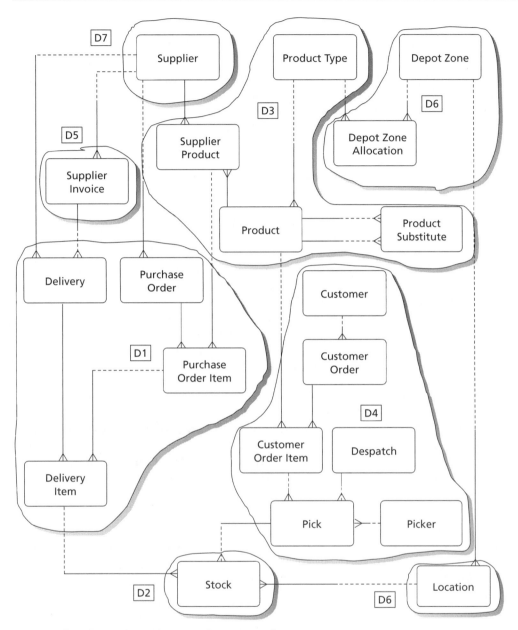

Fig 7.8 The ZigZag Data Store/Entity Cross-reference

we can try to make events and associated processing easy to identify in Function Definition by:

◆ ensuring that each elementary process is *driven* by a single major data flow. If we find that a process is driven by two flows, then each of the flows should be a trigger for mutually exclusive executions of the process;

◆ minimising the number of inter-process data flows. Too many directly linked processes will lead to difficulties in identifying discrete units of processing, and usually indicate that Elementary Processes should be merged.

▶ USER ROLES

The User Catalogue we developed as part of Work Practice Modelling during our investigation of requirements represents a physical view of the current system, with respect to the tasks and activities of users. We can now update the User Catalogue with the new activities and tasks to be supported by the required system. We will then be able to define User Roles, which will effectively provide us with a logicalised view of the users of the required system.

Not all job holders will necessarily come into direct contact with the system. For example, the people maintaining ZigZag's warehouse will never use the required system directly.

Users are those people who interact directly with the system. Subsets of users, requiring access to the same functions and at the same level of authority, are known as *User Roles*. In effect, a user uses the system by 'acting' in his or her 'user roles'.

An extract from the User Role form for ZigZag reveals that some tasks in the new system will be carried out by more than one type of user (or 'Job Title'), and that some types of user can act in more than one role, but that by and large each user has a definable role of their own (Figure 7.9).

> Strictly speaking, the external entities of the Required System DFM should consist of the system's User Roles.

When we start to develop user interfaces in later steps, we will use this list to identify which functions each user role will require access to. Each required access will indicate the need for a user interface. By creating user interfaces for each role, and then assigning these roles to individual users, we can limit the system access of a user to just those functions that they need and are authorised to carry out.

The identification of User Roles marks the point where the influence of the new computerised system on current working practices starts to manifest itself. The business activities that were earmarked as lying inside the system boundary have given rise to DFM processes, which in turn will give rise to user functions dedicated to each User Role. The working practices of these users will therefore be altered because of the system's existence. In some cases the responsibilities of users may shift from one to another.

User Role	Job Title	Activity
Delivery Scheduler	Stock Clerk	Book Delivery Update Delivery Maintain Schedule Overdue Delivery Query Check Available Time Slots Delivery Query
Goods In Clerk	Goods In Clerk	Amend Delivery
Stock Clerk	Stock Clerk	Add New Stock Remove Obsolete Stock Allocate Stock Location
Stock Keeper	Stock Clerk Stock Keeper Goods In Clerk	Confirm Stock Location Amend Stock Location

Fig 7.9 User Roles

▶ SUMMARY

1. A Business System Option describes a potential computerised information system together with its costs and benefits. This description will be both textual and diagrammatic. The final version will usually contain diagrams such as BAMs, DFDs and LDSs, but will usually be supplemented with less formal diagrams whose main purpose will be to illustrate the BSO in order to aid the understanding of the Project Board and its business representatives.

2. Each BSO description, even at the skeleton stage, should include the following non-functional details: approximate cost of the option; development timescale and outline plan; high-level technical description; human–computer interface; organisation of the system; approximate data and transaction volumes; major benefits; and impact on the organisation and other existing systems. It is useful to relate these details to individual functional areas within the BSOs, so that the relative merits of each of those areas can be discussed.

3. A number of outline BSOs are developed, all meeting at least the minimum requirements of the new system, and are short-listed for presentation to the Project Board.

4. The Project Board selects a BSO, which then forms the basis for the rest of the project. This selection has the effect of 'fixing' the Requirements Catalogue, with any further amendments being subject to formal change control.

5. Once the BSO has been selected, the Logical Data Model, Data Flow Model and Requirements Catalogue are updated to reflect the required system.

▶ EXERCISES

7.1. Use Function Point Analysis to estimate the workforce and time needed to complete the ZigZag warehousing system.

7.2. Perform Function Point Analysis on the following data to show that two people working half a year would be enough to complete this project. You may assume this is a straightforward project for an on-line system to be done in a 3GL environment:

	Requirement	Inputs	Outputs	Entity accesses
A1	Requirement 1	10	2	4
A2	Requirement 2	10	3	6
A3	Requirement 3	1	25	1
A4	Requirement 4	10	10	9
A5	Requirement 5	4	10	5
A6	Requirement 6	26	9	2
A7	Requirement 7	5	11	8
A8	Requirement 8	14	4	5
A9	Requirement 9	22	7	4
A10	Requirement 10	6	6	4
A11	Requirement 11	9	9	7
A12	Requirement 12	3	24	5

7.3. For the requirements in Exercise 7.2 show that one person working for four weeks with a second person working for two weeks would complete the Requirements Specification of this project. Also show that Coding and Unit Testing would require about four programmers working for around six weeks.

7.4. Use Function Point Analysis on the data of Exercise 7.2 to estimate the workforce needed to complete each project as depicted in Figure 3.7. How does your answer for the needs of each phase compare to those you obtained for Exercise 7.2? Should the two estimates be the same?

7.5. You are a project manager of an information systems project. You have just completed the Requirements Analysis. From your Data Flow Models, your Logical Data Model and your list of User Requirements you have calculated that the project is worth 352.94 unadjusted function points. You have also calculated that the technical complexity of the project calls for a Technical Complexity Factor of 0.85, and you have established that the implementation will all be on-line. Perform Function Point Analysis to establish the Headcount requirements for a development in a 3GL environment and the Headcount requirements for a development in a 4GL environment.

7.6. If a 3GL programmer costs £600.00 per week, show that a budget of £31,000 would suffice for the Coding and Unit Testing of the project of the previous exercise.

7.7. Develop skeleton Business Systems Options for the Bodgett & Son system (Exercises 3.15–18, 5.10, 6.14) and use them to argue for an appropriate option for the new system.

7.8. Develop skeleton Business Systems Options for the Fresco system (Exercises 3.8–12, 4.3–4, 6.1–2) and use them to argue for an appropriate option for the new system.

PART

3

Specification

During the Specification phase we apply techniques that are more detailed and rigorous than those of Investigation, in order to produce a set of generic programming specifications that can be transported to any technical environment. The products of Investigation will be further transformed in our quest for a system that is maintainable, enhancable and fit for the business's purposes. Of all the Investigation products, the Logical Data Model will prove to be the most important and will be carried forward, becoming the backbone of the new system.

The statement of requirements coming through from Investigation provides the inputs to Specification, but in itself does not specify the new system in enough detail. By the end of Specification we will have started to move into the area of system design, in that the overall shape and structure of the new system will be fully documented. Our objective is to provide sufficient detail for the logical design of user interfaces and internal processing of the new system to take place.

In many ways Specification is the engine room of SSADM. It marks the point at which we move firmly from investigation and analysis to specification and external design, and uses some of the most powerful of SSADM's standard techniques. In contrast to Investigation, where all techniques could be applied to most projects, the techniques of Specification will need careful assessment to confirm whether they are appropriate to the kind of project we are undertaking.

Throughout Specification the Requirements Catalogue retains its pivotal position, as a central repository of functional and non-functional requirements, and continues to be updated with refined or deduced requirements. The Logical Data Model is extended and enhanced using the technique of Relational Data Analysis, and continues to provide the definitive view of the system's data requirements and business rules.

The Data Flow Model, despite being an invaluable aid during the Investigation phase, is less useful as the basis for the specification and design of the new system. In Specification the concept of a 'function' is introduced to tie together more detailed and rigorous models of processing elements. Once the Data Flow Model is updated to reflect the selected BSO it is withdrawn from centre-stage and is only used as a high-level central reference.

In Specification we also introduce the 'third view' of SSADM, representing the effects of events (and therefore time) on data.

Each new function of the system is specified through two sets of techniques, the first ensuring the usability of the function and the second ensuring its utility and efficacy. These two sets of techniques can be tricky to balance and it is down to good project management to guide them through to a successful conclusion. Too much focus on usability may lead to a flashy but vacuous system, while too much attention to the detail of each function's nuance may result in an unfriendly or unwieldy system.

▶ Default structure

As for Investigation, SSADM provides a default structure that can be used as a guide when tailoring the method to a particular environment. This default structure is depicted in Figure P3.1.

Define Work Practice Model

If this has not been done as part of the BSO documentation process, we update the Current Task Models, User Catalogue and User Roles to reflect the selected BSO. We can also define Task Models for any new complex activities within the required system.

Define Required System Processing

Again, if not already completed during the BSO stage, we translate the processing requirements identified in the Requirements Catalogue into Required Data Flow Diagrams.

Develop Required Data Model

The current Logical Data Model is also updated using the chosen Requirements Catalogue. Entity Descriptions are fully completed and the LDM is validated against the required DFM.

Develop users' conceptual models

User Object Models can be developed to illustrate a user-centric view of the required system.

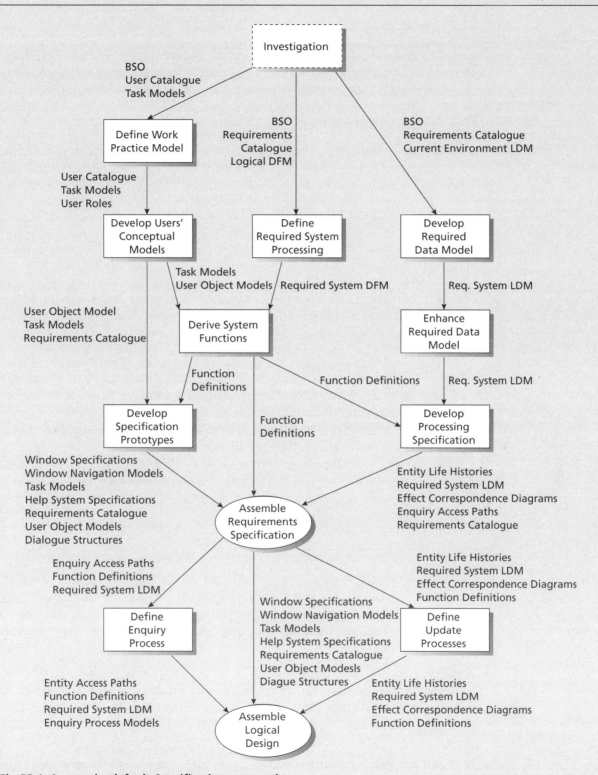

Fig P3.1 A generic, default Specification approach

Derive System Functions

The required DFM is used to draw up an initial set of functions. Function Definition is used to specify units of processing carried out in response to real-world events or enquiry triggers, and to define the structure of system–user interfaces.

Enhance Required Data Model

The technique of Relational Data Analysis is used to analyse data input to and output from the system. The resulting relational data models are then used to validate and enhance the LDM.

Develop Processing Specification

The techniques of Entity Behaviour Modelling and Conceptual Process Modelling are used to specify update processing in detail, and to model its effects on the LDM. SSADM views processing as a set of responses to real-world *events*. Events and their effects are rigorously examined, leading to updates of the LDM, Function Definitions and the fleshing out of processes previously only outlined in the DFM. Enquiries are treated similarly but separately.

Design, prototype and evaluate User Interface Design

The User Interface Design is produced using a variety of tools covering all aspects of the user interface, such as screen layouts, navigation and help systems. Particular attention is required for systems that are accessed directly by customers such as e-commerce and personal banking.

SSADM uses prototyping to verify that user requirements are correctly understood, rather than to incrementally develop the final system. Critical user interfaces are prototyped and the results fed back into Function Definition, User Interface Design and the Requirements Catalogue.

Assemble Requirements Specification

For many projects the Logical Design will now be complete. So the products in hand should be checked for completeness and consistency, and the results presented to the Project Board for review.

It is worth reiterating that the resulting specification should have the following properties:

◆ **It is non-procedural.** In other words, it will state what the system is required to deliver and how it will react to events and enquiry triggers, but will not present procedural algorithms or program specifications (these are developed as appropriate to the chosen technical environment in the Physical Design stage).

◆ **It is not implementation-specific**, i.e. is not tied to a particular technical environment.

◆ **It is capable of re-use.** This is really an extension of the previous point, with emphasis on the fact that the specification can be re-used either as a whole or in parts, by virtue of its logical and modular nature.

For projects where the system is to be built using more traditional procedural, structured and modular approaches to programming (e.g. using a language such as COBOL), we can complete the Logical Design by defining detailed internal processing designs.

Define update processes

If needed, ECDs are transformed into update processing structures, complete with operations governing the effects of each event, integrity error checking and selection and iteration conditions.

Define enquiry processes

Enquiry processing structures are developed with operations and conditions allocated in a similar manner to the update structures.

Assemble Logical Design

All the products are checked again for consistency and completeness before the Logical Design is published for input to Physical Design.

Function Definition

INTRODUCTION

The technique of Function Definition serves a number of different purposes:

◆ to define the services that the computerised system will offer its users;
◆ to help analysts and users towards a shared understanding of what the system will do;
◆ to provide the basis for the system's specification;
◆ to ensure that all design products are traceable to requirements.

To understand properly what a function is we will consider each of these different views or perspectives of Function Definition in turn whilst remembering that no single view captures everything that we might want to say about a function.

The products of Function Definition draw on and draw together work done throughout the Investigation and Specification phases of systems development. This means that it is not possible to have a complete working knowledge of Function Definition without also having understood the techniques of:

◆ Business Activity Modelling and Requirements Definition
◆ Logical Data Modelling and Conceptual Process Modelling
◆ User Interface Design.

Tip

Although we have already covered enough of what we need to know in order to have a useful discussion of Function Definition, we have not yet covered all of the above techniques. So the reader might want to read this chapter again, after having read the relevant chapters.

To visualise what a function is, imagine an information system with which you as a user are familiar. For instance, most of us have used a hole-in-the-wall cash machine (ATM – automated teller machine). As a customer of a bank you have access to *some* of the functions of a cash-dispensing system. Having passed through some security checks – requiring the insertion of a card and inputting of a personal identification number – you are presented with a menu. You use this menu (and any sub-menus) to access the functions of the system. The functions of the system are the things that you can use the system to do. Examples would be 'Withdraw Cash', 'Check Balance', 'Print Mini-Statement', 'Change PIN'.

Note that different types of user (User Roles) have access to different functions. For instance, a person using a cash machine that belongs to their own bank may be offered more functions than the person who banks with a competitor bank and uses that same machine. Obviously, the bank employee who is responsible for loading the cash machine (another User Role) will have access to a completely different set of functions.

Note also that each function will involve the user in some sort of dialogue with the system. This dialogue could be very simple: for instance, if you choose 'Print Mini-Statement' the system may simply respond by printing the statement and then asking if any further service is required; however, if you choose 'Withdraw Cash' there will be a longer dialogue: you will be required to select an amount and choose whether or not to have a printed record of the transaction. The dialogue consists of inputs by the user, e.g. keying in the amount of cash required, and outputs by the system, e.g. showing on the screen the remaining balance available for withdrawal.

All this can be summarised in the following way:

◆ The services offered by a computerised information system to its users are known as 'functions'.

◆ The functions that are available to a given user depend on that user's role.

◆ Functions are usually accessed through a series of one or more menus.

◆ Functions usually involve the user in some sort of dialogue with the system.

> A function is very similar to a Use Case. Use Cases are an increasingly popular way of describing the services offered by a system to its users. They are associated with object-oriented methods, although there is nothing intrinsically object-oriented about Use Cases.

If we have carried out the investigation phase of a project successfully, then both users and analysts should have a similar view about the functionality of the system going into Function Definition. Nevertheless, Function Definition presents an opportunity for a detailed examination of the required system and takes place at a point when it is still not prohibitively expensive to make changes to the scope or requirements of the project (although admittedly this might be undesirable).

During Function Definition, the use of techniques like User Interface Design and Prototyping helps to provide users with a very tangible view of what the final system will be like and contributes to making sure that the 'right' system is ultimately delivered to the users.

In addition to this we have in Function Definition a template or framework that helps ensure that the work of the systems analyst has been carried out thoroughly. Function Definition, in common with many of the techniques in this book, requires the analyst to ask certain relevant questions of the users and focuses the minds of users and analysts on the things that need to be established in order for an effective solution to be developed.

An example is the obligation to document volumetric information and service-level requirements. This data will be used during physical design and implementation when tuning the performance of the system. Whilst database tuning is a specialist activity that is unlikely to be performed by the analyst, it is the analyst's responsibility to agree with the users what the volume of transactions will be and what will be an acceptable level of system performance and availability.

▶ **Learning objectives**

After reading this chapter, readers will be able to:

♦ work with users to organise the system processing to best support their tasks;

♦ identify and document on-line and off-line functions;

♦ use the User Role/function matrix to specify the allowable access to the system;

♦ understand the notation of Jackson structures;

♦ specify the interfaces for off-line functions using Jackson structures;

♦ explain the relationship between Function Definitions and other parts of the specification.

▶ **Links to other chapters**

♦ Chapter 3   The documentation supporting business activities and work practices is used to drive the definition of many functions.

♦ Chapter 6   The Elementary Process Descriptions form the main source for identifying update functions.

♦ Chapter 9   The Used Interface Design provide a definition for the human–computer interface of each on-line function.

♦ Chapter 10   Relational Data Analysis can be performed for the input and output data associated with functions to verify that the data model supports the function.

♦ Chapter 14   Physical Design completes the processing specification of each function.

▶ DEFINING FUNCTIONS

Earlier on, we said that functions define what the computerised information system will do for the user. Clearly it is important to be able to define the things that a system will do for its users. But in order to deliver a satisfactory system we will have to do much more than simply identify the expected functions of that system.

In fact, for each function we will have to:

♦ design the human–computer interface (and, where applicable, interfaces with other computer systems);

♦ specify the database processing that supports each function;

♦ consider ways of preventing erroneous data from ending up in the database;

♦ gather information that can be used to predict and tune the performance of the function;

♦ ensure that the work done on each function is not duplicated elsewhere and make use of work already done elsewhere;

♦ Ensure that the function is included in the user's training manual.

▶ Concepts

Function

We can define the concept of a function in several ways, which together give a good idea of what a function actually is:

◆ A function is a set of processes that users wish to carry out at the same time.

◆ A function is a piece of processing designed to handle the effects of an event (or group of events) on the system.

◆ A function is the basic unit of processing for input to Physical Design (where programs will be specified to implement it).

Each function may encompass several individual update or enquiry processes, each of which responds to a particular event or trigger, but will usually cover just one.

Types of Function

We try to classify each function in three main ways. This is because each of these ways shares certain characteristics that we will need to bear in mind as we design the function.

Enquiry or update functions

First of all we need to decide whether the function has any update element within it. If it does then it will relate to at least one system event for which we will have to specify the update processing later. If the function has no update element than we will classify it as an enquiry function.

A function is classified as an update function if any part of the function can be thought of as inserting, amending or deleting one or more instances of one or more entities in the Logical Data Model. An update function can and often does have an enquiry element.

In many ways enquiries and updates are treated in the same way within Function Definition. However, where an update is taking place there will always be a danger of erroneous data being entered into the database and so particular attention will have to be paid to error handling. This will mean designing the interface to prevent errors from occurring and designing update processes to identify and reject errors.

Update functions may include enquiry elements, but their main purpose will be to update system data.

User- or system-initiated functions

Next we need to establish whether the function is user- or system-initiated.

A function is user-initiated if the user invokes the function. The usual way of doing this would be to select a function from a menu using mouse, keyboard or some other input device.

System-initiated functions are programmed to be invoked at some specific time, or when the system enters some pre-defined state, without any intervention from an end-user.

On-line or off-line functions

An on-line function is user-initiated and involves some sort of dialogue between the system and the user. The database processing (i.e. updates and/or enquiries) of an on-line function happens as soon as the user triggers the processing. Thus, for on-line processing we expect response times of the order of a few seconds (the response time is the time taken to carry out an update once the user has completed the data entry, or the time taken to respond to an enquiry once the user has triggered it).

An off-line function is defined for any piece of update or enquiry processing that takes place at a time that is distinct from the user initiation of the function.

System-initiated functions are by definition off-line.

Components of a function

You should observe that, although from the user's point of view the interface *is* the function, the design of the user interface is only part of what we, as analysts, have to do in order to define a function.

Defining functions will involve the use of design techniques other than Function Definition itself. In this respect the role of Function Definition is to tie together all the different parts of the specification which when taken together allow for the precise implementation of a function.

A useful way of thinking about this 'tying together' role of a function is to consider the Universal Function Model (Figure 8.1).

In this book we discuss a number of different products that help to specify the various components of a function and these are described in Table 8.1.

Because of the desire to satisfy users' requirements and because users tend to perceive the system in terms of its interface it is tempting to emphasise interface design when defining functions. This is reflected in the current trend towards prototyping-based approaches to information systems design.

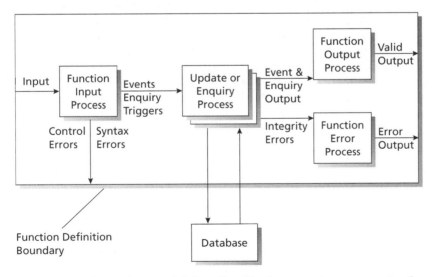

Fig 8.1 The Universal Function Model describes the important components of a function

Table 8.1 How the various components of a function are specified

Universal Function Model Component	Ways of Specifying the Component
Input, Valid Output	If a Required System DFM has been developed, then the I/O Descriptions that form part of this model will indicate the data items that form the input to and output from the function. These can be further described using I/O Structures, Dialogue Design or User Interface Design.
Error Output Function Input Process Function Output Process Function Error Process	These components are usually specified during Physical Design (see the section on the Function Component Implementation Map). However, you will need to build some of the foundations for that work: for instance, identify errors that should be trapped, and define the error messages that the user will see as a result; record the user's requirements for sorting and formatting of the output data.
Update or Enquiry Process	For each event or enquiry we can develop a Conceptual Process Model. A Conceptual Process Model is an implementation-independent description of the database processing and integrity checks that takes place in response to an event or enquiry.

But thinking back to the cash machine example that we started with, it should be obvious that there is much more to the 'Withdraw Cash' function than we can perceive at the user interface. For instance, it will be necessary to check that there is a valid account with enough funds to support a withdrawal. And it will, of course, be necessary to update the customer's balance and/or keep a record of the transaction. At the interface all the user will see is a message saying 'Please take your card and wait for your cash' or a message saying that funds are not available. The user has no idea what is going on behind the scenes – but the designer has to know.

Whilst in no way diminishing the contribution that User Interface Design and Prototyping can make to the validation of the requirements we would, in this section, like to remind readers that there is more to a function than designing the user interface.

In summary, we can say that with Function Definition and the techniques that surround it the analyst has to build two models of the functional requirements, one having a user-centred view and the other a system-centred view.

Function Definition and other techniques

Before we embark on this discussion we would like to make it clear that on any given project it is likely that only a subset of the following techniques will be used. It is up to the project team to be selective in their choice of techniques, choosing the right ones for the problem at hand rather than slavishly and unthinkingly following a standard approach. By discussing quite a large number of techniques in this section we are not suggesting that every technique be used on every project.

The key to understanding a method like SSADM is to appreciate the way in which the different products of the method fit together, how the different techniques interrelate. In the investigative phase they allow the analyst to look at the problem from a number of different views or perspectives and this – it can be argued – gives the analyst a better chance of understanding the problem. In the Specification phase they allow the designer to break the specification into manageable parts whilst at the same time allowing them the confidence that these separate parts will work together to deliver the system as a whole.

Data Flow Modelling

The Required System Data Flow Model shows the update processing that will be needed in the proposed computerised information system in order to maintain (create, update and delete) the data represented in the Required System Logical Data Model. On the Required System Data Flow Model the update processing is represented by Elementary Processes, the data by data stores and the users of the computerised information system by external entities. Each data flow from an external entity to a process represents one of the following:

◆ the notification to the computerised information system of an event together with the data items (event data) needed to carry out the update processing triggered by the event. In other words the data flow says 'Something has happened in the outside world that means that the computerised information system needs updating; here is the data needed to carry out the update';

◆ the triggering of an enquiry upon the computerised information system. The data flow carries any parameters needed by the system in order to carry out the enquiry.

We do not show enquiry processes on the Required System Data Flow Model unless they are an integral part of the update processing, so most of the processes on the Data Flow Diagram are update processes and most of the flows from external entities to processes represent system events.

One way of looking at an update function is that it provides the user with a way of notifying the computerised information system that an event has taken place. Or, put more simply, an update function allows the user to add, amend or delete records in the database. There will typically be at least one function for every event that we identify.

It follows from this discussion that we can use Data Flow Modelling for one of two purposes:

◆ to identify update functions. We use the Required System Data Flow Model to identify system events, and for each event we define an update function;

◆ to illustrate functions. After we have carried out Function Definition we can draw a Required System Data Flow Model as a way of illustrating the update processing for our users. In this case we would be using the Data Flow Model as a validation tool – a presentational device for getting feedback from the users. (It has to be said that the value of this would depend on the user's ability to interpret Data Flow Models. For most analysts the preferred method for validating Functions would be to prototype them. Having said that, we would usually only prototype part of the system, whereas a Data Flow Model can give an overview of the whole system.)

▶

Logical Data Modelling

Each function will handle one or more – often many – data items. Most of these data items will correspond to attributes of entities of the Logical Data Model. We need to ensure that data items are named consistently so that a data item that is used within a function can be related to an attribute where applicable. Where the fields or data items that the user sees at the interface are not given the same name as the underlying attributes to which they relate, then we will have to document this in the data dictionary.

For any enquiry function (or enquiry part of a function) it is important to check that the Logical Data Model can support the function. One way to do this is to perform Relational Data Analysis on the input/output data items of the function. Another way is to sketch an Enquiry Access Path.

Entity Behaviour Modelling

In Chapter 11, Entity Behaviour Modelling, we show how to perform an analysis of the Logical Data Model with a view to both identifying and understanding the effects of events. To simplify a little, what we do with Entity Behaviour Modelling is to name the events that cause:

◆ an instance of an entity to get created or deleted;

◆ attribute values to be assigned or amended;

◆ instances of relationships to be created or destroyed.

The kind of thinking that we employ during Entity Behaviour Modelling usually leads us to think a bit more deeply about the system and find new events that we have not already thought of.

We have already stated that if we can identify events we can identify functions that respond to those events. Thus Entity Behaviour Modelling can be used as a way of identifying functions, or identifying new functions that we might otherwise have overlooked the need for.

Requirements Definition

As discussed earlier, during Function Definition we will take each chosen functional requirement and describe in detail how this requirement is to be met by the system.

The Requirements Catalogue is a primary source for Function Definition.

Function Definitions are an important bridge between the requirements as stated in the Requirements Catalogue and the rest of the Specification. They make the specification traceable – the importance of which has already been discussed.

Work Practice Modelling

Work Practice Modelling offers a more user-centred view of what a function is. With Work Practice Models we emphasise the whole job of the user by modelling all the tasks that the user carries out in response to some business event. The Work Practice Model shows *all* the tasks of the user, not just those supported by the computerised information system. The theory is that we identify the tasks that can be given computerised information system support and this will drive Function Definition (for on-line functions at least). The aim is to develop a system that fits closely with the needs of the user and the way that the user works.

Hence, if we have developed a Hierarchical Task Model for the required system (see Figure 3.17), we can use it to identify on-line functions and produce complete job specifications. If, as part of Work Practice Modelling, we have written scenarios for the carrying out of tasks, then we can use these scenarios to 'walk through' and test or validate each function.

Business Activity Modelling

We should be able to trace each function back to some business activity that it supports. Business Activity Modelling can be used during the investigative phase of a project to drive the definition of functional requirements. These functional requirements form a major input into Function Definition.

It follows that Business Activity Modelling can be used as a way of identifying functions.

The relationship between Business Activity Modelling and Function Definition goes all the way back to our discussion of what an information system is and the statement that an information system exists to support the activities of the users. We could now add to this by saying, a little simplistically, that as far as the user is concerned the functions *are* the information system.

Relational Data Analysis

By taking the input/output data associated with each function and performing Relational Data Analysis it is possible to verify that the Logical Data Model supports the function.

Conceptual Process Modelling

Each function will involve some database processing (update or enquiry or both). During this processing certain business rules will have to be checked and enforced in order to safeguard the integrity of the data in the system.

Conceptual Process Models are a way of specifying the update and enquiry processing that is needed by a function.

User Interface Design

Currently the more popular way of defining the interface of each on-line function (i.e. the human–computer interface or user interface) is to do so in terms of windows and the navigation between windows. 'User Interface Design' is the umbrella term for a set of techniques that help to do this.

Prototyping

The prototyping of windows, screens and reports is a particularly useful and popular way of performing User Interface Design.

From the start of this book we have emphasised the importance (and difficulty) of delivering a computerised information system that meets the user's needs and the business's objectives. In defining a set of functions we are committing ourselves to the shape of the future system. It is obviously desirable to be as sure as possible that that system is what the users want. Prototyping can be used as way of checking that the most critical functions of the computerised information system have been properly understood. (It can be used for other purposes as well, e.g. testing the technical feasibility of a function that relies on untested technology, or testing the performance of the system.)

▶ Identifying functions

During the Investigation phase we develop a number of products that can be used as the basis for identifying functions.

The first point of reference should be the Requirements Catalogue itself. All other techniques feed into the definition of requirements and serve to validate and verify the requirements. Thus, in a well-organised project, the Requirements Catalogue should be the first product that we turn to.

For update functions the key is to identify events that the computerised information system has to handle. These could come from:

◆ the Required System Data Flow Model;

◆ Entity Behaviour Modelling (see Chapter 11);

◆ the Event Catalogue (ideally, system events should be documented as and when they are discovered; if we do this then the resulting event catalogue can be used as a starting point for Function Definition).

Another approach to Function Definition is to consider the users and the computerised information system support that they require. In this case the following would be useful:

◆ Work Practice Model

◆ Business Activity Model.

Alternatively, we could use User Interface Design and Prototyping as a source of Functions, but in this case we should be very careful to document our findings thoroughly.

How big should a function be?

One of the problems that you will face when defining functions is to decide just how much functionality should go into a single Function. It is the same kind of problem that you face when performing Data Flow Modelling or Business Activity Modelling: what is the right level of detail for a process/business activity/function?

Unfortunately, there is no straightforward answer. The scope of a function needs to be defined with reference to the user tasks that the function is going to support.

On the other hand we can give some guidelines.

Typically the smallest amount of work that the user will want to do at the interface is to work with a single event or single enquiry.

For an event, the user will have to enter all the data that is required in order to process that event and will expect as output some confirmation that the update has taken place.

For an enquiry, the user will have to enter any data that is required as a trigger to the enquiry (any parameters for the query) and will expect the results of the enquiry to be output.

Hence, many update functions will map one-to-one with events and many enquiry functions will map one-to-one with enquiries. Your starting point for Function Definition can be to assume that there is one function for each event and enquiry and adjust this view as necessary in the light of discussion with the user, analysis of the user's work practice and consideration of off-line processing.

It will be quite common for a function to package together one or more events and enquiries. When this happens it will often be for two reasons: first, the convenience of the user (saving them from having to navigate around the system to achieve what they perceive as a single task); secondly, because the function needs to be controlled as a single unit (e.g. carrying data forward from one part of the processing to another, allowing the user to backtrack, etc.). Very commonly, users will require some kind of enquiry processing before performing an update. For example, ZigZag is implementing direct-to-customer selling over the Web. When ZigZag's customers use the Web to buy a product they will always start by browsing for the product that they wish to buy (an enquiry) before going on to select a product for purchase, registering as a customer and making their payment (an update). This function is discussed in more detail later on.

Identifying functions from Data Flow Models

Each *major* data flow that crosses the system boundary of the Required System DFM should correspond to at least one system event. So we can identify an initial set of functions by taking the Required System DFM and following each of these data flows into and through the bottom-level processes, tracing all of its data items until they have been either stored or output. We can then identify all of the processes involved in handling the flow (and hence the events) and circle them to form a single function.

For example, by tracing the data flows P.O. Quantities (representing the event Propose Purchase Order) and Availability (representing the event Confirm Order) into the level-1 process Monitor Purchase Order we can create initial functions 1 and 2, as in Figure 8.2. On occasion, when the Data Flow Model contains processes that trigger each other, functions may cover more than one Elementary Process.

More frequently, each function will correspond to a single Elementary Process, particularly if we tied each process to a distinct driving (or triggering) data flow and kept inter-process data flows to a minimum during Data Flow Modelling.

Once all input data flows have been traced, we may be left with some processes that have not yet been allocated to a function. Assuming that these do not represent processes that have been incorrectly modelled, they will represent system-initiated processing. In the ZigZag system process 6, Produce Stock Report, is such a time-initiated process.

Once we have allocated all processes and traced all data flows, the events handled by each function must be identified. Many functions will process a single event, but it is quite permissible for a function to handle batches of different events, which are input in a single data flow. This is especially true for off-line functions where input frequently consists of large amounts of data from different, and often unsorted, forms.

Although we have not covered it yet, the technique of Entity Behaviour Modelling can throw up previously unnoticed events upon which functions should then be defined.

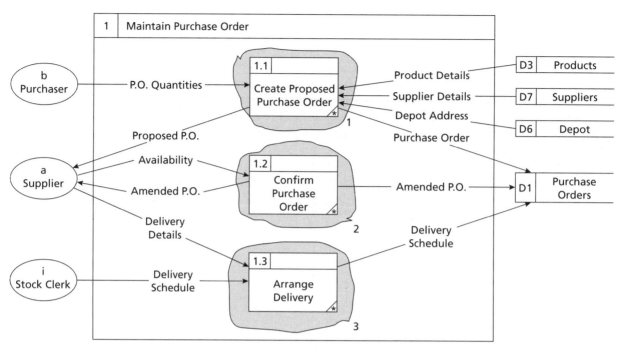

Fig 8.2 Identify functions from the DFD

Identifying enquiry functions

Enquiry functions (other than those which form a major part of the systems functionality) will be documented in the Requirements Catalogue, rather than the Required System Data Flow Model. Each enquiry entry in the Requirements Catalogue will initially be identified as a function in its own right. One such enquiry function from the ZigZag system is the 'Overdue Delivery Report'.

Confirming functions

We should discuss the initial set of functions with users, to verify that they represent the way in which they require processing to be organised. We may find that users wish to split up a function, because they will need to carry out different parts of the function independently, or that they want to combine them to form bigger functions.

For example, the users of the ZigZag system estimate that in about 15% of cases they will need to amend existing deliveries while arranging new deliveries with a supplier. To cater for these occasions we can create two functions in response to elementary process 1.3: one to arrange a delivery for the first time and one to amend an existing delivery. Alternatively, we can retain one function which, when activated, will prompt the user to state whether they wish to arrange a fresh delivery or amend one that has been arranged earlier.

Although we should attempt to satisfy as many of the users' requirements as possible it is important not to get carried away with defining composite functions that will hardly ever be used.

▶ DOCUMENTING FUNCTIONS

As the process of defining functions and confirming them with the users goes on we should take care to document what we have found. A Function Definition is a mainly textual product, the detail of which is progressively refined as work on the project continues. Below we give an outline of what we should record about a function, and why.

The textual part of a Function Definition can be divided into roughly three sections: a descriptive section, a section that provides cross-references to other products that describe some aspect of the function, and a section that relates to the non-functional aspects of the function.

It should be stressed that the exact manner of documenting functions will vary from project to project depending on the CASE tools available, and on the level of documentation required by the local standards for that project. Table 8.2 then can be taken as a guideline, with a warning that not everything will be necessary or applicable to every project.

As an example we have provided a Function Definition for ZigZag's Book Delivery function in Figure 8.3, which uses some of the suggested contents from Table 8.2.

Function Definition	
Function Name: Book Delivery	**Function Id:** 3
Function Type: Update/On-Line/User	
Function Description: Suppliers contact the depot to arrange delivery of goods contained on one or more Purchase Orders. An enquiry will be made to check on free delivery time slots for the suppliers preferred delivery date, if the goods are not to be added to an existing delivery. The delivery is then confirmed for a slot and the relevant Purchase Order Lines (or parts of) are assigned. Suppliers have to state the Purchase Order Line they wish to deliver. The system confirms that the products are indeed expected and we record the quantity expected.	
Business Events: Delivery Arrangement, Delivery Amendment	
Activities: Arrange Delivery, Set Up Delivery Schedule	
System Events: New Delivery Schedule, Updated Delivery Schedule	
User Roles: Delivery Scheduler	
Error Handling: The transaction will be terminated if the goods to be delivered do not match the outstanding Purchase Orders.	
DFD Processes: 1.3 (Arrange delivery)	
I/O Structures: 3.1	
I/O Descriptions: a – 1.3, 1.3 – d1, i – 1.3	
Requirements Catalogue Ref. 10	
Related Functions: None	
Enquiries: Check Available Slots	
Common Processing:	

Fig 8.3 Documentation of the Book Delivery function

Table 8.2 Documenting functions

Function Definition Item	Description
Identifier	A unique key. This can be used wherever you need to refer to the function. For instance, you might want to cross-reference Requirements Catalogue entries to functions by noting the Function Id in the Requirements Catalogue entry. The general aim, of course, is to facilitate the function's traceability. If you are using a CASE tool then it may assign and maintain a unique identifier automatically. Ideally it will also allow you to define links between the function and other related analysis and design products.
Name	Describes the purpose of the function in a few words. The function should be named uniquely. This certainly means that no two functions will be allowed to have the same name. In addition, functions will, ideally, be named differently from requirements.
Type	Classifies the function in three ways: ◆ user- or system-initiated ◆ on-line or off-line ◆ update or enquiry
Description	Gives a detailed account of what the function does
Error Handling	Describes the errors that the function can trap and the error messages that will be displayed when an error occurs
Business Events	The names of the business events associated with the function
Business Activities	A cross-reference to one or more business activities supported by the function
System Events and Enquiries	The names of the events and/or enquiries of which the function is composed
User Roles	The names of all the User Roles that have access to the function
DFD Processes	The identifiers of elementary processes (if any) that relate to the function
I/O Structures	A cross-reference to I/O Structures (if any) that describe the function
I/O Descriptions	A list of all the data items that are input to or output from the function
Requirements	A cross-reference to one or more Requirements Catalogue entries that relate to the function
Functions	A cross-reference to closely related functions (where there are any)
Tasks	A cross-reference to one or more user tasks (in the user Task Model) supported by the function
User Object Model	A cross-reference to objects on the User Object Model (if there is one) that are supported by the function
Dialogues	A cross-reference to Dialogue Designs (if any) for the function
Windows	A cross-reference to graphical user interface designs (if any) for the function
Common Processing	A cross-reference to a process specification for a piece of processing (other than an event or enquiry) that can be used by more than one function
Service-level Requirements	Measurable description of required cost/performance/satisfaction levels, etc.
Volumes	How many times the function will be invoked per hour/day/month/year as appropriate

Ensuring that all products are traceable to requirements

During the investigative phase of systems development we used the technique of Business Activity Modelling as a way of identifying information systems requirements. In particular we used it to define functional requirements by asking the following questions for each Business Activity:

◆ What information could a computerised information system produce that might be useful when carrying out this activity?

◆ What information might we want to record in a computerised information system having carried out this activity?

By starting our investigation with the business activities themselves, we hoped to improve the chances that any new system that we developed would be useful to the business that was paying for it.

Business Activity Modelling throws up many different possible functional requirements (many different possible new systems) and for one reason or another not all of these will be viable or desirable. In the chapter on Business Systems Options we discussed the ways in which the stakeholders in a project will try to come to an agreement on exactly which requirements should be carried forward into any future systems development. The overriding concern here was that the costs of developing and operating a system based on some or other set of requirements should be outweighed by the benefits that that system could reasonably be predicted to bring.

Not all projects will have a distinct stage like Business Systems Options. In projects based on a Rapid Applications Delivery (RAD) approach it is likely that decisions of the sort made in Business Systems Options will be made 'on the fly'. Nevertheless, the same kind of thinking will have to be done, in that the project team will have to be able to make a case that the proposed system will help meet the objectives of the business.

During Function Definition we will take each chosen functional requirement and describe in detail how this requirement is to be met by the system. In other words:

◆ Functional requirements go through some sort of filtering process where we decide if they are carried forward into the development of the new system. We can call these the 'chosen requirements'.

◆ Each function will be based on and traceable to one or more chosen requirements.

◆ We can trace or associate each functional requirement (and therefore each function) with a business activity that it supports.

Because each function is also cross-referenced to the various design products that when taken together *specify* the function, we can be sure that every product of the Investigation and Specification phases of our project is in some way traceable back to the chosen requirements that gave rise to it.

We would argue that the property of traceability is important for at least two reasons:

◆ The project manager is able to check that for any given requirement all the necessary work is taking place (or all the necessary products have been developed) so that the requirement can be implemented.

◆ It is possible to demonstrate that any particular activity of the development team relates in some way to the delivery of a requirement and that the team members are therefore spending their time on the right things.

▶

Without products that are traceable the project manager has to trust to luck that everything necessary is being done, and that is a rather risky strategy. Against this, of course, it has to be said that there is a cost to maintaining the documentation of each product and the links between them.

As an exercise you might like to check the traceability of the 'Record Proposed Purchase Orders' requirement as developed in this book.

The relationship between Function Definition and other techniques is discussed further in the sections that follow.

▶ FUNCTION NAVIGATION MODELS

When working with larger functions it may sometimes be useful to break the Function down into smaller self-contained components. There could be several reasons for wanting to do this:

◆ the Function is very large and it is more manageable to break it down into several parts;

◆ the Function contains more that one event or enquiry – we can define one or more components for each event and enquiry;

◆ there are parts of the function that will be used by other functions too – again, events and enquiries will often feature in more than one function. For instance, the 'Customer Search' enquiry will be used in several different functions.

The notation for the Function Navigation Model used to tie together the components is quite simple (see Figure 8.4). Arrows between boxes on the model show permitted navigation between one component and another. An arrow pointing from a box to nowhere shows that the user can exit the function at that point.

The Web Browse and Buy function allows customers to search ZigZag's product catalogue from a Web browser. If a customer wants to buy an item that they have found, then they can add the item to their 'shopping trolley'. When the customer has finished shopping they register their name, address and e-mail address and make a payment by credit card. Goods are delivered by mail.

The Web Browse and Buy box in Figure 8.5 represents the function and incorporates the enquiry part, i.e. the catalogue search or browse. The function contains a single event ('Place Customer Order'). This is covered by two components in the

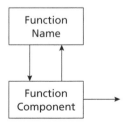

Fig 8.4 Notation for the Function Navigation Model

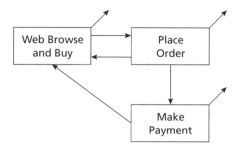

Fig 8.5 Function Navigation Model for Web Browse and Buy function

Function Navigation Model: Place Order and Make Payment. The Make Payment component has been separated from the rest of the order processing because it involves an interface with another system (credit-card details will have to be validated before the order placement can be completed).

▶ IDENTIFYING REQUIRED DIALOGUES

Function Definitions and the User Role documentation can be used to identify exactly who will require access to each function. The easiest way of documenting this is by using a User Role/function matrix (see Figure 8.6).

Each cross indicates that a dialogue between a User Role and a system function is required. We must, of course, check the matrix with users to verify that we have correctly understood which functions each User Role will need access to.

Once the matrix is complete we should identify all dialogues that are critical to the success of the new system (it is possible that they all are). This will help us to make a short-list of the most important dialogues for prototyping, and might be useful in assessing the suitability of various technical system options later on. It will also help us define the users' work practices.

There are a number of questions we can ask when selecting critical dialogues, but any decisions we make must be agreed with users.

User Role \ Function	1. Propose P. Order	2. Confirm P. Order	3. Book Delivery	4. Update Delivery	5. Maintain Schedule	6. Receive Delivery	7. Store New Stock	8. Setup Dispatch	9. Match C. Order
Delivery Scheduler			X	X	X				
Goods In Clerk				X		X	X		
Purchaser	X								
P.O. Clerk	X	X							
Stock Keeper							X		

Fig 8.6 User Role/function matrix

User Role \ Function	1. Propose P. Order	2. Confirm P. Order	3. Book Delivery	4. Update Delivery	5. Maintain Schedule	6. Receive Delivery	7. Store New Stock	8. Setup Dispatch	9. Match C. Order
Delivery Scheduler			(X)	X	(X)				
Goods In Clerk				X		(X)	(X)		
Purchaser	(X)								
P.O. Clerk	(X)	(X)							
Stock Keeper							(X)		

Fig 8.7 User Role/function matrix with critical dialogues

◆ Will the dialogue be used very frequently? Any that are used only once or twice a year are unlikely to be critical.

◆ Does the underlying function carry out tasks essential to the success of the business?

◆ Are large numbers of entities accessed or updated, and are the access paths likely to be complex?

◆ Is the dialogue associated with new system functionality?

◆ Does the dialogue represent processing that is likely to be high-profile or politically sensitive?

Once we have identified the critical dialogues we should annotate the User Role/function matrix by circling the appropriate crosses (or replacing them with the letter C) as shown in Figure 8.7.

Jackson structures

Jackson structures are named after their originator Michael Jackson. This is not the Michael Jackson to be found in the ZigZag catalogue, but the author of *Principles of Program Design*, published in 1975.

Jackson structures offer an elegant diagrammatic way of showing sequence, selection and iteration in program or data structures.

Selection

In Figure 8.8, X is the *parent* of A, B, C and D; so A, B, C and D are the *children* of X. In Figure 8.9, Y is the child of Z and the parent of B, C and D. Note that a parent box can have one or more children but a child box has one and only one parent.

A Jackson structure always has one *root node*. This is the box that appears at the top of the diagram and it carries the name of the thing whose structure we want to describe. A root node always has at least one child.

A Jackson structure always has at least one *leaf*. A leaf is a box that does not have any children.

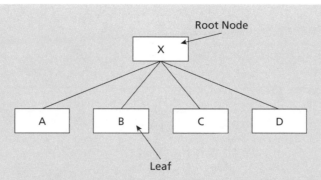

Fig 8.8 **X is a sequence of A followed by B followed by C followed by D**

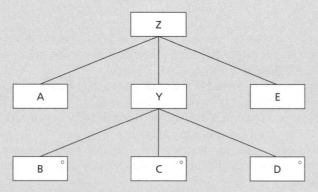

Fig 8.9 **Y is a selection of B or C or D**

The diagram of Figure 8.8 says that X is a sequence of A, B, C and D. Note that we read the diagram by the leaves from left to right: A is followed by B is followed by C is followed by D.

Selection

There is also a *sequence* in Figure 8.9: Z is a sequence of A, Y and E. But we have a further type of structure in Figure 8.9: a *selection*. Y is a selection of B or C or D. In other words, when we reach Y in the sequence we get one and only one of B, C and D.

This means that there are a number of allowable 'routes' through the Jackson structure for Z: e.g. ABE, ACE, ADE. By contrast, ABCE would not be a permitted structure for Z (for any given Z we can have only *one* of B, C and D).

A selection node has two or more children. Each of these children (each optional element of the selection) has a small 'o' in the top right-hand corner to show that it is an option.

As we can see from Figure 8.9, a Jackson structure can have a number of levels of parents and children with a mixture of sequences and selections.

Iteration

In Figure 8.10 we once again have a sequence: T is a sequence of A followed by W, followed by C. We say W is an *iteration*. In other words it represents a repeating element. The

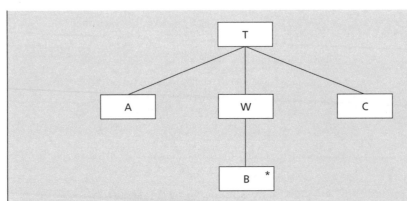

Fig 8.10 W is an iteration of Bs

repeating element can occur 0, 1 or many times. In the example above this means that we can have 0, 1 or many Bs. Hence four of the infinite number of possible structures for T would be AC, ABC, ABBBC and ABBBBBC.

The repeating element is marked with an * in the top right-hand corner to indicate that it can occur 0, 1 or many times.

An iteration has one and only one child, but the child can be the parent of a whole sub-structure (you will see lots of examples in this book).

We can combine sequences, selections and iterations into a structure of any size according to our needs, but to ensure that the structures are always unambiguous you should observe the following rules:

◆ all the children of a sequence must be plain boxes;

◆ all the children (two or more of them) of a selection must be boxes with an 'o' in the top right-hand corner;

◆ the child of an iteration (one and only one child) must be a box with an * in the top right-hand corner.

Thus the types of structures shown in Figure 8.11 are not legal because they break the above rules and cannot be clearly and consistently interpreted.

Jackson structures, or Jackson-like structures, appear in this book in a variety of guises and these are summarised in Table 8.3.

If you find the discussion in this section a little too abstract then you might want to return to it after having read the section on I/O Structures.

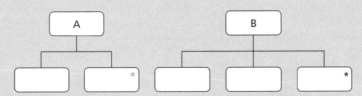

Fig 8.11 Is A a sequence or a selection? Is B a sequence or an iteration?

Table 8.3 Uses of Jackson structures

Product	Purpose
Enquiry Access Paths	Show how the Logical Data Model is navigated to support an enquiry
Effect Correspondence Diagrams	Show how the effects of a single event are co-ordinated
Update/Enquiry Process Models	Provide procedural but implementation-independent specification of database processing
Entity Life Histories	Model the behavioural aspects of the system by showing – for each entity – the effects of events and changes of state
I/O Structures and Dialogue Designs	Model the interface of a function (or part of a function) by describing the structure of the input and/or output data.

▶ I/O STRUCTURES

The reason that we have paused at this point to discuss Jackson structures is that they can be used to describe the input and output data of functions (we call the resulting diagrams *I/O Structure Diagrams*).

The argument for using Jackson structures for this purpose is that they give a detailed, logical, implementation-independent view of the input and/or output data.

Against this it has to be said that they are not very user-friendly and when used to describe on-line functions in general, and graphical user interfaces in particular, they can become excessively unwieldy without offering many clues as to the actual look and feel of the interface.

In this book we have included some I/O Structures for on-line functions in the section on User Interface Design but for the reasons mentioned above few analysts would recommend them and several more popular alternatives are discussed.

SSADM in its latest version (4.3) suggests that I/O Structures be used only for off-line functions, including reports and system-to-system interfaces (which may be one of the largest components of some systems), and that is what we will illustrate in this section.

We will explain the notation of I/O Structure Diagrams using the Proposed Purchase Order function as an example. This function has two parts. First, purchasers suggest products to be bought. Then, overnight, all the suggestions are sorted by preferred supplier and the system produces proposed purchase orders. Clearly, purchasers work on-line while the production of each proposed purchase order is done off-line. The function therefore consists of two parts, one on-line and one off-line.

The off-line part is illustrated in Figure 8.12. The diagram says that each time this function runs there will be an output of zero, one or many purchase orders. Each purchase order will consist of an order heading and zero, one or many purchase order items.

On an I/O Structure Diagram the leaves (i.e. those boxes that have no other boxes below them) are known as *I/O Structure Elements* and these represent groups of data

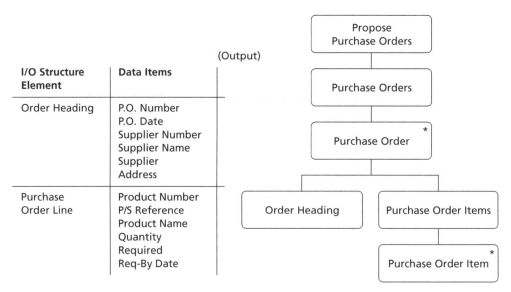

I/O Structure Element	Data Items
Order Heading	P.O. Number P.O. Date Supplier Number Supplier Name Supplier Address
Purchase Order Line	Product Number P/S Reference Product Name Quantity Required Req-By Date

Fig 8.12 Off-line I/O Structure Diagram and I/O Structure Description for Propose Purchase Order

items that are either input to or output from the function. Each element is labelled as either input or output and will only include logical data items (not system control messages, operator ids, etc.).

Note that in Figure 8.12 we do not need to label the I/O Structure Elements as input or output as the entire diagram is output only.

For each I/O Structure Diagram we complete an I/O Structure Description detailing the contents of the I/O Structure Elements. For elements that consist of a single data item this is a trivial exercise, but usually each element represents more than one data item.

We complete the documentation of I/O Structures by assigning a unique numeric identifier to each I/O Structure (preferably consisting of the function identifier, plus a unique suffix), and adding this to the Function Definition.

Functions that are largely update in nature are likely to have substantial inputs but only minor confirmation output, whereas enquiry functions will generally have a small amount of triggering input but a large amount of output (after all, it is the output that interests us in enquiry functions).

If you have drawn a Data Flow Model then you can use the I/O Descriptions from that model as the basis for developing I/O Structures.

Enquiry functions will be documented mainly in the Requirements Catalogue, so we will not have I/O Descriptions to use as a starting point for drawing I/O Structure Diagrams. Instead we can draw them in direct consultation with users.

▶ SUMMARY

1. Function Definition plays a central role in the specification of the new system and defines what the computerised information system will do for the user.

2. As a technique Function Definition requires analysts to work at their understanding not only of the user's information system requirements but also at their understanding of how the new system will fit in with the user's tasks.

3. Defining Functions involves the use of design techniques other than Function Definition itself, and in this respect the role of Function Definition is to tie together all the different parts of the specification, which when taken together allow for the precise implementation of a Function.

4. The primary source of Functions is the Requirements Catalogue. Other sources include: the Required System Data Flow Model, Entity Behaviour Models, the Work Practice Model and the Business Activity Model.

5. When working with larger functions it may sometimes be useful to break the Function down into smaller self-contained components.

6. I/O structures can be used to detail the input and output data of, in particular, off-line Functions.

▶ EXERCISES

8.1. Produce an I/O Structure (diagram and description) for the following Natlib off-line function description:

Every Friday at 11.00 a.m. the system should produce a report of all overdue loans. The report should be broken down by due return date (longest-overdue first), and provide details of the reader, book title and copy for each loan.

Assume that the following attribute names are all that are needed:

Due Return Date	Loan No.
Book ISBN	Book Title.
Reader No.	Reader Name
Reader Address	Reader Tel. No.
Book Copy No.	

8.2. Identify the main functions of the West Munster Hotel system described below:

West Munster Hotel was built in 1867 by the great-great-grandfather of the current owner. It has 30 rooms of three types: Bridal Suite, Double with Bathroom and Double with Shower.

A system analyst has done some analysis for the development of a billing system for West Munster Hotel. In what follows the analysis results are summarised.

Room prices vary according to season. Currently two seasons are recognised: 'Low', from 20 September to 20 May, and 'High', from 21 May to 19 September. Prices for each room type per season are updated every year.

Customers, some of whom belong to companies, book rooms throughout the year. Each booking is for one room. A name of a customer is always required when a booking is made.

Bookings are either confirmed at the time of booking or are provisional. A provisional booking has to be confirmed within three weeks. If it is not confirmed in three weeks, the booking lapses, and is deleted from the system.

Upon arrival, customers who propose to pay their bill by credit card have to inform the hotel by providing the relevant credit-card number. West Munster accepts three methods of payment: Cash, Cheque with Supporting Cheque Card, and Credit Card.

Cheque and credit cards are credit-checked while customers are shown to their room. If the credit check shows the card to be unacceptable West Munster staff inform the customer

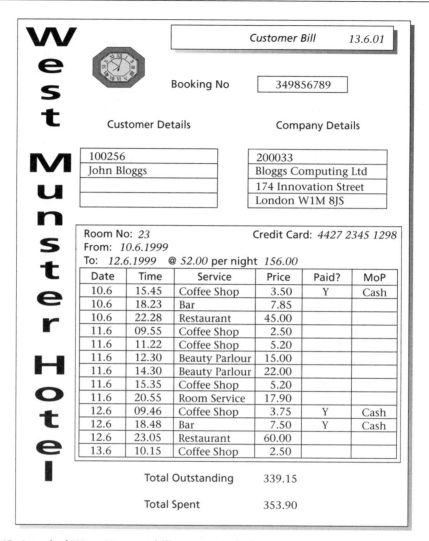

Fig 8.13 A typical West Munster bill. MoP stands for 'Method of Payment'

and request another card or another method of payment. Until the new method of payment is acceptable, the customer cannot use the hotel's services without paying in cash.

During their stay, guests can use the hotel's services, which are: Bar, Restaurant, Coffee Shop, Room Service and Beauty Parlour. These services are either paid for directly or passed-on to the room. The total of each service used is recorded in the system.

When a booking is taken, whether provisional or confirmed, a Booking No is allocated, a Room No is earmarked, a Customer Number and/or a Company Code No are attached to the booking, and the expected start and end dates are recorded.

West Munster has 50 companies and 300 customers on record. Each room has an average of 30 bookings per year. When at West Munster, guests use the hotel's services 10 times on average.

A typical West Munster Hotel bill is shown in Figure 8.13. The Logical Data Model for the West Munster Hotel system is shown in Figure 8.14.

8.3. Identify the main User Roles of the West Munster Hotel system and use them to produce a User Role/Function Matrix

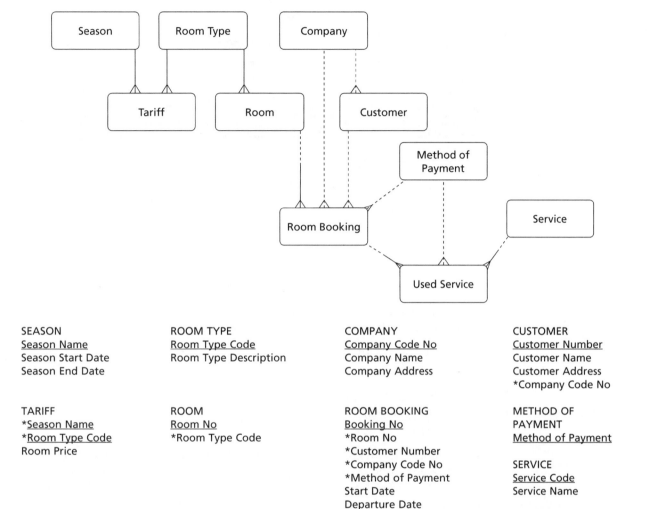

Fig 8.14 Logical Data Model for the West Munster Hotel system

SEASON
Season Name
Season Start Date
Season End Date

ROOM TYPE
Room Type Code
Room Type Description

COMPANY
Company Code No
Company Name
Company Address

CUSTOMER
Customer Number
Customer Name
Customer Address
*Company Code No

TARIFF
*Season Name
*Room Type Code
Room Price

ROOM
Room No
*Room Type Code

ROOM BOOKING
Booking No
*Room No
*Customer Number
*Company Code No
*Method of Payment
Start Date
Departure Date
Provisional (Y/N)
Arrival Time
Bill Date
Card Number
Card Accepted (Y/N)

METHOD OF
PAYMENT
Method of Payment

SERVICE
Service Code
Service Name

USED SERVICE
*Booking No
*Service Code
Service Date
Service Type
Amount Charged
*Method of Payment

8.4. The systems analyst of Exercise 8.2 has not done any Data Flow Modelling. Write a report stating the advantages and disadvantages of not producing a DFM for the West Munster Hotel system. Finish off your report with a statement as to whether a DFM is essential for the success of this project.

8.5. From the scenario in Exercise 8.2 deduce the system's requirements and use them to produce complete set of Function Definitions for the West Munster Hotel system.

8.6. Study the suggested answer to Exercise 6.4 and use it to produce Function Definitions and a Function Navigation Model for the functions associated with the business event of a book's return.

9

User Interface Design

INTRODUCTION

In this section we discuss a number of techniques and products that can be used to help in the design and specification of the user interface, namely:

◆ Window Navigation Models

◆ Window Specifications

◆ Prototyping

◆ User Object Modelling

◆ I/O Structures

Each of the above has its unique strengths and weaknesses and it is unlikely that all of them will be used on a single project. Where a technique is chosen for use it may well be applied selectively, i.e. not necessarily to all the on-line functions of the system.

User Interface Design will encompass the design of many types of interface (e.g. character, windows-based) for many types of user (e.g. customer, occasional, frequent, expert). In each case we will have to emphasise different aspects of the design. For instance, the customer-facing parts of the interface will have to prioritise ease of use (probably at the expense of speed of data entry). Contrast this with a function that is being used by the same person many times per day for high-volume data entry: here we will want to design the interface to minimise the length of time taken to complete each transaction.

User Interface Design involves the following activities:

◆ **Agree style guide:** the purpose of the style guide is to define a standard for the look and feel of the interface. The starting point for this will probably be the commercial style guide, e.g. Microsoft's style guide. To this will be added any organisational or application-specific guides. It is possible to define more than one interface for each function to cater for different platforms or for different user needs.

◆ **Design windows in outline:** the product of this step can be sketches done on paper, or using a drawing tool. A popular approach is to work with users on a whiteboard (preferably the type from which we can make photocopies). Alternatively you could use a large sheet of paper and Post-it notes. Outlines will probably be produced for all the windows in the user interface. For simple Functions they will suffice as a specification and it will not be necessary to produce Window Navigation Models or detailed Window Specifications.

◆ **Design Window Navigation Model:** a useful starting point can be the Function Navigation Model (if you have developed one) but you should realise that a function component could be implemented by several windows or a window could cover several components, so there will not necessarily be a straightforward mapping between the two.

◆ **Specify windows in detail:** the level of detail will depend on the importance and complexity of the Function. For simple Functions a sketched outline of the window should suffice, perhaps annotated with comments. For more complex Functions there will need to be a detailed description of all the window controls and actions and the dependencies between windows.

▶ Learning objectives

After reading this chapter, readers will be able to:

◆ model the interface of an on-line Function in terms of its input and output data;

◆ model the way in which users will be able to navigate around a Function;

◆ model the way in which users will be able to access Functions and move from one function to another;

◆ make an informed choice between a number of alternative techniques for User Interface Design;

◆ plan, run and report on prototyping sessions.

▶ Links to other chapters

◆ Chapter 3   The Work Practice Model for the required system will be developed in conjunction with the User Interface Design, leading to user job specifications.

◆ Chapter 7   The technical environment outlined in the BSO will guide the approach adopted by User Interface Design.

◆ Chapter 8   User Interface Design will be carried out for each function.

◆ Chapter 14   Physical Design will complete the specification of all User Interface Design.

▶ WINDOW NAVIGATION MODELS

In a graphical user interface (GUI) environment, the interface of each on-line Function will be implemented using one or more windows. Before designing or building these it may be useful to get an overview of the different windows that will support a Function and how those windows interact.

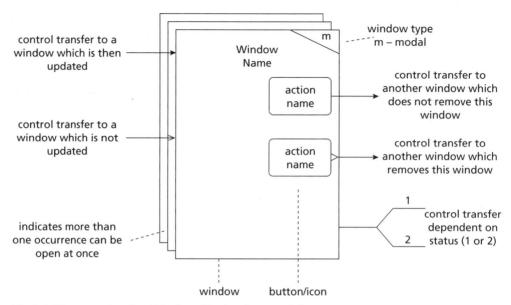

control transfer to a window which is then updated

control transfer to a window which is not updated

indicates more than one occurrence can be open at once

Window Name

action name

action name

window type m – modal

control transfer to another window which does not remove this window

control transfer to another window which removes this window

1

control transfer dependent on status (1 or 2)

2

window   button/icon

Fig 9.1 The notation for Window Navigation Models

You can use a Window Navigation Model to do this. Its notation is shown in Figure 9.1 and a Windows Navigation Model for the Amend Delivery function is shown in Figure 9.2.

When developing Window Navigation Models we should bear in mind the following points:

◆ Users will often want to quit or suspend a transaction before completing it: we should define exit points that allow the user to do this.

◆ The Window Navigation Model should be checked for consistency against the relevant Function Navigation Model and Task Model (Work Practice Model).

◆ We should try to minimise the amount of navigation that users have to do in order to complete a task. In particular users should not have to switch back and forth across different windows.

◆ Where possible the structure of dialogues should be consistent across different Functions.

Note that the Window Navigation Model does not show any detail as to the content of a window. It is really only concerned with the navigation between windows, i.e. the main components of the dialogue for the Function in question. Further information about the content of windows could be added but this will make the model quite cluttered. We prefer instead to produce a separate specification for each window (see next section).

By the end of this chapter we will end up with several representations of the external design of the Amend Delivery function: the I/O Structure of Figure 9.12, the Dialogue Structure of Figure 9.15 and the Window Navigation Model of Figure 9.2 (as we have said several times, it is unlikely that you would want to develop all of these for a particular Function). Note that each model is concerned with the way that a user navigates through a particular Function. Note also that the Window Navigation Model does not indicate which data items are involved in the Function. It will

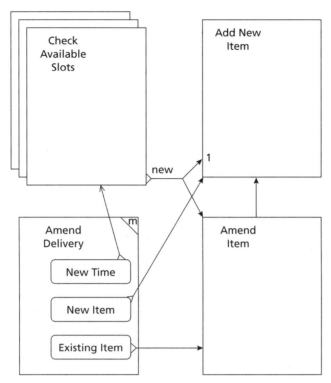

Fig 9.2 A Window Navigation Model for the Amend Delivery function

therefore need to be accompanied by some form of window specification containing this information (unless you happen to have produced I/O Structures as well).

▶ WINDOW SPECIFICATION

It will be important for every on-line function to define the windows that form the interface to that function (assuming that a windows-based interface is being used). The type of things that we have to document for each function include:

◆ data entry fields: names, size, optionality, validation, format and defaults (where this has not been specified elsewhere);

◆ the use of list boxes, their contents, sort order, etc.;

◆ the behaviour of the window, e.g. whether the window is modal (i.e. the user has to finish with that window before they can move to any other window);

◆ tab sequences (the default order in which we move from control to control);

◆ help: identifying the things that the user will require on-line help for.

Window Specification is a useful technique to use within Prototyping to sketch an outline of the window to an appropriate level of detail.

We can bring together the User Interface Design techniques to demonstrate the Web Browse and Buy activities of a Web-based retail customer. We started our study of this new type of ZigZag customer during Business Analysis and further developed the required function during Function Definition. In what follows we present the

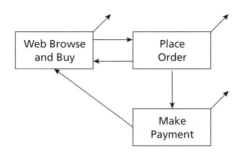

Fig 9.3 Function Navigation Model for Web Browse and Buy function

Function Navigation Model we produced earlier together with the Window Navigation Model and Window Specification for the same function.

In Figure 9.3 we have reproduced the Function Navigation Model for the Web Browse and Buy function. This is described in the chapter on Function Definition (Chapter 8). The browse, order and pay components of this function are to be implemented across a series of web pages. These pages are a subset of the ZigZag's website and are described by the Window Navigation Model of Figure 9.4. Outline Window Specifications for some of these windows are shown in Figure 9.5.

▶ PROTOTYPING

Prototyping provides us with one of the most valuable opportunities to demonstrate and validate our understanding of system requirements by presenting users with a model or example of what the system will actually look like.

Prototyping has become an increasingly popular technique for a number of reasons: perhaps the availability of relatively quick and easy-to-use visual development tools, and perhaps also because a large number of projects are concerned with adding improved interfaces to existing systems.

It has also been recognised for its important contribution towards the validation of users' requirements. What better way to confirm that you have understood what the users want than to actually show them what the system will be like?

Caution

Some systems development approaches rely very heavily on prototyping to the point where there is little other documentation of the design. In some approaches a prototype is built and iteratively improved until it is refined enough for delivery. Sometimes this approach sacrifices maintainability and correctness of specification in favour of speed of delivery.

In projects where off-line processing forms a substantial part of the system, where there are complex calculations, or where financial transactions are involved, it is extremely risky to base anything other than design of user interfaces on prototyping alone.

In many ways specification prototyping can be considered as a mini-project in its own right and as such it requires its own project management procedures.

In this section we shall assume that we are building the prototype to learn from it before 'throwing it away'. Thus we typically use prototyping to correct and

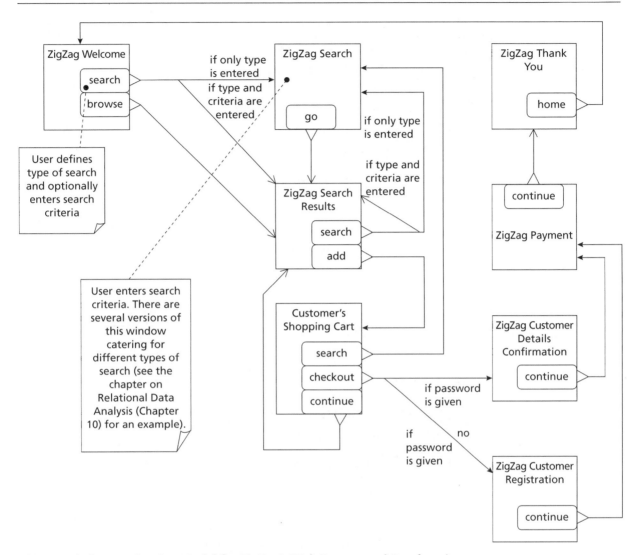

Fig 9.4 Window Navigation Model for ZigZag's Web Browse and Buy function

enhance our specification of user requirements, *not* to incrementally develop the final system. We call this type of prototyping 'specification prototyping'. If any of the prototypes developed at this point can be used in Physical Design and implementation then this should be regarded as a bonus and not our specific objective. Like most of the techniques covered in this book, the use of prototyping is not mandatory – the project team will have to decide if it is worthwhile for the project in question – but when used properly it can lead to enhancement of a wide range of products and to increased user commitment.

▶ Specification prototyping issues

Once a decision to undertake prototyping has been made and management procedures set up, the activities involved are quite easy to understand and satisfying to

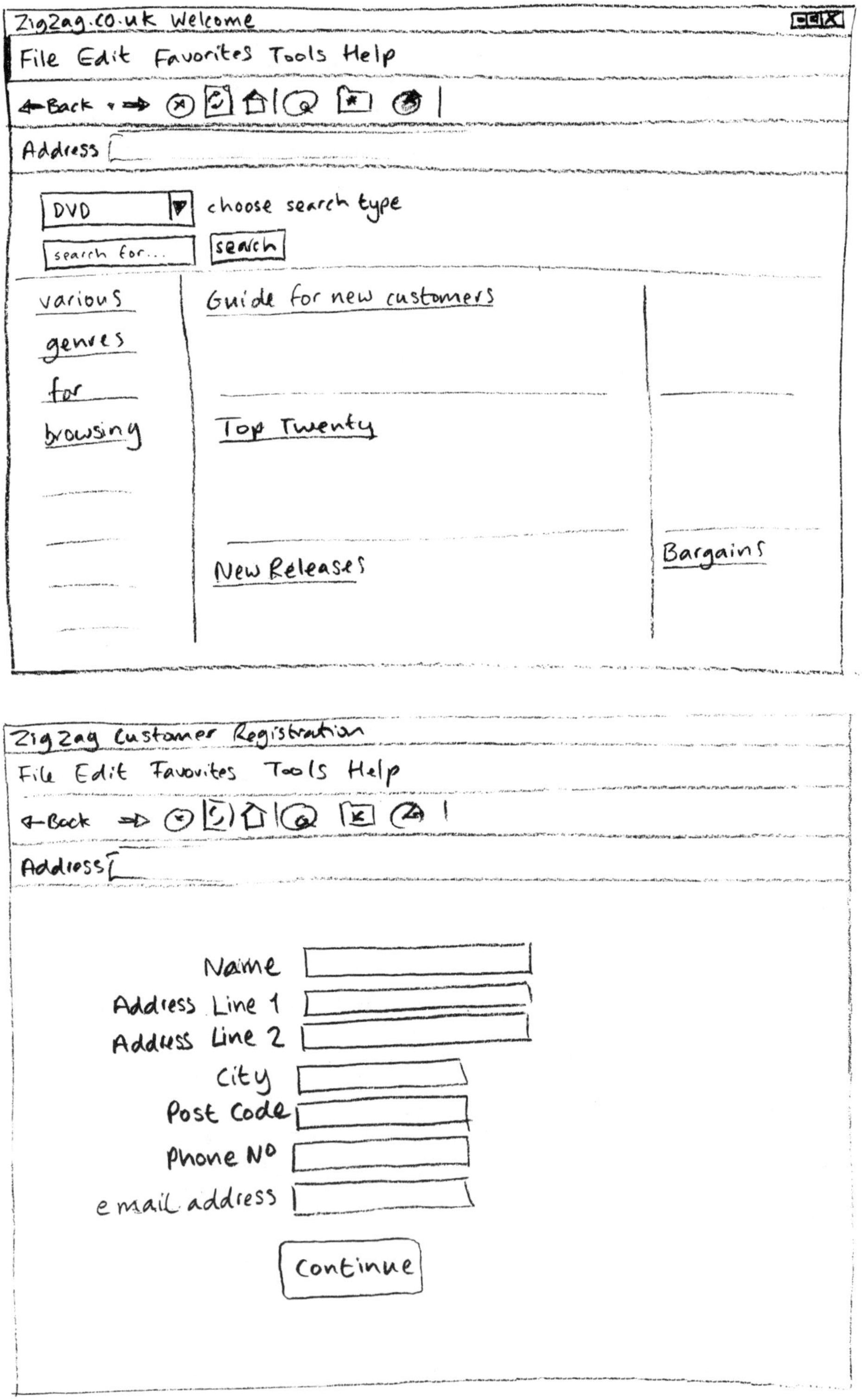

Fig 9.5 Sample Window Specifications

carry out. However, before we can proceed there are a number of issues that need addressing.

Project suitability

Specification prototyping can involve a great deal of time and effort, so for each project we should assess its suitability before committing the necessary resources.

Projects that are likely to benefit from specification prototyping will generally fall into one or more of the following categories:

◆ high-risk projects;
◆ high cost projects;
◆ projects that are likely to result in large-scale changes to working practices;
◆ politically sensitive projects;
◆ projects involving users with little or no experience of computer systems, or analysts with little experience of the business area;
◆ projects that involve large elements of new functionality, or for which there is no existing system.

Projects that are unlikely to benefit substantially from specification prototyping will generally fall into one or more of the following categories:

◆ projects that merely aim to replace existing systems, with little or no extra functionality (although some of these projects will be particularly concerned with improving the user interface, in which case prototyping will be useful, perhaps essential);
◆ low-cost projects, where specification prototyping would have significant impact on cost or benefit;
◆ projects where user requirements are very precisely and rigidly defined.

Note

Our discussion of prototyping follows closely on the heels of Function Definition, as this is one of the most useful points at which to introduce the topic. But it is important to remember that prototyping can be used at many points in a systems development project, in order to enhance understanding of functional requirements.

Risks of prototyping

Virtually all of the risks involved with prototyping are associated with its management and presentation:

◆ **False user expectation:** when users are shown a superficially working system they may believe that final implementation is imminent. Users must be made aware that prototypes are just highly visual and sophisticated simulations or models of systems.

◆ **Limits of prototyping tool:** users may receive a misleading impression of the final solution's likely appearance. In many cases the prototyping tool will not use the same technical platforms as the final system and may be less sophisticated, or alternatively offer more facilities than the implementation environment (for example where a PC-based 4GL is used for prototyping, giving a colourful and a highly graphical user interface, while a mainframe-based 3GL will be used to build the final system, with monochrome screens, etc.).

◆ **Uncontrolled system design:** prototyping can be difficult to control. An atmosphere can develop where a prototype is demonstrated to users, changes are made and then it is re-demonstrated to users and so on, almost indefinitely. The rather informal physical design of prototypes can also lead to non-standard designs and constant uncontrolled changes to the project's scope.

◆ **Lack of documentation:** it is extremely important that the results of prototyping be fully documented and that appropriate updates to function definitions, the LDM, the Requirements Catalogue, etc. are applied. If this is not done then the purpose of prototyping will have been missed.

◆ **Losing sight of the business objectives of the project:** it may be tempting to follow through each idea for improving and enhancing the system but the costs of doing so should be borne in mind and measured against the business benefit that the change or enhancement might bring.

Benefits of prototyping

For suitable projects there are a number of potential benefits that may justify the use of specification prototyping:

◆ **Improved communication:** users can immediately understand and relate to actual screens or reports, in a way which is just not possible with some of the more abstract models in the specification. Although some training may be necessary in the prototyping tool, this is likely to be far less than in the interpretation of diagrams.

◆ **Validation of user requirements:** users will be able to provide feedback quickly and effectively on which aspects of the system requirements have been correctly interpreted.

◆ **Assessment of system capabilities:** users may not be familiar with the capabilities of computer systems (probably less true these days than in the past). Using an appropriate prototyping tool to present a mock-up or simulation of a system may help users to more fully understand these capabilities. This in turn may lead to additional or modified user requirements.

◆ **Increased user commitment:** in large projects users can feel alienated or even forgotten as time progresses without anything concrete for them to get involved with. The central and highly practical role of users in specification prototyping often leads to a greater sense of commitment and project ownership in users.

◆ **Improved project morale:** analysts and users usually find prototyping a satisfying and enjoyable activity, leading to improved morale and a greater feeling of teamwork.

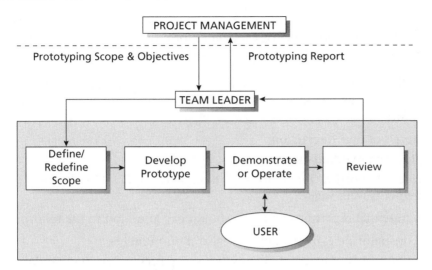

Fig 9.6 Suggested prototyping cycle

▶ Management of prototyping

The management of specification prototyping requires careful handling. Prototyping products are less rigorously defined than most other SSADM techniques, but there are some basic guidelines for us to follow.

The exact method of managing the prototyping process will vary from project to project, but the prototyping cycle illustrated in Figure 9.6 suggests a number of phases that should be borne in mind.

Planning

The overall scope and objectives of prototyping should be defined before the iterative part of the cycle. The main aim here is to know when prototyping should stop by defining what it is expected to achieve.

On the first iteration of the prototype the scope and objectives can be confirmed and on subsequent iterations they will be refined or redefined in the light of the review that has taken place.

The planning stage will involve choices about the prototyping tool and the nature of the prototype used. If the final implementation environment for the project is known then the prototype should conform to the characteristics of that environment. This does not mean that it has to be developed using the same tools. In fact, the prototype could consist simply of hand-drawn sketches on paper.

Define/redefine scope

The overall scope of the prototyping exercise will have been agreed with management already. This is done by examining the outputs Function Definition and Work Practice Modelling to assess which specific on-line dialogues and reports should be prototyped.

The scope of the prototype could vary from a single function or part of a function through to the whole of the user interface.

The User Role/function matrix will be annotated with detail of critical dialogues, and these will be the obvious first candidates for prototyping. We should also consider any other dialogues that users wish to view, to confirm that their requirements are properly understood.

For each on-line dialogue we will have one or more potential screen prototypes. To help us decide which of the candidate dialogues are suitable for prototyping we can look for:

◆ high levels of user interaction;

◆ complex data manipulation;

◆ dialogues supporting functions that have been difficult to define, or where requirements were vague.

For each potential report prototype we should pay attention to the following:

◆ legal constraints, e.g. disclaimers or registration numbers;

◆ the requirements of external organisations, e.g. Inland Revenue standard forms.

Development

The development of the prototype may be guided by any organisational or vendor-specific style guides. It might also include advice from experts in the field of HCI.

The prototype should use realistic data and there should be sufficient variety to demonstrate that the prototype can support different scenarios.

A mixture of types of prototype can be used: starting with low-fidelity, paper-based prototypes and moving on to the use of tools for building functional models of the system. Care should be taken that any functionality that is simulated for the users is actually feasible.

Preparing for the prototype demonstration

Before we demonstrate prototypes to users, we should draw up specific objectives and agendas for the session as it is all too easy in the rather informal atmosphere of prototyping to get side-tracked and to waste valuable time on trivial issues.

To help in this task we can draw up a Prototype Demonstration Objective Document for each pathway. As well as listing specific discussion points for each dialogue, we can use this document to note down general tasks, such as an explanation of procedure, in the form of an agenda for the entire prototyping session.

The Prototype Demonstration Objective Document for the Book Delivery Function is shown in Figure 9.7.

As part of preparing for the demonstration we should also design some test data. This will often include both input and output data, as the prototype may not possess any true processing capabilities, but merely simulate the correct responses to pre-specified inputs. We should make test data comprehensive enough to illustrate all the relevant user requirements, and to simulate a full range of system responses (including major error processing or validation).

Demonstration

We will demonstrate each pathway to one or more users belonging to the appropriate User Role. Ideally, demonstrations should be conducted by two analysts from the project team.

> You may consider it more appropriate to let the users 'drive' the prototype rather than simply observe it.

Prototype Demonstration Objective Document		
Pathway No: 001	**Function Name:** Book Delivery	**User Role:** Delivery Scheduler
Agenda: 1. Discuss prototyping aims and procedures. 2. Explain operation of prototyping tool. 3. Carry out demonstration. 4. Discuss feedback and possible re-demonstration.		
Component No.	**Discussion Point**	
5	Check input data items if Id unknown.	
6	Check field sizes.	
6	Check output details are sufficient.	
6	What if stated quantity exceeds expected?	

Fig 9.7 A sample Prototype Demonstration Objective Document

The 'lead' analyst will be the designer of the prototype and will be the one who actually demonstrates the prototype. The second analyst will act as note-taker and help to keep the demonstration to the preset agenda. The designer of a prototype will often feel defensive about any criticisms of it, so the second analyst should also act as an objective observer, ensuring that the wishes of the *user* are fully recorded.

We document the results of the demonstration in a Prototype Result Log (Figure 9.8). This will detail each request made by the user during the demonstration, which we will later annotate with a change grade. There are seven grades suggested by SSADM:

N. No change needed.

C. Cosmetic. This refers to change requests associated with layout and format, not content. If the only change requests logged are in this category, we would not carry out any further iterations of the prototyping cycle.

D. Dialogue-level, i.e. changes which affect the content of the dialogue only.

P. Pathway changes. These will generally refer to requests regarding the sequence of the pathway. They may lead to changes in the I/O Structure for the function.

S. This indicates a possible need to change installation standards.

Prototype Result Log			
Pathway No:	Function Name:		User Role:
Component No	Result No	Result Description	Change Grade

Fig 9.8 A sample Prototype Result Log

A. Analysis errors. Any changes in this category will indicate that errors have been made in systems analysis. If the errors are very great they may lead to repeats of earlier steps.

G. Global change requests. Some requests may have implications outside the business area under investigation.

We pass the completed Pathway Result Logs, along with our recommendations, to the team leader, who will decide on what will happen next. This may involve carrying out further demonstrations using refined prototypes, or halting the cycle if there is likely to be little benefit gained from continuing. If any major problems have arisen from the demonstration then these may need reporting to management.

The team leader will also ensure that any relevant SSADM products are updated. Any entirely new requirements should be added to the Requirements Catalogue. We should review changes of all types other than N or C to make sure that all of their implications are understood, and that they are practical when the requirements of the system as a whole are taken into account.

Review

Once all of the prototype demonstrations have finished the team leader should prepare an overall report of their results. This should include comments on the prototyping exercise as a whole, e.g.:

◆ Were the objectives of the step met?
◆ Was the original scope adhered to?
◆ Was the exercise a success?

It will also summarise any changes to the requirements specification, and refer to any products that have been updated or amended as a result.

▶ USING I/O STRUCTURES TO SPECIFY USER INTERFACES

Functions are units of work that users wish to carry out using the new system. To do so they will need to hold a dialogue with the system, i.e. to interface with each function. From a logical point of view Jackson structures lend themselves well to describing the dialogue between user and system, many analysts and nearly all users consider them too abstract for specifying on-line user interfaces.

> **Note**
>
> We should remember that in many systems the bulk of the work will be in the design of off-line interfaces, a purpose to which Jackson structures are well suited.

Most update functions will already have a number of I/O Descriptions associated with them, as identified from the Required System Data Flow Model. These provide a list of all data items input to the function and of the major data items output from the function, thus providing us with an initial, albeit limited, view of the user interface. However, output elements of I/O Descriptions contain only those data items

I/O Description				
From	To	Data Flow Name	Data Content	Comments
b	1.1	P.O. Quantities	Product Number, Qty Required, Req-By Date, Req-By Time-Period.	The purchaser may state a date by when the product should be delivered or a time-period, e.g. 3 weeks from now.
1.1	a	Proposed Purchase Order	P.O. Number, Supplier Number, Supplier Name, Supplier Address, Req-By Date, Product Number, P/S Reference, Product Name, Qty Required, Product Price.	Each Purchase Order will contain several lines (usually up to about 12). Product No is the unique number of a product, e.g. the bar code P/S Reference is the reference number suppliers use for that product.

Fig 9.9 I/O Description of data flows from Figure 8.2

that make up the final output from the function, and so omit any outputs designed to verify or confirm user input. We can illustrate this by looking at the function Propose Purchase Order, which was originally identified using the Data Flow Model (see Figure 8.2). The I/O Descriptions for this function are given in Figure 9.9.

The data flows do not indicate whether any output is received by the purchaser (or the Purchase Order Clerk). The data flows appear to be showing that the purchaser inputs a group of data items (on-line) without response from the system, which ultimately results in the output of proposed purchase orders (off-line) which are sent to suppliers. This cannot be true though, and we expect a friendlier system which, when a purchaser inputs a product number, will respond by displaying the product name so that the purchaser can verify that the product number is correct.

This kind of response is not part of the required final output of the process, but merely acts as an input confirmation. For this reason it is not included in the Data Flow Model (the model would get very cluttered if you tried to show all these flows), but it *will* be documented as part of a Function Definition.

Looking again at the I/O Descriptions in Figure 9.9, we can see that inputs and outputs are shown as separate unstructured groups of data items, with occasional comments that certain sub-groups will be repeated within the overall group (e.g. *product number, quantity required*, etc. in I/O Description 1.1-a). This would imply that we input a continuous stream of data that results in a continuous stream of output data.

However, in reality users will require input and output data to be interleaved, with some data items being repeated or omitted.

For example, let us take the on-line interface between users and the function Propose Purchase Order, represented at the moment by I/O Description b-1.1.

Purchasers will enter data in batches throughout the day, for a range of products. For each product there will be entered a product number, each of which will be confirmed by the output of relevant product details.

This textual description is not sufficiently precise or rigorous enough to act as a specification of the users' required interface, but it does point to the need to model the structure, as well as the content, of such interfaces. We can do this by using I/O Structures. These were introduced earlier in the chapter on Function Definition (Chapter 8) when we used them to describe the off-line part of the Propose Purchase Order Function (see Figure 8.12).

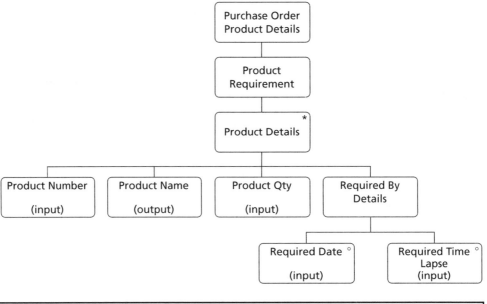

I/O Structure Description		
I/O Structure Name: Purchase Order Product Details		**Id**
I/O Structure Element	**Data Item**	**Comments**
Product Number	Product No	
Product Name	Product Name	The product's name is output to provide a visual check
Product Qty	Quantity Required	
Required Date	Req-By Date	
Required Time Lapse	Req-By Time-Period	

Fig 9.10 I/O Structure for the on-line proposal of purchase orders

An I/O Structure for an on-line function or part of a function is drawn to reflect the kind of dialogue users will expect, i.e. with inputs paired with appropriate system responses, as in the case of Purchase Order Product Details in Figure 9.10.

Tip

Some functions will include both enquiry and update components, e.g. Amend Delivery (which is a function related to Book Delivery, see Figure 8.3). In these cases we can produce a separate I/O Structure for each component. The I/O Structure for the enquiry component Check Available Slots is shown in Figure 9.11, and the remaining update component of Amend Delivery is shown in Figure 9.12. I/O Structures can be made more manageable in this way.

The interface of the Book Delivery function is not straightforward since it involves two components and an element of decision making on the part of the delivery scheduler. For this reason it is a candidate for prototyping.

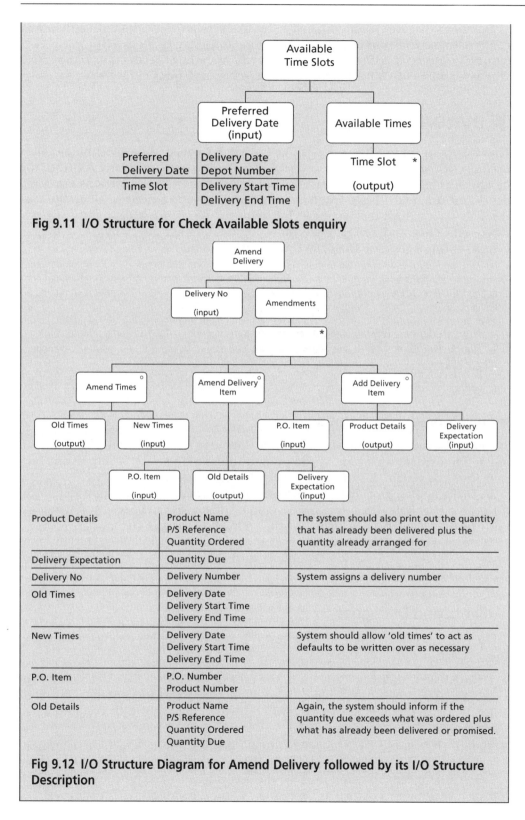

Fig 9.11 I/O Structure for Check Available Slots enquiry

Fig 9.12 I/O Structure Diagram for Amend Delivery followed by its I/O Structure Description

Functions that are system-initiated are unlikely to have any input other than user confirmations. For example, the enquiry function Produce Stock Listing will contain only outputs as it is triggered by date and time, while the function Remove Old Purchase Orders will only require yes/no-type confirmations of deletions.

▶ DIALOGUE DESIGN

The work done on I/O Structures can be further developed using a technique called *Dialogue Design*. Like I/O Structures, Dialogue Designs are particularly suited to the design of character-based interfaces. Then again, even where Functions are being developed using GUI-based interfaces there will *often* be a need to guide the user through a prescribed dialogue, e.g. for frequent use, or mass data entry screens, where efficiency and accuracy are more important than any initial ease of use. Wherever this is the case Dialogue Design can be useful.

Note

Dialogue Designs are logical, abstract representations of the interface. In some ways this is an advantage (for example it makes them portable) but users will want to see something more tangible: screen designs, for instance.

▶ Developing user dialogues

To many users the internal workings of a computer system are entirely unknown and to some extent irrelevant. The important things to them are the inputs and outputs from the system, i.e. their dialogues with the system.

In Dialogue Design, a dialogue provides the 'front-end' to a Function Definition for a given User Role. The process of developing dialogues can be seen as having two fairly distinct elements: Dialogue Identification and Dialogue Design.

Dialogue Identification begins early in the project life-cycle and is typically completed by the end of Function Definition with the creation of the User Role/function matrix.

▶ Identifying Dialogues

To help place Dialogue Design in context it is worth recapping briefly on the activities of Dialogue Identification:

◆ **Define User Catalogue:** early on in the project we create a User Catalogue, containing details of all job holders who require access to the system. Thus each job holder is a potential user who will engage in dialogues with the automated system.

◆ **Define User Roles:** User Roles are usually defined during Function Definition. Each User Role represents a collection of users who will carry out similar (or the same) automated tasks, and thus require access to the same functions. Thus each user belonging to a particular User Role will use a common set of dialogues.

◆ **Create User Role/function matrix:** following the definition of functions we create a User Role/function matrix. This documents which functions each user role will require access to, thereby identifying all required dialogues. The responsibilities of each User Role can be defined using Work Practice Modelling.

◆ **Identify Critical Dialogues:** we then identify which of these required dialogues are regarded as critical to the success of the system. In particular this helps us to select dialogues for prototyping.

▶ Designing dialogues

The design of dialogues at this stage is concerned with logical issues, not physical ones such as screen layout, colour usage, etc. The technique is built on the concept of logical screen components, which may eventually be implemented over more than one physical screen or window, or form just one part of a single physical screen/window.

Our main aim in designing dialogues is to support the way in which users wish to interface with the system. Clearly the best way to achieve this is to involve users in the process as much as possible.

Although the technique does not take account of many physical considerations at this point, we should take into account some style issues, such as the maximum number of data items on a screen. Information on this sort of thing will usually be available from an organisation's standards manual or 'style guide'.

Some dialogues may have already been prototyped. However, our primary concern in Specification Prototyping is to validate user requirements, not to design final dialogues. We can use the dialogues or menu structures produced as a result of prototyping, but we should always review them before doing so (and in any case they will usually need further expansion and integration with other dialogues or menus).

Figure 9.13 illustrates the activities involved in Dialogue Design.

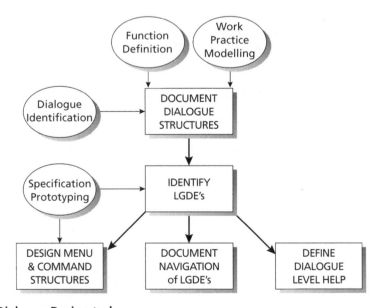

Fig 9.13 Dialogue Design tasks

Dialogue Element Descriptions			
Dialogue Element	Data Item	Logical Grouping of Dialogue Elements	Mandatory/ Optional LGDE
Product Details	Product Name P/S Reference Quantity Ordered		
Delivery Expectation	Quantity Due		
Delivery No	Delivery Number		
Old Times	Delivery Date Delivery Start Time Delivery End Time		
New Times	Delivery Date Delivery Start Time Delivery End Time		
P.O. Item	P.O. Number Product Number		
Old Details	Product Name P/S Reference Quantity Ordered Quantity Due		

Fig 9.14 Dialogue Element Description Form for Amend Delivery

Document dialogue structure

We will now discuss Dialogue Design by taking the I/O Structure of Figure 9.12 and developing it a little further:

The I/O Structure for Amend Delivery is carried forward to form the basis for the Dialogue Structure. We then begin to fill out a Dialogue Element Description form (Figure 9.14) by copying all of the data items from the I/O Structure Description, each Dialogue Element being equivalent to an I/O Structure Element.

Note that the Dialogue Structure does not replace the I/O Structure – it is merely based on it.

Identify Logical Groupings of Dialogue Elements (LGDE)

We now examine the Dialogue Structure in order to identify groups of Dialogue Elements that logically belong together, as components of the dialogue. The views of users are paramount at this point, but individuals may well differ on the question of what constitutes a logical grouping for their purposes. Representatives of different User Roles, all requiring access to the same function, could require different paths or navigation through the dialogue, reflecting the ways in which they operate. It is quite acceptable to produce more than one set of LGDEs for a given Dialogue Structure, leading to more than one Logical Dialogue for the function.

In the case of Amend Delivery we will assume that only one dialogue is required for all User Roles with the LGDEs shown in Figure 9.15.

Identifying which elements should be logically grouped together is really a matter of discussing with users which elements 'feel right' together. Elements that are always input as a set, which form a group of inputs and resultant outputs, or reflect

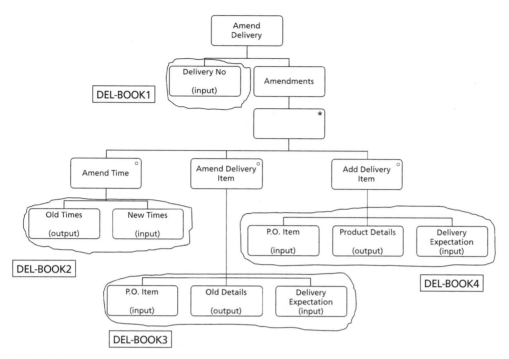

Fig 9.15 Dialogue Structure with LGDEs

tasks that are always carried out in sequence, are candidates for grouping. Clearly there should be a close relationship between LGDEs and the components of the relevant Function Navigation and Window Navigation Models.

A few other points to bear in mind when looking for LGDEs are:

◆ Never group together elements that have elements outside the group between them.

◆ Each element must belong to one and only one LGDE.

◆ Avoid grouping elements together that represent mutually exclusive selections.

Once the LGDEs have been added to the Dialogue Structure we should update the Dialogue Element Description form as shown in Figure 9.16.

Complete Dialogue Control Table

The LGDEs provide us with logical screen components. Whenever a user interacts with the function they will navigate through a number of these components.

For example, the most frequently used or default path through the Amend Delivery dialogue will be to amend the agreed delivery time, i.e. DEL-BOOK1, DEL-BOOK2.

However, this is not the only path. For example, we may begin by entering an existing delivery number to add an item to the delivery, i.e. DEL-BOOK1, DEL-BOOK4. This happens about 15% of the time.

The default path and any alternatives can be documented using a Dialogue Control Table, as in Figure 9.17.

As well as indicating possible pathways we can use the table to document the frequency (or 'percentage path usage') of each path, and to show the minimum,

Dialogue Element Descriptions			
Dialogue Element	Data Item	Logical Grouping of Dialogue Elements	Mandatory/ Optional LGDE
Delivery No	Delivery Number	DEL-BOOK1	
Old Times	Delivery Date Delivery Start Time Delivery End Time	DEL-BOOK2	
New Times	Delivery Date Delivery Start Time Delivery End Time		
P.O. Item	P.O. Number Product Number	DEL-BOOK3	
Old Details	Product Name P/S Reference Quantity Ordered Quantity Due		
Delivery Expectation	Quantity Due		
Product Details	Product Name P/S Reference Quantity Ordered	DEL-BOOK4	

Fig 9.16 Adding LGDEs to the Dialogue Element Descriptions

Dialogue Control Table							
Function: Amend Delivery							
LGDE	Occurrences			Default	Alternative pathways		
	min	max	ave	Pathway	alt1	alt2	alt3	alt4
DEL-BOOK1	1	1	1	X	X	X	X	X
DEL-BOOK2	0	1	2	X			X	X
DEL-BOOK3	0	20	6		X	X	X	X
DEL-BOOK4	0	20	8			X		X
% path usage				40	15	15	15	15

Fig 9.17 Dialogue Control Table

maximum and average number of occurrences of each LGDE in a typical execution of the dialogue. Clearly, any LGDE with a minimum number of occurrences of greater than zero will be present in all alternative pathways, and is known as *mandatory*. Where the minimum number of occurrences is zero the LGDE is *optional*. The Dialogue Element Description form should be updated to reflect whether LGDEs are optional or mandatory (Figure 9.18).

Note that any LGDEs consisting of sequence boxes only must be mandatory. If an LGDE covers just part of a selection, or a whole selection containing a null box, it must be optional. Iteration LGDEs can be mandatory or optional depending on whether the possibility of zero iterations is allowed in the particular dialogue under scrutiny.

Dialogue Element Descriptions			
Dialogue Element	**Data Item**	**Logical Grouping of Dialogue Elements**	**Mandatory/ Optional LGDE**
Delivery No	Delivery Number	DEL-BOOK1	M
Old Times	Delivery Date Delivery Start Time Delivery End Time	DEL-BOOK2	O
New Times	Delivery Date Delivery Start Time Delivery End Time		
P.O. Item	P.O. Number Product Number	DEL-BOOK3	O
Old Details	Product Name P/S Reference Quantity Ordered Quantity Due		
Delivery Expectation	Quantity Due		
Product Details	Product Name P/S Reference Quantity Ordered	DEL-BOOK4	O

Fig 9.18 Complete Dialogue Element Descriptions

Design Menu and Command Structures

Once we have completed the documentation of possible pathways within each dialogue, we turn our attention to pathways through the system as a whole (i.e. between different dialogues and menus).

Menu Structures provide a good idea of how the system fits together from a user's point of view, but do not strictly specify all the alternative ways of accessing any given dialogue. This function is fulfilled by Command Structures, which provide details of where a user may go to in the Menu Structure once a dialogue is finished.

Define dialogue-level help

While designing logical dialogues we will also consider the question of where dialogue-level help might be required.

By 'dialogue-level help' we mean help with navigation problems, i.e. how to exit, where to go next, etc. The level of help needed will depend largely on the complexity of the system and will be driven by the needs of users.

Once again we are operating at a logical level, so we will not consider the format of help screens or the procedure for obtaining them. The main thing at this point is to identify where help is required.

Fig 9.19 Menu

Fig 9.20 Dialogue

Menu Structures

Menu Structures are fairly straightforward. We represent menus by 'square-cornered' boxes as in Figure 9.19, whereas individual dialogues are represented by round-cornered boxes as in Figure 9.20. We then combine these basic elements to represent a hierarchy of menus and dialogues for the user role (see Figure 9.21 for example).

SSADM suggests that a Menu Structure should be constructed for each User Role. The bottom level in a Menu Structure hierarchy will then consist of the dialogues required to interface with all of the functions identified for the appropriate User Role in the User Role/function matrix. For example, taking the User Role 'Delivery Scheduler' the required dialogues are:

◆ Book Delivery
◆ Amend Delivery
◆ Maintain Schedule
◆ Overdue Delivery Query
◆ Check Available Slots
◆ Delivery Query
◆ Allocate Stock Location

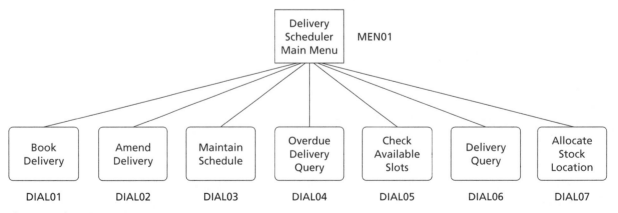

Fig 9.21 Flat Menu Structure

Dialogue names will normally be taken from the name of the function they represent. If a function is accessed by more than one User Role, then it may be represented
by more than one dialogue. In this case the dialogue name for each User Role will
consist of the function name plus a qualifier indicating the User Role to which it
belongs.

Once the bottom level of the Menu Structure is known we can construct intermediate levels in the hierarchy by grouping dialogues that belong together logically
under sub-menus. There are no strict rules for doing this, and in the extreme case
we could hang all of the dialogues from the top of the User Role main menu as in
Figure 9.21.

Alternatively, we could group dialogues together under intermediate levels of
menus. As a general guide any groupings of dialogues should aim to support the
way in which users carry out activities. In the absence of firm requirements from
users we might consider groupings based on the following:

♦ functions of a similar type (e.g. group all updates under one menu, and all
   enquiries under another);

♦ functions that access the same data;

♦ the grouping of processes on the Required System DFM.

In the case of the Delivery Scheduler we will produce three main groups based on
delivery data, Schedule Maintenance and Enquiries (Figure 9.22).

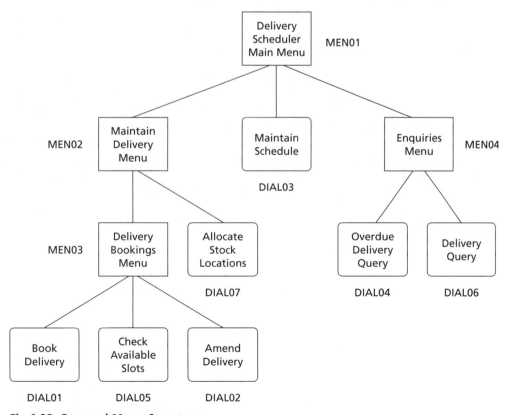

Fig 9.22 Grouped Menu Structure

Command Structures

The Menu Structure gives us a fair idea of how we can navigate to the dialogue we require by following the menu hierarchy. However, this does not give us the whole picture. We may frequently wish to jump from one dialogue to another dialogue, or back to the beginning of the same dialogue without returning to a menu.

For example, on completion of the dialogue DIAL01 (Book Delivery) of Figure 9.21 we may wish to do any of the following:

◆ book another delivery
◆ check slot availability
◆ return to the main menu
◆ return to the Delivery Bookings menu

We document this information using a Command Structure. Each dialogue in the Menu Structure will have its own Command Structure detailing all of the places that control can pass to upon completion of the dialogue.

Figure 9.23 shows the Command Structure for DIAL01 (Book Delivery).

Command Structure		
Dialogue Name: Book Delivery		**Dialogue Id.** DIAL 01
Option	**Dialogue or Menu**	**Name/Id.**
Book another delivery	Dialogue	Book Delivery DIAL 01
Check Slot Availability for new delivery	Dialogue	Check Available Slots DIAL 05
Quit to Main Menu	Menu	Delivery Scheduler Main Menu MEN 01
Quit to Maintain Delivery	Menu	Maintain Delivery MEN 02

Fig 9.23 Command Structure for DIAL01 (Book Delivery)

Command Structures include no details of how each completion option is to be implemented (e.g. using a function key or mouse button) – this is an issue for Prototyping and Physical Design. If the technical environment is known at this point – and it probably will be – we should take any constraints into account regarding possible navigation, e.g. it may not allow us to jump directly to dialogues below other sub-menus.

▶ USER OBJECT MODELLING

User Object Modelling is a technique that attempts to represent the computerised information system as the user might think of it. In other words, it is used to represent the user's mental model of the information system.

It can be argued that Function Definition leads analysts to take a too system-centred view of their work, concentrating on the way in which update and enquiry processing will be triggered and hence emphasising a view of the system that is based around the Logical Data Model, events and enquiries, rather than the user.

The intention of User Object Modelling is to redress the balance by taking a user-centred view of the system or its interface, forcing the analyst to think about the system as the user perceives it.

It should be noted that the system-centred view captures the analyst's understanding of the business rules and data, and in most cases it will be appropriate for this to drive the Specification phase, with user-centred design techniques used as complementary tools for systems design, rather than as alternatives to a system-centred approach.

Tip

User Object Modelling is probably most useful as an informal requirements gathering tool: for instance, you might try developing one with users during a workshop as a way of discussing and clarifying functional requirements prior to prototyping.

We would not recommend User Object Modelling as a framework for detailed design: we believe that there are already more than enough techniques at your disposal and that User Object Modelling does not bring enough unique benefits to the design process to really justify its use in this way.

There are four basic User Object Model (UOM) concepts:

♦ User Objects
♦ User Object Model attributes
♦ actions
♦ associations

A User Object is a part of the user interface that the user can interact with. It is useful to distinguish between those objects that present data for viewing and editing and other types of object like printers, drives and folders.

The data-presenting objects will mainly consist of data that is related to the Logical Data Model, although they could include data from elsewhere; examples include data imported from other applications and data provided by links to web sites.

User Object Model (UOM) attributes are not unlike attributes of the Logical Data Model, i.e. they are data items. However, there is no neat one-to-one mapping of UOM attributes to LDM attributes. For instance, we may, at the interface, see a single field called 'address', but this may map to several LDM attributes.

> It is important to realise that we are modelling things as the user sees them, so there is no expectation that our User Object Model will be normalised (see Chapter 10).

Bear in mind also that not every data item that is visible at the interface relates directly to the persistent data that is modelled by the Logical Data Model. Examples include derived data, e.g. a total that is calculated from data that is stored rather than being stored itself. Another example would be a web-page link.

Actions are the things that users can do to objects. Some actions will be the triggers to updates or enquiries, but again these are a subset of the actions that can appear on a User Object Model. For instance, we could include actions to cut, copy and paste data or to sort in ascending or descending order or to undo a previous action. These actions could all change the way that data is presented at the interface without impacting at all on the data stored in the database.

Associations are relationships between objects that are provided by the system. Typically they represent the ability to navigate from one object to another (but they are not really formally defined, unfortunately).

A User Object Model consists of a User Object Structure plus User Object Descriptions. The latter should describe each attribute and action and cross-reference them (where applicable) to LDM attributes, and to events and enquiries

---

Note

Version 4.3 of SSADM introduced the User Object Model and presented it as a major tool in the specification of the system. We do not believe that this is justified. We think that the User Object Model might have some value as a vehicle for discussing requirements, but we probably would not make a stronger case for it than that. Please note that this has nothing to do with a bigger debate about the role that object orientation can play in information systems design, and that the User Object Model is *not* the same thing as a Class Model or Object Model in an object-oriented approach. In downplaying the User Object Model we are not rejecting object-oriented ideas.

▶ Developing User Object Models

We can build a single User Object Model for the whole application, for each User Role. Two suggested ways of developing user object models are:

◆ From the Required Task Model. For each task or sub-task, ask which objects the user want to work on. Another way of looking at this is that many low-level tasks will involve actions on objects.

◆ Directly from interviews with the users. In this case the discussion will focus directly on the appearance of the proposed system's interface.

In the case of arranging a delivery for ZigZag, delivery schedulers perceive that the activity entails the use of a diary, which consists of many time slots. During the activity, a delivery note is created. This delivery note is associated with one or more half-hour slots and with one or more purchase orders. Figure 9.24 depicts an Object Model for the delivery scheduler.

Note

User Object Models have superficial similarities to object-oriented design models. Indeed, the User Object Model notation can be extended to incorporate any object-oriented (OO) notational convention depending on the standards of the organisation. However, it should be noted that most object-oriented approaches use object-oriented models to describe the underlying business data and business rules (for which we have used the Logical Data Model, Entity Behaviour Model and Conceptual Process Models). It is not so common to use an object-oriented model to describe the interface, although, as we have already shown, this is perfectly feasible.

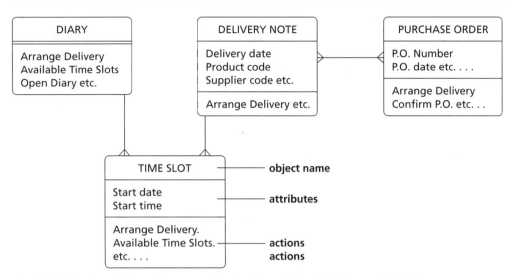

Fig 9.24 The User Object Structure for the User Role of Delivery Scheduler

▶ SUMMARY

1. I/O Structures and Dialogue Designs are not the most fashionable of products for the specification of user interfaces, but they offer a number of important features that should not be lightly discarded:

   ◆ They force the analyst to specify precisely the data contents and structure of each screen or window.

   ◆ They require careful consideration of the nature of the dialogue between user and machine.

   ◆ They encourage the analyst to collect useful volumetric data.

2. In a graphical user interface environment, the interface of each on-line Function will be implemented using one or more windows. Before designing or building these it may be useful to get an overview of the different windows that will support a Function and how those windows interact. You can use a Window Navigation Model to do this.

3. Window Navigation Models and Window Specifications are perhaps more suited to the development of modern WIMP-based graphical user interfaces, and if developed with care can deliver all the features outlined in point 1 above, whilst also providing the user with a more concrete impression of what the system will look like.

4. For many people prototyping has become an essential part of systems development. Its benefits are indisputable and it is one of the few truly user-centred techniques. But it does need to be carefully managed and used, both to control the costs of performing it and also to ensure that the results are properly recorded and acted upon.

5. A User Object Model consists of a User Object Structure plus User Object Descriptions. The latter should describe each attribute and action and cross-reference them (where applicable) to LDM attributes, and to events and enquiries.

6. User Object Modelling is a technique that attempts to represent the computerised information system as the user might think of it. In other words, it is used to represent the user's mental model of the information system. User Object Modelling is probably most useful as an informal requirements-gathering tool: for instance, you might try developing one with users during a workshop as a way of discussing and clarifying functional requirements prior to prototyping.

▶ EXERCISES

9.1. Produce a Window Specification for the following Treebanks enquiry function description:

When a playing court number is entered into the system it should return a list of all the sessions for which the court has been booked, giving details of the member or team that placed each booking.

Assume that the following attribute names are all that are needed:

Court No.
Session No.
Session Start Time
Session End Time
Member No.
Member Name
Team No.
Team Name

9.2. Produce an I/O Structure (diagram and description) for the Treebanks enquiry of Exercise 9.1. Compare the two approaches and discuss the merits of I/O Structures when compared to Windows Specifications.

9.3. Produce a Window Navigation Model and a Window Specification for the Bodgett & Son function 'Record Estimate Decision':

The estimate number should be entered and the customer's details displayed for confirmation. The user should then have a choice of entering an acceptance or rejection for each individual job within the estimate. When decisions are entered there will often be a mix of acceptances or rejections, and job details should be displayed for each job number for confirmation. Details of any jobs for which no decision has been recorded should then be displayed.

The following attribute names are derived from the I/O Description associated with this function along with appropriate system responses:

Estimate No.
Customer No.
Customer Name
Customer Address
Job Decision
Job No.
Job Description

9.4. Produce an I/O Structure (diagram and description) for the Bodgett & Son function of Exercise 9.3. Compare and contrast the production of I/O Structures and Window Specifications for this function.

9.5. Translate the 'Customer Invoice' and 'Customer Request Form' physical layouts of Figure 9.25 into I/O Structures. These two reports are required by the current system of the Fresco ticket agency of Exercise 3.8. You may assume that all Fresco's customers are allocated a customer number, each performance has a number, e.g. 2341, and a performance may be requested at more than one venue.

9.6. Identify the User Role within Fresco that will use the functions of Exercise 9.5. Produce a Menu Structure for that User Role and use it to plan a prototyping session to test the specifications of these two functions.

```
                    FRESCO TICKET AGENCY
                       1 High Street
                         London
                         W1 1AA
                    Tel: 0171 911 5000

Invoice No: 12345
Date: 29/02/00

Customer: 113557
                    F. Bloggs
                    25 Low Rd
                    Hightown NE4 8QT
                    0155 23675

Performance:    Mozart Symphony 40, LCO      (No: 2341)
Venue:   London Concert Centre
                    25 South St
                    London

Date: 14/03/00    Time: 8.00 p.m.

Tickets:  2 at £10
                    2 at £12

Performance:    Macbeth                       (No: 2452)
Venue:   London Theatre
                    1 The Parade
                    London

Date: 19/04/00    Time: 7.00 p.m.

Tickets:  4 at £20

                              Total Cost:  £124
                              Payment by: Credit Card
                              Number:    6319 0621 9914
```

```
                CUSTOMER REQUEST FORM

Customer:   113557
                    F. Bloggs
                    25 Low Rd
                    Hightown NE4 8QT
                    0155 23675

Performance:     Mozart Symphony 36, LCO
Venue:   London Concert Centre
                    25 South St
                    London

Preferred Dates:    07/03/00   14/03/00   21/03/00
Price Range:        £10–12
Number of Tickets: 4

Performance:     Haydn Symphony 104, LCO
Venue:   London Concert Centre
                    25 South St
                    London

Preferred Dates:    13/04/00
Price Range:        £20–25
Number of Tickets: 6
```

Fig 9.25 Fresco 'Customer Invoice' and 'Customer Request Form' physical layouts

9.7. You are a member of a team who is a bit concerned with some of the functions identified during Function Definition. Write a brief report explaining to the team why Specification Prototyping should be undertaken and explain how you will proceed to avoid the possibility of the prototyping session overwhelming the development.

9.8. Use the West Munster User Role/function matrix from Exercise 8.3 to produce Menu Structures for the new system.

9.9. Use the User Role/function matrix of Exercise 9.8 to pick two West Munster critical functions. Plan and run a prototyping session for these two functions and report your findings in the proper manner.

10

Relational Data Analysis

INTRODUCTION

In Chapter 7 we completed the top-down analysis of data, and produced the Required System LDM (see Figure 7.6). Logical Data Modelling is a very useful and powerful technique for analysing data, as it concentrates on the concepts that are of significance to the business, and allows for subsequent refinement of the model as our knowledge increases. It also allows us to create a system-wide view of information needs right from the very start of analysis.

We now turn our attention to bottom-up data analysis using the technique of Relational Data Analysis (RDA). Our aim is to create data model extracts from collections of individual data items, which we can then use to enhance or confirm the Required System LDM.

To carry out RDA on all known system data in order to build a complete LDS would be very time-consuming and complicated, especially for large systems. Instead we select some of the most important or most complex functions (using the kind of criteria mentioned earlier for identifying critical dialogues) and carry out RDA on the data items that are input and output as part of that function.

Tip

We have chosen here to demonstrate RDA principles after Function Definition, but RDA can be used informally at any time to validate or enhance the Logical Data Model as we go along. In these cases our sources of data items might be actual forms or reports we may have picked up when analysing current data, or entity descriptions themselves from when we are creating the Required System LDM.

> ▶ Learning objectives
>
> After reading this chapter, readers will be able to:
> - ◆ understand the process of normalisation;
> - ◆ perform Relational Data Analysis;
> - ◆ recognise the importance of normalised databases;
> - ◆ recognise First, Second and Third Normal Forms;
> - ◆ augment and formalise their understanding of Logical Data Modelling;
> - ◆ decide when in the development cycle to perform normalisation;
> - ◆ evaluate the integrity of a data structure.

> ▶ Links to other chapters
>
> - ◆ Chapter 5 (and Chapter 7) Data models to be compared and contrasted with the data model produced using Relational Data Modelling are produced.
> - ◆ Chapter 8 The formal Function Definitions that are used as input to Relational Data Analysis are developed.

▶ RELATIONAL DATA ANALYSIS CONCEPTS

RDA is based on material published in the 1970s by Edgar Codd of IBM, proposing the application of mathematical set theory and algebra to the organisation of data. Up to that time data tended to be stored in a relatively *ad hoc* fashion, in line with the way in which computer systems had been developed. Computer files frequently mirrored paper documents or the way in which data was entered into the system. Data duplication (and hence data redundancy) was rife, leading to problems with maintenance and flexibility. Codd's aim was to solve many of these problems by applying mathematical principles to the storage and manipulation of data, thereby reducing redundancy and increasing flexibility.

A detailed discussion of the work of Edgar Codd in setting out his theory of the 'relational model' of data is well beyond the scope of this book (for a definitive view see Date, 2000). Instead we will look briefly at some basic concepts, then move on to the main technique of RDA, that of 'normalisation'.

▶ Relations or tables

In the relational model, data is viewed as consisting of two-dimensional tables (or more formally, *relations*), consisting of rows and columns of attributes (see Figure 10.1). For a table to be properly relational it must obey a number of rules:

- ◆ There must be no duplicate rows. This means that each row will be uniquely identifiable. We use the concept of the primary key, as discussed in Chapter 5, to describe the group of attributes that together identify each row.

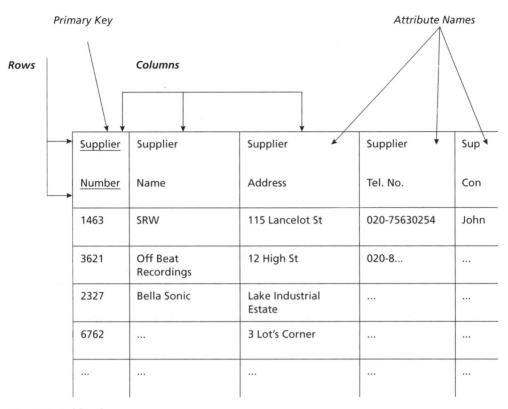

Fig 10.1 Table elements

◆ The order of the rows is not significant. If a sequence is required to add meaning to the data (for example in a table that holds data on positions in a race) then a new column should be added that stores values for the position attribute.

◆ The order of the columns is not significant. It makes tables easier to read if the primary key is given as the first column or columns; this is for presentational purposes only.

◆ Attribute names are unique throughout the system.

◆ For a given value of the primary key there must be no more than one value for each attribute in the row, i.e. attributes must be 'atomic'. Collections of attributes that can have more than one value for a given primary key are known as 'repeating groups', and a table that contains them is termed 'unnormalised'. We will look at how we can resolve repeating groups by creating new tables to hold their information a little later in this chapter.

▶ Tables and entity types

In essence a table is the relational equivalent of an entity type in a Logical Data Model. In SSADM, tables are represented by listing their attributes under the table name, with the primary key underlined:

SUPPLIER

Supplier Number

Supplier Name

Supplier Address   or

Supplier Tel. Number

Supplier Contact Name

SUPPLIER

Supplier Number
Supplier Name
Supplier Address
Supplier Tel. Number
Supplier Contact Name

The same information is sometimes depicted in books as:

SUPPLIER (Supplier Number, Supplier Name, Supplier Address, Supplier Tel. Number, Supplier Contact Name).

▶ Repeating groups and levels

Tables that include repeating groups of attributes for a single value of the primary key (i.e. unnormalised tables) are documented with the repeating group indented. For example, consider a list showing all the suppliers of a particular product, like the one in Figure 10.2.

If we place the contents of the whole list in just one table for the moment, we will represent it as:

PRODUCT'S SUPPLIERS

Product Number

Product Name

Product Type Code

Product Type Name

    Supplier Number

    Supplier Name

    S/P Ref. Number

    Cost Price

    Main Supplier Y/N

The indentation of the attributes below Product Type Name indicates that this group of attributes will repeat (with different values) for each value of Product Number (the table's primary key). The problem with indenting repeating groups is that it is possible to have numerous levels of repeating group within repeating group and so on, which leads to presentational problems. An alternative to indenting repeating groups is to assign a 'level' number to each repeating group:

PRODUCT'S SUPPLIERS	level
Product Number	1
Product Name	1
Product Type Code	1
Product Type Name	1
Supplier Number	2
Supplier Name	2
S/P Ref. Number	2
Cost Price	2
Main Supplier Y/N	2

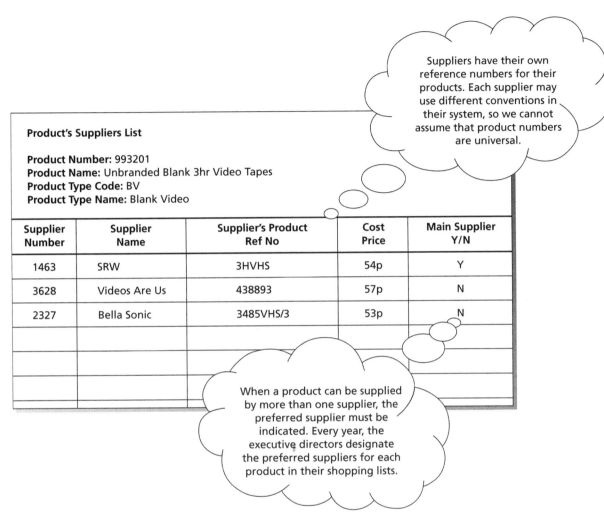

Fig 10.2 Sample list of a product's suppliers

▶ Functional dependencies

An attribute X is said to be *functionally dependent* on an attribute Y if each value of Y is associated with only one value of X.

For example, suppose each product is of one product type. This means that each Product Number is associated with only one Product Type Code. Product Type Code is therefore functionally dependent on Product Number. The opposite is not true as each Product Type Code may be associated with many Product Numbers.

Another way of phrasing functional dependency is to say that the value of X can be *determined* from the value of Y, or that Y *functionally determines* X.

So Product Number functionally determines the value of Product Type Code, or the value of Product Type Code can be determined from the value of the Product Number, i.e. given the value of a Product Number we can always establish the value of the associated Product Type Code.

The concept of functional dependencies is an important one in RDA, as we shall see in the next section.

▶ NORMALISATION

The process of normalisation involves applying a series of refinements to groups of data items in order to produce tables that conform to specified standards, known as *normal forms*.

Unnormalised tables are converted to First Normal Form by removing repeating groups into separate tables. Second and Third Normal Forms are achieved by reducing and splitting tables so that the only functional dependencies that exist are between the primary keys and the remaining non-key attributes.

There are further normal forms which are beyond the scope of this book and are not part of SSADM. They address problems with data organisation that are relatively infrequent (for a full discussion see Date, 2000).

Advantages of normalisation

Before describing normalisation in detail it is worth mentioning some of its advantages briefly. Data in Third Normal Form (3NF) consists of tables of closely associated attributes which are entirely dependent on 'the key, the whole key, and nothing but the key'.

This has the effect of minimising data duplication across different tables, thereby resolving many of the problems associated with data redundancy. In particular it should reduce the incidence of 'update anomalies'.

'Update anomalies' is the collective term for problems with modifying, inserting and deleting data in an unnormalised environment. These can be illustrated by considering the unnormalised contents of the list of a product's suppliers again.

If we were to implement this data structure as it stands, and to use it as the only place in which product details were stored we would encounter the following problems:

1. **Insertion anomalies.** No new suppliers could be added to the system without adding a product.
2. **Deletion anomalies.** If the last remaining product for a given supplier were deleted, then all information on that supplier would be lost.
3. **Amendment anomalies.** Any change to a supplier's details (e.g. to the telephone number) would mean that every product for that supplier would need amending to keep it in line.

(The same anomalies also apply to the supplier information if we were to keep the one data structure alone.)

Many of the problems with poorly organised data arise from failures to deal with these anomalies, especially amendment anomalies, which would occur even if some details were also duplicated in a separate table.

The other major advantage of well-normalised data is its flexibility or ease of expansion. If new groups of attributes are added to the system they can usually be accommodated with minimal knock-on effects.

There are also disadvantages, of course. The principal one is that maintaining and manipulating a large number of tables is often physically slow. However, these are problems of physical design and implementation and should be ignored at this point.

▶ Unnormalised Form (UNF)

The first step in normalisation is to identify the group of data items to be normalised and to pick a key that will uniquely identify each occurrence of the group (it may be composite or compound); this gives us a UNF table.

Our source of data items will be the screen designs, input and output designs, or whichever means was used to describe the external design during Function Definition. The UNF tables will include all data items, whether input or output. Repeating groups should be identified quite easily and need to be indicated in the UNF table by indenting or level numbers. Once again, we will use as an example the Product's Suppliers List in Figure 10.2, the contents of which provide us with the UNF table show in Figure 10.3.

UNF	level	1NF	2NF	3NF	Table Name
Product Number	1				
Product Name	1				
Product Type Code	1				
Product Type Name	1				
Supplier Number	2				
Supplier Name	2				
S/P Ref. Number	2				
Cost Price	2				
Main Supplier Y/N	2				

Fig 10.3 UNF and levels formally documented using an RDA 'working paper'

▶ First Normal Form (1NF)

A table is in *First Normal Form* if it contains no repeating groups.

If our UNF table does include a repeating group then we remove it to form a table of its own, and give it a primary key consisting of the primary key of the parent UNF table *plus* an additional attribute or attributes to enable unique identification of its rows (Figure 10.4). The additional attributes required to make the primary key of the new table unique are those that uniquely identify occurrences of the repeating group within the UNF table as a whole.

In Figure 10.4, the Supplier Number is chosen as the additional attribute that uniquely identifies occurrences of the repeating group. Choosing this additional attribute from among the repeating group is not always straightforward and great care should be taken when deciding which of the repeating group attributes are sufficient to uniquely identify the other attributes of the repeating group. Once it has been identified, it is prudent to underline the chosen attribute(s) to emphasise its significance.

UNF	level	1NF	2NF	3NF	Table Name
Product Number	1	Product Number			
Product Name	1	Product Name			
Product Type Code	1	Product Type Code			
Product Type Name	1	Product Type Name			
Supplier Number	2				
Supplier Name	2				
S/P Ref. Number	2	Product Number			
Cost Price	2	Supplier Number			
Main Supplier Y/N	2	Supplier Name			
		S/P Ref. Number			
		Cost Price			
		Main Supplier Y/N			

Fig 10.4 Moving from UNF to 1NF

It is important to note that the inclusion of the primary key of the parent table in the new table is what maintains the link between the two tables. Without this link we would not be able to use the resulting data model to reconstruct the original Product's Suppliers List.

▶ Second Normal Form (2NF)

A table is in *Second Normal Form* if, in addition to being in 1NF, it contains no non-key attributes that are dependent on only *part* of the primary key.

Any such attributes in the 1NF tables are removed to form a new table with the relevant partial key as its primary key.

In our example we scan the four attributes that hang under the key of the second 1NF table. We immediately notice that Supplier Name is dependent solely on Supplier Number, which is only part of the key of the second table in Figure 10.4. In other words, given the Supplier Number we can always determine the relevant Supplier Name, regardless of the Product Number. To achieve 2NF we will need to remove any partially dependent attributes to form a new table whose key is the part of the original key on which these attributes depend (see Figure 10.5). The other three attributes, S/P Ref. Number, Cost Price and Main Supplier Y/N all need the whole of the key to be determined, so they remain attached to that compound key. (This last point may need some time to sink in. What we are saying here is that if, for example, we want to know how much it will cost us to buy a particular product we need to state *both* the product we are talking about and the supplier from whom we may purchase this product – since different suppliers may charge different prices for the same product. In other words, a product's 'cost price' depends on *both* Product Number and Supplier Number. Similarly, S/P Ref. Number and Main Supplier Y/N depend on both those attributes too.)

Note that we could be left with a table without any non-key attributes. This is quite allowable and fairly common. The table concerned may represent the equivalent of

UNF	level	1NF	2NF	3NF	Table Name
Product Number	1	Product Number	Product Number		
Product Name	1	Product Name	Product Name		
Product Type Code	1	Product Type Code	Product Type Code		
Product Type Name	1	Product Type Name	Product Type Name		
Supplier Number	2				
Supplier Name	2				
S/P Ref. Number	2	Product Number	Product Number		
Cost Price	2	Supplier Number	Supplier Number		
Main Supplier Y/N	2	Supplier Name	S/P Ref. Number		
		S/P Ref. Number	Cost Price		
		Cost Price	Main Supplier Y/N		
		Main Supplier Y/N			
			Supplier Number		
			Supplier Name		

Fig 10.5 Moving from 1NF to 2NF

a link entity, or an entity for which we will discover additional attributes when we tackle other system functions.

The essential question to ask at this point for each non-key attribute is:

"Does this attribute depend on the *whole* of the primary key?"

If it does not then it should be removed and placed in a new table. One thing to take care over is that the primary key of the parent table is left intact – there are usually other attributes that *do* depend wholly on it.

In other words, *there is no way* that the keys of 1NF will disappear. When moving from 1NF to 2NF to 3NF, new keys may be identified while existing keys will remain even if they have no other attributes associated with them.

The first 1NF table in our example is automatically carried forward as 2NF, since it has a simple key that cannot be subject to partial dependencies.

▶ Third Normal Form (3NF)

A table is in *Third Normal Form* if, in addition to being in 2NF, there are no non-key attributes that depend on *other* non-candidate key attributes.

Any such dependent attributes are removed from the 2NF table to form a new table having the attribute that determines them (their determinant) as its primary key. The determinant is left in the parent table as a foreign key in order to maintain links between the data.

If we look at our example, the table with a primary key of Product Number contains the Product Type Name, which is dependent directly on Product Type Code rather than Product Number. So we need to remove this attribute to form a table

UNF	level	1NF	2NF	3NF	Table Name
Product Number	1	Product Number	Product Number	Product Number	PRODUCT
Product Name	1	Product Name	Product Name	Product Name	
Product Type Code	1	Product Type Code	Product Type Code	*Product Type Code	
Product Type Name	1	Product Type Name	Product Type Name		
Supplier Number	2				
Supplier Name	2			Product Type Code	PRODUCT
S/P Ref. Number	2			Product Type Name	TYPE
Cost Price	2				
Main Supplier Y/N	2	Product Number	Product Number	Product Number	SUPPLIER
		Supplier Number	Supplier Number	Supplier Number	PRODUCT
		Supplier Name	S/P Ref. Number	S/P Ref. Number	
		S/P Ref. Number	Cost Price	Cost Price	
		Cost Price	Main Supplier Y/N	Main Supplier Y/N	
		Main Supplier Y/N			
			Supplier Number	Supplier Number	SUPPLIER
			Supplier Name	Supplier Name	

Fig 10.6 A complete RDA normalisation

with Product Type Code as its primary key, while at the same time leaving Product Type Code behind as a foreign key in the 'parent' table (Figure 10.6).

Dependencies of this sort are sometimes more difficult to spot than partial key dependencies, as they involve asking for each non-key attribute:

"Is this attribute dependent on *any other* non-key attribute(s)?"

The other 2NF tables in our example are already in 3NF, so the process of normalisation is complete for the purposes of RDA.

As each normalisation is completed, each 3NF table is given a meaningful name so that we can refer to it easily later on.

▶ Partial inter-key dependencies

The following example exposes a possible RDA pitfall: suppose we wish to list all the purchase orders sent to a particular supplier. (This function does not exist as such in our case, but it would be part of the function that provides a monitor of suppliers' performance. Supplier monitoring did not make it through BSO.)

A rough report design for this 'List all Purchase Orders of a Supplier' function is shown in Figure 10.7.

Moving into 1NF, two data groups appear. The first has *supplier number* as its key. The second appears to have the pair

Supplier Monitoring

Supplier Purchase Order Status Report

Supplier Number: 1463

Supplier Name: SRW

Report Start Date: 1/1/01

Report End Date: 1/1/02

These two report dates are not stored by the system, so they are not involved in RDA

P.O. Number	P.O. Date	P.O. Status	Delivery Date	% Delivered
12299	16/1/01	Delivered	22/1/01	100
12976	29/2/01	Delivered	14/3/01	97
13812	6/4/01	Cancelled		
15002	3/7/01	Delivered	14/7/01	83
15557	20/9/01	Delivered	2/10/01	92
17113	8/11/01	Delivered	16/11/01	89

Notes:
Delivery Date is the date of the last consignment delivered
P.O. Status is entered by the P.O. Clerk
% Delivered is the % in monetary terms delivered

Fig 10.7 Supplier monitoring report design

Supplier Number
Purchase Order Number

as its key. Before we unquestionably accept this compound key we have to scrutinise it by enquiring whether any part of it depends on any other part of it. Careful consideration reveals that Supplier Number *depends* on Purchase Order Number since a purchase order is directed to a specific supplier. In other words, given a purchase order number we can, without a shadow of a doubt, pinpoint the supplier for which the purchase order was raised or, to use the proper terminology, Purchase Order Number *determines* Supplier Number. This means that the key of the second 1NF group is simple. Figure 10.8 shows the full RDA working for this function where we also see that the first-normal-form groups are in third normal form too.

So, before moving to 2NF we should look at all composite or compound keys to see if any of the attributes in the primary key are redundant, i.e. are not needed

UNF	**level**	**1NF**	**2NF**	**3NF**	**Table Name**
<u>Supplier Number</u>	1	<u>Supplier Number</u>	→	→	SUPPLIER
<u>Purchase Order Number</u>	2				
Purchase Order Date	2				
Purchase Order Status	2	*Supplier Number	→	→	
Delivery Date		<u>Purchase Order Number</u>			PURCHASE
% Delivered		Purchase Order Date			ORDER
		Purchase Order Status			
		Delivery Date			
		% Delivered			

> Note that since the 1NF groupings have simple keys, they are automatically in 2NF too.

Fig 10.8 RDA of the report of Figure 10.7, which contains an inter-key dependency

to uniquely identify the rows in the table. This is equivalent to asking whether *all* attributes (including potentially redundant attributes in the primary key) are dependent on the *same part* of the key.

▶ Additional normalisation examples

We will now look at some more examples of normalisation, partly to illustrate one or two additional aspects of the process and partly to provide us with a reasonable set of tables with which to build an LDS extract which we will then compare with the Required System LDM.

Record Customer Order

With the advent of e-commerce, ZigZag sees an opportunity to expand its customer base to include retail customers. Since this is a new requirement we need to be doubly careful in implementing it.

It is envisaged that retail customers will access the ZigZag web page, browse the on-line price list, and pick the products they wish to buy.

A printed version of a typical customer order is displayed in Figure 10.9 from which we can identify the following data items:

Customer Order Date	Customer Order Number	Customer Number
Customer Title	Customer First Name	Customer Surname
Customer Address	Customer City	Customer Postcode

All of these occur once only on each order. These are followed by the data items, which form a repeating group:

Product Description	Product Reference Number	Quantity Ordered
Retail Price	Format	Line Value

Order Date: 2/3/01
Your Order Number: 0103020079

Your Customer Number: 10007375
Mrs Joanna Smith
983A Gloucester Road
London
SW7 9JJ

DESCRIPTION	REF	FORMAT	QTY	PRICE	TOTAL PRICE
Beatles 1	232702	CD	2	14.99	29.98
Unbranded Blank 3hr Video Tapes	993201	BV	5	0.80	4.00
Never on Sunday	351223	CD	1	15.99	15.99
Beethoven's 2nd	104673	VHS	1	6.99	6.99

Thank you for your order

Total Items 9   56.96
Postage and Packing   2.95
Total Amount Due   59.91

These values can be calculated so they don't need to be stored.

Method of Payment   VISA
Card Number   ************

Registered in England 1234567890
VAT Number 999999999

ZigZag only accepts credit card payments. The credit card number is not printed for security reasons

Company Registration Numbers and VAT Numbers are not stored in the data model

Fig 10.9 A typical retail customer order printout

Beneath this repeating group we then find the following data items, each of which occurs once only:

Order Total Items	Order Subtotal	Postage and Packing
Total Amount Due	Method of Payment	Customer Card Number

Naming conventions

The attribute names we have identified do not map exactly to the names found on the customer's order. We have instead taken care to name the attributes in a manner that is relevant to the computerised information system we are striving for. Thus 'Your Order Number' has become 'Customer Order Number', 'DESCRIPTION' has become 'Product Description', 'REF' has become 'Product Reference Number', 'QTY' has become 'Quantity Ordered', and so on. The names we have chosen for the system are more precise but it would be cumbersome for the customer, a human being, to have to acknowledge these names.

In a similar vein we have broken 'Mrs Joanna Smith' up into three attributes, aptly named Customer Title, Customer First Name and Customer Surname.

Synonyms

Finally, the astute reader will have noticed that 'Product Reference Number' has been labelled just 'Product Number' in Figure 10.2. This is an example of a *synonym* or alias. Synonyms should be carefully stated and recorded in the data dictionary.

Attributes should be named with care and consistency as soon as possible. All alternative names, including abbreviations, should be recorded with diligence.

Out of those 21 attributes we choose to ignore Line Value, Order Total Items and Order Subtotal because they can be recalculated every time we wish to reconstruct the customer order. We also choose to ignore Postage and Packing since this is a constant value throughout the system.

We are therefore left with the following attributes on which we will perform RDA:

CUSTOMER ORDER	level
Customer Order Date	1
Customer Order Number	1
Customer Number	1
Customer Title	1
Customer First Name	1
Customer Surname	1
Customer Address	1
Customer City	1
Customer Postcode	1
Product Description	2
Product Reference Number	2
Quantity Ordered	2
Retail Price	2
Format	2
Total Amount Due	1
Method of Payment	1
Customer Card Number	1

Note how it is possible for 'level 1' attributes to be identified even after the higher levels have been detected.

It now remains for us to identify the key attributes for each level before we proceed with RDA.

Looking at the level-1 attributes we can identify four candidates for the primary key. These are Customer Order Number, Customer Number, Customer Postcode and Customer Card Number.

Customer Card Number is dismissed in our case because it is liable to change and we cannot have a primary key that changes all the time.

Customer Postcode is tempting since many credit-card clearing houses seem to use it as a check when authorising a credit-card transaction. A moment's thought here would dismiss this choice since a postcode, while well and tightly defined, does not apply to only one customer's orders (e.g. customers from the same block of flats would share the same postcode).

We are thus left with the two serious key candidates: Customer Order Number and Customer Number. In order to dismiss Customer Number we need to remember the reason for considering the original set of attributes, namely the fact that they have come together as part of a customer's order, not a customer's *registration*. It is therefore the Customer Order Number that our system will be generating to record the customer's new order, since the Customer Number will probably pre-exist to be used by the customer in many orders. In other words, one Customer Number applies to many of this customer's orders, which means that given the Customer Number we will not be able to *uniquely* identify the above order. On the other hand, given the Customer Order Number we will be able to pinpoint the right order. Thus Customer Order Number is the only choice for a key to the level-1 attributes.

We now turn our attention to the five level-2 attributes of the customer's order. Here our task is simpler since only Product Reference Number is a plausible key for this repeating group. (The fact that it is not the first attribute in the list should not bother us at all.)

With the keys in hand we proceed with the normalisation of the customer order, as shown in Figure 10.10 overleaf.

Having chosen the keys for each level of the customer order, the 1NF is straightforward.

To move to the 2NF we note that Product Description, Retail Price and Format apply to the product only, and so depend on the Product Reference Number alone. This leaves only Quantity Ordered which clearly depends on both elements of the compound key.

The move to 3NF needs some further elucidation as far as the Method of Payment and the Customer Card Number are concerned: we have chosen to associate these two attributes with the Customer Order table instead of the Customer table. We have done so to retain a record of all the different means a customer may have chosen when ordering. If we place the customer's card number with the other customer information we run the risk of losing some historical information that may be of use.

Search for title

ZigZag is currently a business-to-business (B2B) supplier. It is now expanding to sell directly to retail customers through the Internet. By changing from a B2B to a B2C (business-to-customer) operation, careful consideration has to be given to the new functionality and usability expected by the information system. While previously the users of the system were employees of ZigZag who could be trained in the

UNF	level	1NF	2NF	3NF	Table Name
Customer Order Date	1	Customer Order Number		Customer Order Number	CUSTOMER
Customer Order Number	1	Customer Order Date		Customer Order Date	ORDER
Customer Number	1	Customer Number		*Customer Number	
Customer Title	1	Customer Title		Total Amount Due	
Customer First Name	1	Customer First Name		Method Of Payment	
Customer Surname	1	Customer Surname		Customer Card Number	
Customer Address	1	Customer Address			
Customer City	1	Customer City		Customer Number	CUSTOMER
Customer Postcode	1	Customer Postcode		Customer Title	
Product Description	2	Total Amount Due		Customer First Name	
Product Reference Number	2	Method Of Payment		Customer Surname	
Quantity Ordered	2	Customer Card Number		Customer Address	
Retail Price	2			Customer City	
Format	2			Customer Postcode	
Total Amount Due	1				
Method Of Payment	1	Customer Order Number	Customer Order Number	Customer Order Number	CUSTOMER
Customer Card Number	1	Product Reference Number	Product Reference Number	Product Reference Number	ORDER
		Product Description	Quantity Ordered	Quantity Ordered	ITEM
		Quantity Ordered			
		Retail Price	Product Reference Number	Product Reference Number	PRODUCT
		Format	Product Description	Product Description	
			Retail Price	Retail Price	
			Format	Format	

Fig 10.10 Normalising a retail customer order

nuances of the system, now there is no telling who will use the Web to access the system. Extra care should therefore be given to new functions that address the B2C requirements.

Having identified the requirements during the Investigation phase of the development, we can use a mini-project-cycle of Function Definition, normalisation and prototyping to ensure that the resulting system caters for the needs of retail customers too (see Figure 10.11).

A new requirement that is a direct consequence of the decision of ZigZag to extend its operations by branching out into e-business is to allow a more user-friendly search of artists and their work.

In the original system the stock clerk would use the product reference numbers to locate products. While we wish to retain this functionality, we clearly cannot expect customers visiting ZigZag's website to have to do the same. When customers enter the system through the Net they should be able to search for a product using more natural means such as product titles or artists' names.

For example, a customer may be looking for a film or a song but can only remember the name of an artist involved in the piece. We should be able to design a sys-

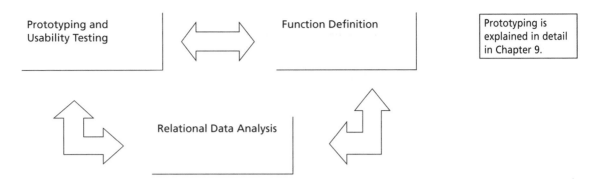

Fig 10.11 A development mini-cycle for e-commerce applications

Search for an artist

Please enter the artist's name and one or more keywords in the appropriate boxes below.

Then press the Go button.

Artist	
Genre	All genres ▾
Involvement	Everything ▾
Format	All Formats ▾ Go

Fig 10.12 Extract from search page for the work of an artist

tem that is sympathetic to such a normal occurrence by providing a search facility that can accommodate as many search variations as we can think of (Figure 10.12).

Emulating the ideas of the better web pages, ZigZag has decided to allow customers to locate products by using artists' names as well as product names. A user should be able to enter an artists' name and get all the works of that artist listed. It should further be possible to limit the search by:

Genre: classical, pop, blues, jazz, action, comedy, etc.
Format: CD, DVD, VHS, vinyl, single, LP, hardback, paperback, etc.
Involvement: actor, director, composer, lead singer, drummer, orchestra, author, scriptwriter, etc.

For example, a search with no restrictions on genre, format or involvement for products where Michael Caine is involved may look like the one depicted in Figure 10.13.

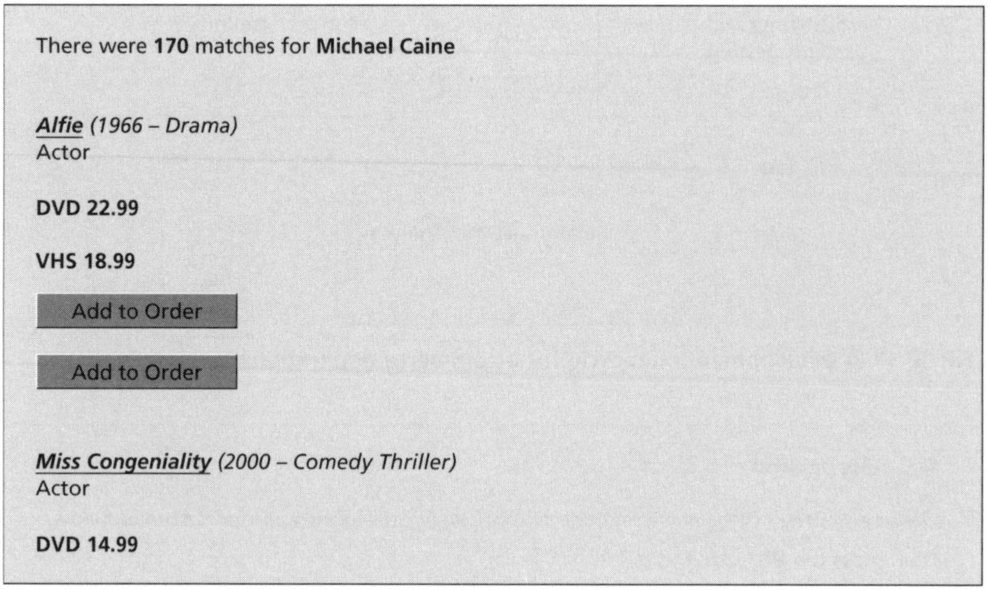

There were **170** matches for **Michael Caine**

Alfie (1966 – Drama)
Actor

DVD 22.99

VHS 18.99

Add to Order

Add to Order

Miss Congeniality (2000 – Comedy Thriller)
Actor

DVD 14.99

Fig 10.13 An imaginary search result of Michael Caine's work

Before we try to normalise the content of our search, some pause for thought is required for us to understand the interdependency between the information being output.

Under the artist's name, Michael Caine, we have a list of titles. For each title we have the various formats this title is available in (e.g. *Miss Congeniality* is available on DVD and VHS). For each title we also have the types of involvement the artist has in the title. Since some artists have more than one involvement in a single title, we see that 'involvement' is a repeating group within 'title'.

It takes some time to detect that *two* separate pieces of information have been combined and are shown for each title. The first concerns a title and the format this title is available in. The second concerns the artistic involvement of a person in a title. These two pieces of information are independent of each other: The fact that *And Not Many People Know This, Either* can be obtained in hardback and paperback does not affect the fact that Michael Caine wrote the book. Conversely, the fact that Michael Caine wrote *And Not Many People Know This, Either* does not change according to whether we can buy the title in hardback or paperback.

The contents of Figure 10.13 lead to the normalisation of Figure 10.14. The normalisation is not straightforward and requires some elaboration.

First, we consider the levels. From the arguments we presented earlier, we have 'title' as a repeating group within artist name (one artist, many titles). We then have for each title two independent repeating groups: one of involvement round a title (an artist can have various involvements per title), and one of format round title (a title can come out in many formats). We have chosen to spread the UNF attributes to extenuate the independence of the level-3 repeating groups.

Looking now at the 1NF column we note that we effectively have two independent normalisations here. The first concerns the artistic involvement of a person in a title, which is catered for in the first three groupings of the 1NF column.

UNF	level	1NF	2NF	3NF	Table Name
<u>Artist Name</u>	1	<u>Artist Name</u>	→	→	ARTIST
<u>Title</u>	2	<u>Artist Name</u>	<u>Artist Name</u>	→	ARTIST TITLE
Genre	2	<u>Title</u>	<u>Title</u>		
		Genre			
			<u>Title</u>	→	TITLE
			Genre		
<u>Involvement</u>	3	<u>Artist Name</u>			INVOLVEMENT
		<u>Title</u>	→	→	(ROLE)
		<u>Involvement</u>			
<u>Format</u>	3	<u>Title</u>			TITLE FORMAT
Product Price	3	<u>Format</u>	→	→	(PRODUCT)
		Product Price			

Fig 10.14 RDA of the output shown in Figure 10.13

The second concerns the format of a title and is a bit trickier to deal with. This is because the format of a title is not dependent on the artist. Looking at the format repeating group (which repeats independently of the involvement group) we appear at first sight to have the following:

<u>Artist Name</u>	1
<u>Title</u>	2
<u>Format</u>	3
Product	
Price	3

However, if we give the candidate key a little more thought, we see that we can drop Artist Name from the key as the Format repeating group does not relate to the artist at all but is present in relation to the title only (in effect we have two independent enquiries in one here: one dealing with the films of the requested artist, and one dealing with the formats and prices of the titles listed in the previous enquiry. We thus end up with:

<u>Title</u>	2
<u>Format</u>	3
Product	
Price	3

Produce Stock Report

The smooth running of ZigZag's operations depends heavily on efficient stock control. It is vital to be able to locate a product easily, to find empty slots, to produce stock reports and to facilitate stock-taking.

A Stock Report is a lengthy report that shows the location and a breakdown of the quantity in stock of every product currently held by ZigZag. A portion of such

Product Reference Number	Product Name	Format	Stock Id	Location	Quantity in Location	Quantity in Stock
104673	Beethoven's 2nd	DVD	3459247	070603	20	
			3489335	070604	24	
			3444987	110519	24	68
104674	Beethoven's 2nd	VHS	3329573	021023	42	
			3329574	020923	12	64
104680	Captain Corelli's Mandolin	DVD	4295818	010310	48	
			4295819	010311	44	92
. . .						

Fig 10.15 A section of a ZigZag Stock Report

UNF	level	1NF	2NF	3NF	Table Name
Product Number	1	Product Number			PRODUCT
Product Name	1	Product Name			
Format	1	Format	→	→	
Stock Id	2	Qty in Stock			
Location	2				
Qty in Location	2				
Qty in Stock	1	*Product Number	→	→	STOCK
		Stock Id			
		Location			
		Qty in Location			

Fig 10.16 Normalisation of a ZigZag Stock Report

a Stock Report is contained in Figure 10.15. The normalisation of the ZigZag Stock Report is shown in Figure 10.16.

There is one tricky bit in the normalisation of the Stock Report which happens when moving from UNF to 1NF. It is not straightforward to recognise that Product Number depends on Stock Id. The reason why Stock Id determines Product Number is that Stock Id is an artificial number that the system plants for convenience. The purpose of defining a Stock Id is so that we can attach to it information about how many items of a particular product we hold in a particular location. Thus Product Number (as well as Location and Qty in Location) depends on Stock Id.

Once the dependency of Product Number on Stock Id is discovered, the rest of the normalisation is uneventful.

(If we wished we could have avoided storing Qty in Stock since it contains a derivable value – if we add the relevant Qty in Location values we get the value for Qty in Stock – but we chose to store it in any case. Later, when we discuss denormalisation, we will revisit this decision.)

▶ RATIONALISING 3NF TABLES

Once we have carried out normalisation on a number of functions we will have several sets of tables in 3NF, which we now rationalise into a single, larger set. Any tables that share a primary key should be merged, as should tables with matching candidate keys. We will also look for attributes that now act as foreign keys when compared with primary keys in other 3NF sets. A little care is needed to ensure that any synonyms are identified, as failure to do so could lead to missing or spurious merges.

The list of tables resulting from our four examples is as follows:

PRODUCT'S SUPPLIERS LIST:

1. PRODUCT	2. PRODUCT TYPE	3. SUPPLIER PRODUCT	4. SUPPLIER
Product Number	Product Type Code	Product Number	Supplier Number
Product Name	Product Type Name	Supplier Number	Supplier Name
*Product Type Code		S/P Ref. Number	
		Cost Price	
		Main Supplier Y/N	

RECORD CUSTOMER ORDER:

5. CUSTOMER ORDER	6. CUSTOMER	7. CUSTOMER ORDER ITEM	8. PRODUCT
Customer Order Number	Customer Number	Customer Order Number	Product Reference Number
Customer Order Date	Customer Title	Product Reference Number	Product Description
Customer Number	Customer First Name	Quantity Ordered	Retail Price
Total Amount Due	Customer Surname		Format
Method Of Payment	Customer Address		
Customer Card Number	Customer City		
	Customer Postcode		

SEARCH FOR TITLE:

9. ARTIST	10. ARTIST TITLE	11. INVOLVEMENT (ROLE)	12. TITLE	13. PRODUCT
Artist Name	Artist Name	Artist Name	Title	Title
	Title	Title	Genre	Format
		Involvement		Product Price

PRODUCE STOCK REPORT:

14. PRODUCT	15. STOCK
Product Number	Stock Id
Product Name	*Product Number
Format	Location
Qty in Stock	Qty in Location

An examination of the above tables indicates that 11 of them appear once only, so we can accept them for the time being as they are. The only table that appears more than once is PRODUCT, which appears with various guises four times (as tables 1, 8, 13 and 14). We will use this table to demonstrate how to rationalise the 3NF tables:

◆ **Merging tables:** Tables 1 and 14 have the same key but different attributes which all contain information that we need to store about products. We therefore merge the two tables to get:

16. PRODUCT
<u>Product Number</u>
Product Name
*Product Type Code
Format
Qty in Stock

◆ **Dealing with synonyms:** If we now consider tables 8 and 16 we see that, while they both contain generic product information, the key for table 16 is Product Number while the key for table 8 is Product Reference Number, but we have already discussed that these are synonyms. We therefore merge the contents of these product tables into one that also absorbs all the different attributes we spot:

17. PRODUCT
<u>Product Number</u> (or <u>Product Reference Number</u>)
Product Name
*Product Type Code
Product Description
Retail Price
Format
Qty in Stock

◆ **Resolving candidate key issues:** We now need to consider table 13, which has also been christened 'product', and compare it with table 17. While both tables 13 and 17 contain product information, their keys are quite different. Table 17 has a simple, artificially generated, key while table 13 has what appears to be a compound key made of natural values (a title and a format, both of which exist in their own right). This is a common occurrence in the world of databases where for convenience we plant numbers on things – such as order numbers, student id numbers, driving licence numbers, bank account numbers, all of which have good reason to exist. In the case of our two product tables, while we understand that a product is actually a manifestation of a title in a given format, we decide to go for the more convenient Product Number key. In doing so we will not obliterate the Title and Format pair from the entity because it provides us with some useful, in business terms, information. Taking this into consideration gives the following content to Product:

17. PRODUCT
<u>Product Number</u> (or <u>Product Reference Number</u>)
Product Name
*Product Type Code

Product Description
Retail Price (or its synonym Product Price)
Title
Format
Qty in Stock

The astute reader may now spot that we have introduced a synonym into this entity: in the current system Product Type Code had such values as VHS, DVD, CD. These are precisely the values we associate with the attribute Format. We thus conclude that Product Type Code and Format are synonyms:

17. PRODUCT
<u>Product Number</u> (or <u>Product Reference Number</u>)
Product Name
*Product Type Code (or *Format)
Product Description
Retail Price (or its synonym Product Price)
Title
Qty in Stock

- **Recognising new master entities:** From the preceding arguments it appears that, in the context of the information held in the ZigZag system, 'genre' is an important notion in its own right, especially as customers can easily relate to different genres such as classical, pop, comedy, drama, biography. We therefore decide to elevate Genre into an entity type that, for the time being at least, is an all-key entity whose key is also called 'Genre'.

- **Recognising new primary keys:** Sometimes an output does not contain primary key values. The output of Figure 10.13, for example, contains full titles. These titles, while imperative when directed to humans, in database terms constitute clumsy and cumbersome keys. We might therefore replace the attribute Title with Title Id as the key to the Title table. Having done so we need to revisit all tables that had Title as a foreign key and readjust them to include Title Id instead. (Providing an artificial key for each title helps resolve some issues that the reader may be grappling with since titles are not unique and they transcend formats and genres. We have, for example, two films called 'Psycho', a symphony and a film called 'Beethoven's Second', and a book and a film called '1984'. By giving each of these six titles a separate, artificial, title id we avoid the duplication problem.)

- **Annotating newly discovered foreign keys:** Now that Genre has been recognised as the primary key of an entity we should scan our tables again and mark any occurrences of Genre in other entities as foreign keys. Such a scan on our tables reveals that Title inherits Genre as a foreign key to become:

  12. TITLE
  <u>Title</u>
  *Genre

- **Dropping spurious entities:** Finally, we revisit tables 9, 10 and 11, which contain what is effectively a cascading key, with a view to questioning whether we need all three tables to produce a list of an artist's work. Table 9 contains the base artist information and is therefore necessary. Similarly, table 11 indicates exactly what an artist's involvement was on a piece of art, allowing for more than one contribution per title. Table 10 on the other hand links an artist with a title but does not indicate what it is that the artist did. We soon realise that table 10, the Artist Title table, adds nothing to the information, and we decide to drop it.

All the other tables appear only once so we just carry them forward intact.

Caution

When we merge tables formed during normalisation there is a danger that we may introduce new functional dependencies that take us out of 3NF (this is especially true for merges involving large numbers of non-overlapping attributes). To check that the resulting tables are still in 3NF we can apply a couple of tests, known as the '3NF tests':

1. For a given value of the primary key, is there one and only one possible value for each attribute in its row?

2. Are all attributes in the table dependent on the key, the whole key and nothing but the key?

All merged tables that result from rationalising the ZigZag tables pass the 3NF tests.

▶ CONVERTING 3NF TABLES TO LDSs

The prime objective of RDA is to enhance the Required System Logical Data Model using the results of the close examination of data items enforced by normalisation. In order to compare our 3NF tables with the Required System LDM we will convert them into an LDS extract, which we hope will represent a subset of the entire Required System LDM we have produced earlier in the development cycle.

The conversion process is a fairly mechanistic one, following four basic steps:

1. Represent each table as an entity type box and list the primary and foreign key attributes of each entity inside the box. Remember to clearly annotate the foreign key elements of hierarchic keys with an asterisk (Figure 10.17).

2. For each element of every compound key, check that an entity exists which has that element as its primary key. If any such entities are missing they should be created with their sole attribute being the required primary key.

For example, the key of Involvement, the entity that records the role, be it acting, authoring, directing, composing, etc., an artist has had in a piece of art, contains the attribute Involvement. We have not formally identified a table with this as its primary key but it appears reasonable to create such an entity here. This entity will record the recognisable types of involvement one can have with a piece of art. We therefore include this new entity, Involvement Type, in our diagram, as in Figure 10.18.

Note that we would also need to look for non-key attributes in other relations, to check if the new key, as they would now become foreign keys.

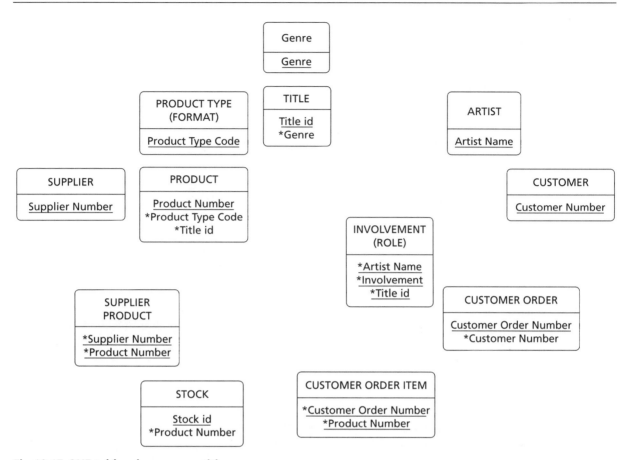

Fig 10.17 3NF tables drawn as entities

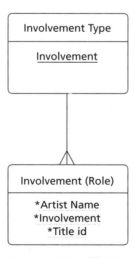

Fig 10.18 The entity Involvement Type is identified because one element of the compound key of Involvement (Role) is a foreign key that does not belong to any of the entities of Figure 10.17

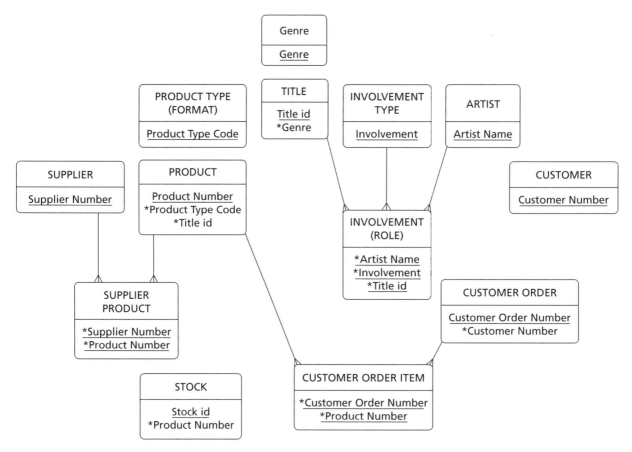

Fig 10.19 Small keys grab large keys

3. Add a master–detail relationship between every pair of entities where the *entire* primary key of one (the master) is an attribute or collection of attributes in the *compound* key of the other (the detail) (see Figure 10.19).

4. Add a master–detail relationship between every pair of entities, where the primary key of one (the master) is a foreign key of the other (the detail), see Figure 10.20.

The resulting LDS will not have relationship names, or include details of optionality or exclusivity. Don't worry about this as these features are not required for the purposes of comparison with the Required System LDM, and the LDS extract will not be carried forward beyond this comparison.

▶ COMPARING LDSs

We now compare our two data structures (the extract we have just produced using RDA and the Required System LDS we produced earlier in the development – see Figure 7.6) and decide whether any discrepancies are because of errors in Logical Data Modelling or whether they arise from the fact that RDA examines only an extract of the system.

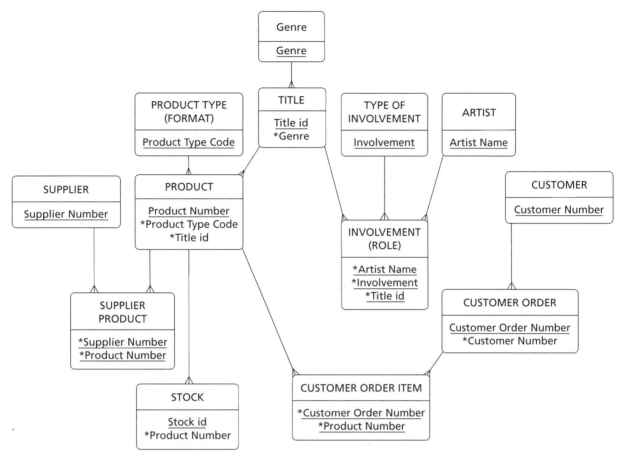

Fig 10.20 Crows feet grab asterisks, to complete RDA

In practice there may be large numbers of entities involved in the comparison, so a fair amount of time is likely to be spent in identifying corresponding entities in the two models. Probably the best starting point is to look for common attributes, in particular common primary keys or candidate keys.

When comparing the two models we should look for:

◆ **Differences in attributes.** RDA will frequently uncover additional attributes as it concentrates much more on data item detail than Logical Data Modelling does. It may also allocate some attributes to different entities than those in the Required System LDM. These differences should be discussed with users in order to decide on the correct interpretation.

◆ **Additional entity types.** Logical Data Modelling may result in entities that are not in 3NF. This may lead to additional entities in the data model extract produced using RDA, which should be incorporated in the Required System LDM. On the other hand, RDA may result in extra entities that are redundant. Each entity has to be carefully scrutinised and its inclusion justified with care. When comparing the data model produced by RDA for ZigZag we note that the entities

Artist, Involvement (Role) and Type of Involvement have been recognised for the first time. We choose to accept them, as they are a direct consequence of expanding the ZigZag operations to include e-commerce retail customers. These three entities were not important in the current system where we only needed to identify products without being concerned about artists and other things artistic. The advent of e-commerce has changed all this since now we have to deal directly with customers who *do* care very much about artists and therefore need a system that will accommodate this necessity.

An important new entity that RDA has identified is Title. Title is a new entity that recognises the distinction between a title and the formats in which this title may be offered. We made this distinction when arguing how to normalise the Search for Title function. Again, the reason why Logical Data Modelling appears not to have identified this entity is because the current system did not have a need for it.

Given the appearance and acceptance of Title, it is clear that the entity Genre, which is a master of Title, should also be accepted.

Note

The reader should be aware that many entities identified here using RDA could have been identified using Logical Data Modelling and that many entities identified using Logical Data Modelling could have been identified using normalisation if we had chosen the appropriate functions to normalise. The reader should be clear that we are only talking about new or spurious entities in the areas where a comparison is viable. For example, normalisation confirmed the Supplier Product entity that Logical Data Modelling had spotted during Business Systems Option. Similarly, entities like Supplier Invoice, Product Supplier, Depot Zone, Picker, which were identified through Logical Data Modelling remain safe. If we were not sure about them we could have picked an appropriate function and normalised it to check the viability of the data model.

◆ **Additional relationships.** Any additional relationships in the RDA model should be looked at carefully to check that they have not been missed out in the Required System LDM intentionally, to reduce redundancy.

In the RDA model we have a relationship between Product and Stock. This relationship is maintained in the LDM by the links from Stock to Delivery Item to Purchase Order Item to Supplier's Product to Product. These links are much stronger because it enforces the fact that we order only what is on a supplier's catalogue and we only accept deliveries of what we have ordered. We therefore drop the RDA relationship from our enhanced model. (It is possible that during Physical Design this relationship will be resurrected for reasons of efficiency, and when we may be forced to perform denormalisation. It is important to understand why it actually is redundant in *logical* terms.)

▶ SUMMARY

1. Relational Data Analysis (RDA) is based on material published in the 1970s by Edgar Codd of IBM, proposing the application of mathematical set theory and algebra to the organisation of data.

2. RDA is used to create data model extracts from collections of individual data items, which can then be used to enhance or confirm the Required System LDM and to provide the basis for database design.

3. The process of normalisation involves applying a series of refinements to groups of data items in order to produce tables that conform to specified standards, known as normal forms.

4. Data in Third Normal Form (3NF) consists of tables of closely associated attributes which are entirely dependent on 'the key, the whole key, and nothing but the key'. This has the effect of minimising data duplication across different tables, thereby resolving many of the problems associated with data redundancy.

▶ EXERCISES

10.1. Normalise (to 3NF) the Treebanks function described in Exercises 9.1 and 9.2.

10.2. Normalise (to 3NF) the Natlib function described in Exercise 8.1.

10.3. Normalise (to 3NF) the Bodgett & Son function described in Exercises 9.3 and 9.4.

10.4. Normalise the functions of Exercise 9.5. Produce a Logical Data Structure from them.

10.5. Normalise the West Munster Hotel bill from Exercise 8.2.

10.6. In the context of the SSADM default structure, explain why Function Point Analysis performed after Function Definition and Relational Data Analysis would be more accurate than Function Point Analysis done during Business Systems Options.

Entity Behaviour Modelling

INTRODUCTION

Up to this point we have developed both process-oriented and data-oriented views of the system, the former through Business Activity Modelling, Data Flow Modelling and Function Definition, the latter through Logical Data Modelling and Relational Data Analysis. At every stage the work on one view of the system has been co-ordinated and cross-referenced with the work on the other so that the process and data perspectives are truly consistent with each other.

In Entity Behaviour Modelling we present a technique that can be used to bring the process and data views of the system together in a meticulous way. In so doing the analyst is forced to consider the ways in which the data changes over time, or, put another way, to define the different states that entities can have.

The importance of this will not become clear until you have read both this chapter and that on Effect Correspondence Diagrams, but, put simply, it enables you, the analyst, to specify important integrity checks on the update processing.

Entity Behaviour Modelling is one of the most rigorous techniques that SSADM has to offer. Unfortunately it is also difficult to master and time-consuming to apply, and therefore is not the most popular or widely used of techniques.

Entity Behaviour Modelling is both an analysis and a design technique. In analysis mode it should help to identify the more obscure or less obvious events that other techniques have failed to reach. It should also help to clarify in your mind the exact effects that each event has on the system's data.

In design mode Entity Behaviour Modelling provides the basis for the specification of all the update processing. If used correctly the process specifications (Effect Correspondence Diagrams) – complete with integrity checking – simply 'fall out' of the Entity Behaviour Model with very little extra effort on the analyst's part (it has to be said that using Entity Behaviour Modelling in this way will require a lot of commitment and many practitioners will question whether all the extra effort is worthwhile).

The view of the system represented by Entity Behaviour Modelling is often known as the behavioural view.

There are other non-SSADM techniques that can be used to model this view. They include Statecharts and State Transition Diagrams.

Although this is not a technique for the faint-hearted it *can* be applied in a light-handed manner and if you want to do this you should follow this chapter up to and including the section on basic behaviour. In this way you will be able to get *some* of the benefits of the technique without too much of the pain.

> **Tip**
>
> If modelling even the basic behaviour of entities seems too heavy-handed, then we would suggest that you at least develop the Entity Access Matrix, as it provides a very useful overview of the system processing in a very concise manner.

Entity Behaviour Modelling involves two related techniques: Event Identification and Entity Life History Analysis. It is closely bound up with Conceptual Process Modelling – an umbrella term for the development of Enquiry Access Paths, Effect Correspondence Diagrams, Update Process Models and Enquiry Process Models.

Entity Life Histories are diagrammatic representations of how entity types are or can be affected by events. They detail the allowable sequence, iterations and optionality of events for each entity, and thus capture and record further business rules. As we have already said, Entity Life History analysis often requires a great deal of effort and discipline in order to reduce ambiguity and ensure completeness. The rigour of the technique should repay this effort by producing a greater understanding of system needs and revealing flaws or omissions in our specification to date. It will also provide the basis for detailed design of system processing in the steps that follow.

The involvement of users at this point is essential to the success of the exercise, as only they are likely to have a full and detailed enough understanding of events and their effects. In particular we will rely heavily on users to provide details of exceptions to the normal lives of entities, the specification of which will frequently take far longer than that of standard processing.

Although it would be extremely unlikely that we could show an Entity Life History to a user and expect useful feedback, the process of developing Entity Life Histories will generate many important questions. You can put these questions to the user and use the answers to improve your understanding of the requirements.

The related activity of developing Effect Correspondence Diagrams places the project firmly in the design phase. Having modelled the effects of events from the perspective of each entity, we then take the opposite view and model the effects on entities from the perspective of each event.

Entity Behaviour Modelling and Effect Correspondence Diagrams are only concerned with the processing of events, i.e. with updates. Using the Entity Access Matrix we will also begin the detailed specification of enquiries, by formally documenting the required entity accesses of each enquiry. These can then be further described using Enquiry Access Paths.

The level of detail and rigour involved in Entity Behaviour Modelling is likely to lead to a number of modifications and additions to the Required System Logical Data Model and Function Definitions as we uncover new or revised user requirements. For this reason Function Definition cannot be considered to be complete until Entity Behaviour Modelling is finished.

▶ Learning objectives

By the end of this chapter you will be able to:

- ◆ describe the concepts and purpose of Entity Behaviour Modelling;
- ◆ identify and document system events;
- ◆ chart the effects of events on entities using an Entity Access Matrix;
- ◆ develop simple Entity Life Histories;
- ◆ refine simple Entity Life Histories to take account of alternative and exceptional lives;
- ◆ use Entity Life Histories to define the important states of an entity.

▶ Links to other chapters

- ◆ Chapter 5   Entity Behaviour Modelling examines the effects of events on the entities of the Logical Data Model.
- ◆ Chapter 12   The Entity Behaviour Model and the Conceptual Process Model provide cross-checking views of the effects of events on the Logical Data Model.

▶ ENTITY LIFE HISTORIES

As we have discussed, Entity Life Histories document all of the events that can affect (i.e. cause a change to or constrain the life of) an entity. They model the business rules applicable to the processing of a particular entity, and thus specify allowable sequences and combinations of events. It is important to realise that an Entity Life History must allow for all the possible lives of all the legal instances of that entity type.

▶ Concepts

Before looking at the technique in detail, it is worth reconsidering the concepts of events and effects.

Events

An event is a real-world action that causes an update to the data held within the system. Therefore it acts as a trigger for a defined set of processing, and does not represent the processing itself.

If the reader thinks of the information system as storing facts about some aspect of the real world, then it follows that when that part of the real world changes, the stored data must change too.

The information system does not get to know about the changes in the real world by magic. The users of the system have to constantly feed it with new data ('event data') that describe the changes that have taken place.

This is the purpose of an update function: to provide an interface by which the users of a system can inform the system of a real-world event.

In this chapter, when we talk about an event, we mean a conceptual model event (or *system* event). This is a real-world happening whose effects can be described in terms of changes to the data represented by the Logical Data Model.

Tip

The systems analyst will encounter many events. Many of these may be relevant to the running of the organisation, but not all of them will be reflected in the conceptual model. Thus the event 'purchaser goes sick' is of no relevance to the system we are developing, although it could of course be an important event in some other system – a system that was supporting human-resource activities, for instance.

Once again the skill of the systems analyst is to work with the users to clarify which events are to be considered relevant in the context of the current project. In general you will find it useful to ask two questions of an event:

♦ Do the users have a requirement to record the fact that the event has taken place?

♦ Is the event different (in its effects on the data model) from other events that have already been identified?

Effects

The change that an event causes to an entity is called an *effect*.

An effect could result in the creation or deletion of one or more entity instances, a change in one or more attribute values, the creation or destruction of a relationship.

For instance, when a purchase order is first proposed there are four effects: one on Purchaser Order, one on Purchase Order Item, one on Supplier Product and one on Supplier. These effects cause the following changes: a new instance of Purchase Order is created; one or more instances of Purchase Order Item are created; values are assigned to Purchase Order Number and Date, and to the Quantity Required and Required-by-Date attributes of each new instance of Purchase Order Item; relationship instances are created between Purchase Order and Supplier, Purchase Order and Purchase Order Item, Purchase Order Item and Supplier Product.

It is quite possible for a single event to have two or more mutually exclusive effects on an entity occurrence, dependent upon the state of that entity occurrence. For example, the effects of the event Archive Stock on the entity Delivery Item depend on whether or not we are deleting the last occurrence of Stock of the Delivery Item. If it is the last one, we delete the Delivery Item as well as the Stock occurrence; if there are other Stock occurrences belonging to the Delivery Item, then the Delivery Item is not deleted. It is not possible to know which effect is going to take place without reading the database to see what the situation is.

It is also possible for a single event to simultaneously affect more than one occurrence of an entity, but in differing ways. For example, the event Merge Stocks will cause one stock of a product to be added to another. The quantity of the first stock will thus be reduced to zero, while the quantity of the second stock will be increased by the original quantity of the first. In cases such as this, the entity is said to be *taking on different roles*.

States

States are important because there are rules that determine how an entity moves from one state to another. The definition of these rules is an important part of ensuring the integrity of the data in the database.

Note

In large multi-user databases users will make mistakes when entering data – it is inevitable that erroneous data will get into the system. By identifying the important states of an entity, and the events that cause these states to change, we can build rules into the update process-ing of our system that will identify and prevent *some* of these errors.

The current state of an entity defines which events can happen to an entity next. For instance, if the cancellation of a purchase order has been recorded on the sys-tem then we would not expect the 'Purchase Order Confirmed' event to occur. If a user tries to record the confirmation of a purchase order that has been cancelled then the system should stop them – producing an error message and advice on what to do – instead of allowing the update to go ahead.

From our point of view the easiest way of thinking about a state is to say that a state is the point that an entity has reached in its life. In fact, we can actually define for each entity an extra attribute called a *state indicator*. If an entity has states 'pro-posed', 'confirmed' and 'cancelled' then these will be the three values that can be assigned to the state indicator attribute.

▶ Notation

With one or two minor differences, Entity Life Histories use the same Jackson-like structure notation as I/O Structure Diagrams. This requires us to issue a word of warning to the novice reader: the *notation* is the same, but the *context* of the dia-grams and therefore their meaning are completely different.

On the Entity Life History the root node (the top box) contains the name of the entity being described. The leaves of the diagram (boxes that have no children) rep-resent effects and each effect box carries the name of the event that causes that effect.

On the I/O Structure Diagram the root node contains the name of a function (or function component) and each leaf represents a data item, or a group of data items, that is input to, or output from, that function.

Sequence

Entity Life Histories comprise a sequence of events read from left to right, i.e. from birth to death. A birth event is an event that causes new instances of an entity to be created. A death event is one that marks the point in an entity's life where no more events will occur (except, possibly, deletion). In the most trivial of cases an Entity Life History may include only an insertion (creation or birth) event, but the majority will include some form of mid-life and end-life as well.

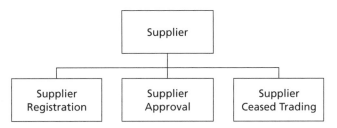

Fig 11.1 Sequence of events in the life of Supplier

The Supplier Entity Life History of Figure 11.1 is not complete. It is a first-cut diagram, which shows only the main events in the life of the Supplier entity. It is usual to develop Entity Life Histories in a number of passes, with each pass adding greater detail to the life.

In Figure 11.1, the Entity Life History for Supplier, we see that the Supplier entity has three events in its life, and that each occurrence of Supplier can experience each event just once.

The system is notified of a new Supplier when the Purchasing Department identifies a new company from which they want to buy products. This is the Supplier Registration event.

Suppliers have to undergo a number of checks to make sure that they can meet ZigZag's standards (relating to timeliness of deliveries, quality of goods, ability to deliver in sufficient quantities, etc.). When these checks have been completed the Supplier record will be updated. This is the Supplier Approval event.

It is possible that the Supplier will go out of business. If this is the case, then the Supplier record has reached the end of its life and may be deleted, or marked for deletion. Hence the event Supplier Ceased Trading.

Note

There will be many events in the life of the real-world Supplier that will not be reflected in the Supplier Entity Life History. For example: Supplier Celebrates 50 Years in the Trade may be an important event in the history of the Supplier in question, but it is not of any great interest to our purchasers.

Selection

Some effects are options: each time that a selection is reached, one and only one of the optional effects of the selection can occur. Thus in Figure 11.2, we see two selections: Supplier Product Birth and Supplier Product Death (these are structure boxes, and it is not necessary to name them, but it is often convenient to do so).

The diagram tells us that an occurrence of Supplier Product is created either by the Supplier Registration event (when the details of the Supplier are first recorded), or by the New Supplier for Product event (when a decision is taken to source a product from an existing supplier). Similarly, each occurrence of Supplier Product is killed off by either the Product Discontinued event or the Supplier Ceased Trading event. In other words, an occurrence of Supplier Product will die when one of its masters dies, whichever dies first.

Selections sometimes represent a small opportunity for re-use. The code that we need to write for Supplier Product Death may be the same for both of its optional death effects (see the section on super-events).

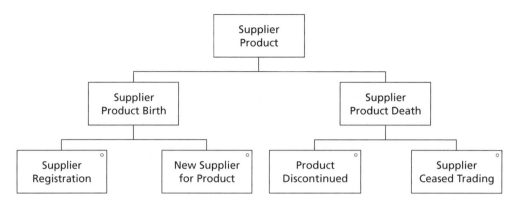

Fig 11.2 Selection: not every instance follows exactly the same life

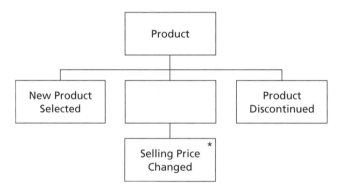

Fig 11.3 Iteration: a product can change its price many times

Iteration

Iterations are used to illustrate cases where an event can affect an entity instance more than once. Figure 11.3 uses the first-cut Entity Life History for the Product entity to illustrate iteration. The Selling Price Changed event can affect an occurrence of Product more than once because a given Product can have its Selling Price increased or reduced many times: one week we reduce the price of a CD as part of a special promotion; the next week we put it back to its original price.

This leads us to a further point. The *business events* Sales Promotion and Price Change both have the same effect on our system: an update to the Selling Price attribute of Product. Because their effects are the same, we only need to define one conceptual model (or system) event for these two business events.

As ever, the precise requirements of the users have to be considered. We could imagine that the Sales Promotion event requires us to record two dates between which a lower price is to be charged for a product, whilst the Price Revision event requires us to record only the date from which the price increase is to take place. Then the two events would be different and we would need to name two different events in our Entity Behaviour Model. The Required System Logical Data Model for ZigZag does not support these requirements and we will consider them no further, but the reader might like to consider the ways in which the Logical Data Model and Entity Life Histories might be changed in order to support them.

> Although the 'New Product Selected' event can occur many times within a system it can occur only once *per occurrence* of Product. Thus this effect is not iterated. By definition birth effects are not iterated.

It is important to remember that by definition an iteration occurs *zero*, one or many times, which means in this case that an occurrence of Product might experience its death event without ever having been affected by the Selling Price Changed event. This is correct in business terms: some products will be discontinued without ever having had their prices changed.

Jackson structures revisited

The rules for drawing Jackson structures have already been discussed in the section on I/O Structures and Dialogue Design. The same rules apply to the drawing of Entity Life Histories as to the drawing of I/O Structure Diagrams:

◆ An iteration has just one child (but remember that that child can be the parent of an entire substructure: see Figure 11.6 for example).

◆ All the children of a parent must be of the same type (by type we mean that some sort of box must be used: plain or with an 'o' or with an '*').

▶ DEVELOPING ENTITY LIFE HISTORIES

The development of Entity Life Histories is best carried out in a number of stages:

1. Identify events and document them on the Entity Access Matrix.

2. Analyse the basic behaviour of the entities.

3. Perform an *up-pass* analysis of entity behaviour: starting from the bottom of the Required System Logical Data Model (i.e. with those entities that are details only) and working upwards through the structure, document the life of each entity. Try to identify mutually exclusive behaviour patterns. For iterated patterns of events identify the way in which the iteration stops.

4. Perform a *down-pass* analysis of entity behaviour: starting from the top of the Required System Logical Data Model (i.e. with those entities having no masters), work down through the data model reviewing each Entity Life History, looking in particular for death events, and examining the way that the death events of masters and details interact. Identify super-events; see box later in this chapter.

5. Consider deletion strategy.

6. Add operations and state indicators.

▶ Identifying events and documenting their effects

Event identification

There are a number of ways in which events can be identified.

During Function Definition the Required System Data Flow Model is examined for updates, which are indicated by the presence of an arrow pointing towards a main data store. These updates are responses to events (remember that an event is a trigger to a set of update processing), and these events are named and documented as part of the Function Description.

In the section on Function Definition we discussed the identification of the events Purchase Order Proposal and Purchase Order Confirmation from the Required System Data Flow Model. The reader might like to review the Required System Data Flow Model and try to identify further events.

Another good way of identifying events is to examine the Required System Logical Data Model. For each entity we ask the following questions:

◆ Which events cause occurrences of this entity to be created?

◆ Which events cause occurrences of this entity to be deleted?

◆ Which events set the values of non-key attributes?

◆ Which events create and delete occurrences of relationships?

◆ Which events cause changes to the state of an entity?

When we ask of Delivery 'Which event creates occurrences of Delivery?' we find that these are created when the supplier phones to confirm the delivery date and time: the Delivery Confirmation event.

Having identified an entity's birth event, a further question that we should always ask is whether the entity shares the birth event with any of its masters or details.

A clue here is the presence of a fully mandatory relationship between master and detail. This suggests that the master and one or more detail occurrences will be created at the same time, i.e. in response to the same event. It also suggests that they will be deleted together.

A further question that can be asked when identifying a detail that shares a birth event with its master is 'Does the detail have its own independent birth as well?'. Again, a similar question can be asked of entities that share a death event.

An example is Supplier Product, which shares a birth event with Supplier: when suppliers are first registered the purchaser can also record any products that they are thinking of using that supplier for, hence the sharing of the Supplier Registration event. However, purchasers often decide to extend the range of products that they buy from an existing supplier. When this happens the system must be updated by creating new Supplier Product records. Hence the alternative birth event for Supplier Product: New Supplier for Product.

Mandatory attributes (those that are defined as 'not null' in the attribute descriptions) will have their values set by the birth event, but we should still ask if there are any other events that cause those values to change.

We have seen one example of this already: the Selling Price Changed event (see Figure 11.3). The price of a product is recorded when the details of the product are first recorded, but prices change, and there is an event to reflect this.

This does not mean that we have to have a separate event for each non-key attribute that can change its value. Often there are a number of attributes, e.g. name, address, telephone number, which the user might want to edit from time to time. We can define a single change of details event for this situation.

Some attributes are optional: they may be set to null when an entity is created. Again, we should look for the event that leads to a value being given to that attribute. In ZigZag, Purchase Order has the attributes Quantity Requested: a mandatory attribute input with the Purchase Order Proposal event; and Quantity Confirmed: set to null at first and then updated as part of the Purchased Order Confirmation event.

The relationship between two entities is represented by a foreign key attribute in the entity that is at the detail end of the relationship. When an occurrence of a relationship is created this foreign key attribute is given a value. The value ties the detail occurrence to the master entity that has that value as its primary key.

In Entity Behaviour Modelling we talk about masters gaining details and details being tied to masters. This is another way of saying that a relationship occurrence is being created, that a foreign key is being given a value.

We should look for events that cause gains and ties. When an end of a relationship is mandatory, there will be a gain (for a master) or a tie (for a detail) as part of the birth effect of the entity at the mandatory end of the relationship. Thus Supplier Product is tied to Supplier and tied to Product at the time of its creation.

When an end of a relationship is optional, we can expect the event that causes a gain or tie to occur in the mid-life of the entity at the optional end of the relationship. Thus the Invoice Arrival event will be a mid-life event in the Entity Life History for Delivery (when a delivery is set up, it has no related invoice because the invoice is not sent by the Supplier until after the delivery has been made).

Another point to consider with relationships is whether they are fixed or transferable. In our system most of the relationships are fixed: once we create a relationship between a Purchase Order Item and a Purchase Order, those two occurrences remain linked together, i.e. the Purchase Order Item will never be linked to a different occurrence of Purchase Order.

An exception is the relationship between Stock and Location. Stock can be moved, or merged with other stocks (of the same delivery). When this happens, the stock location changes, e.g. stock that was in aisle 3, shelf 4, rack 1 is now in aisle 3, shelf 5, rack 2. We have to cut the Stock occurrence from one Location occurrence, and tie it to a different one. The relationship between Stock and Location is said to be transferable, because the detail can be attached to more than one master during its lifetime (but only one master at any one time, of course).

Event Identification can start early on in a project, perhaps as early as Business Activity Modelling. However, we cannot fully identify and analyse events until we have a stable Logical Data Model and a well-defined set of requirements, and it is not until Function Definition and subsequently Entity Behaviour Modelling that events are identified and documented in earnest.

Events are informed to the system by users, via functions. It follows that every event must have at least one related Function Definition, and every update function should be related to at least one event. Events and Function Definitions should be cross-checked to ensure that this is the case.

It is likely that previously unidentified events will be discovered, so we will have to define new functions for these events because there are unlikely to be any existing Function Definitions that cover them.

Documenting events in the Event and Enquiry Catalogue

By now it should be obvious that there are many ways in which events can be identified and that events can be identified at many points in a project. Whenever and however events are discovered we should document the following information:

♦ event name and id (if necessary);

♦ event description;

◆ related business activities or business events;

◆ average and maximum occurrences;

◆ event data;

◆ entry point entity;

◆ entity accesses.

The cross-referencing of events to the Business Activity Model is good practice because it reminds us that the purpose of our work is always to support business activities. It lends our project further traceability from the early stages of investigation through to the later stages of design.

The volumetric information is important because in Physical Design and Technical Systems Options we will need to have an idea of the volumes of transactions that the Technical Systems Architecture must support.

The event data is the data that must be provided by the user (input via a function), or by another process, in order for the processing of the event to take place (see the sections on Conceptual Process Modelling).

The entry point entity is the first entity accessed by the event (again, see the sections on Conceptual Process Modelling).

The entity accesses are the effects of the events and the type of those effects (see the section on the Entity Access Matrix).

Similar information should be recorded for each enquiry.

Entity Access Matrix

Note that the Entity Access Matrix is a useful product in its own right and you should consider developing one even if you do not intend to draw Entity Life Histories (although in that case you cannot expect to get the same level of understanding of the system's processing).

As events are identified we should document their effects on the Entity Access Matrix. This is a grid that lists all the entities of the Required System Logical Data Model along the top, and all the known events and enquiries down the side.

Each column–row intersection of the matrix can be used to record the effect of an event on an entity. Each effect can be described with one or more letters each of which gives an indication of the type of effect taking place (see Table 11.1). The matrix can also be used to show which entities are read by each enquiry.

Tip

The Entity Access Matrix can help in identifying events. We look down each column of the matrix and ask:

◆ *Does the entity have at least one creation and one deletion event?* If not, then the reasons should be investigated and if the omission is valid they should be documented in the Required System Logical Data Model. If the omission is not valid we should try to identify the missing event and we will have to alter Function Definition and possibly the Required System Data Flow Model (if we think it is important to keep it up-to-date).

Similarly we can look across each row of the matrix and ask:

◆ *Does every event affect at least one entity?* If it does not then we have definitely made an error, as events only exist to trigger update processing.

Table 11.1 Suggested keys for the Entity Access Matrix. Note that the above letters can only be used for events, except for R and K, which can be used for enquiries as well as events

I	Insert	Indicates that one or more occurrences of the entity are created by an occurrence of the event.
M	Modify	Indicates that one or more occurrences of the entity have their attribute values modified by an occurrence of the event. We include the possibility that the state indicator itself is the attribute being modified (sometimes this might be the only discernible effect of an event).
D	Death	Indicates that one or more occurrences of the entity are 'killed' by an occurrence of the event (the occurrences remain, but only for enquiry purposes).
B	Bury	Indicates that one or more occurrences of the entity are deleted by an occurrence of the event (the occurrences are physically deleted, archived or marked for deletion; the user can no longer see them).
T	Tie	Indicates that one or more occurrences of the entity are linked to a master entity, via a relationship. (The corresponding effect on the master is shown on the matrix using a G.)
C	Cut	Indicates that one or more occurrences of the entity have the relationship with their master deleted. (The corresponding effect on the master is shown on the matrix using an L.)
G	Gain	Indicates that one or more occurrences of the entity have their relationships with one or more detail entities created. (The corresponding effect on the detail is shown on the matrix using a T.)
L	Lose	Indicates that one or more occurrences of the entity have their relationships with one or more detail entities deleted. (The corresponding effect on the detail is shown on the matrix using a C.)
S	Swap masters	Indicates that one or more occurrences will be cut from one master and tied to another. (The corresponding effect on the master is shown on the matrix using an X.)
X	Swap details	Indicates that one or more occurrences of an entity lose their details and that one or more occurrences of the entity gain them. (The corresponding effect on the detail is shown on the matrix using an S.)
R	Read	Indicates that one or more occurrences of the entity are read to access attribute values, or to navigate the Logical Data Model.
K	Check	Indicates that there is no discernible effect on the entity, but that the effect is constrained to happen at a specific point in that entity's life.

▶ Entity Life Histories for ZigZag

In the following sections we will develop Entity Life Histories for some of the entities in the case study. As we do so we will introduce the further concepts and notation of Entity Life History Analysis: effect qualifiers, entity roles, parallel lives, super-events, and Quits and Resumes.

Entity Event	Product	Supplier	Supplier Product	Purchase Order	Purchase Order Item
Ad Hoc Purchase Order Raised				I	I
Purchase Order Proposal				I	I
Purchase Order Confirmation					M
Purchase Order Cancellation				D	D
Supplier Ceased Trading		D	D	D	D
New Supplier for Product			I		
Supplier Registration		I	I		
New Product Selected	I				
Product Discontinued	D		D		
Selling Price Changed	M				
Supplier Approval		M			

Fig 11.4 Incomplete Entity Access Matrix for ZigZag

We do not have the space to develop the full set of Entity Life Histories, so we will concentrate on the section of the data model containing the Purchase Order, Purchase Order Item, Supplier, Supplier Product and Product entities.

Entity Access Matrix for ZigZag

The Entity Access Matrix for the part of the case study we are concerned with is shown in Figure 11.4. In practice it is not possible to entirely complete the Entity Access Matrix before starting to draw the Entity Life Histories. Rather, an initial Entity Access Matrix is drawn up, the drawing of Entity Life Histories commences, and the Entity Access Matrix is kept up-to-date as the behavioural analysis progresses.

Hence, in Figure 11.4 we show the Entity Access Matrix that is consistent with the examples of Entity Life Histories that we have discussed so far.

Some will prefer to use the Entity Access Matrix as a working document – a rough starting point for Entity Behaviour Modelling and nothing more. Others will want to keep the Entity Access Matrix entirely consistent with the Entity Life Histories as they develop. A good CASE tool should do this automatically, but CASE tools tend to be poor in their support for the Entity Access Matrix. Maintaining consistency by hand can be a tiresome task.

Basic behaviour

We will start by reviewing the Entity Life Histories of Figures 11.1, 11.2 and 11.3. These were somewhat simplified. For instance, the mid-life of Product contained no *gain* effects. Product is a master of two entities: Supplier Product and Product Substitute. The birth events of these entities should be placed in the mid-life of Product, because relationships with Product are created each time one of these events takes place. This pattern (birth event of detail, iterated in mid-life of master)

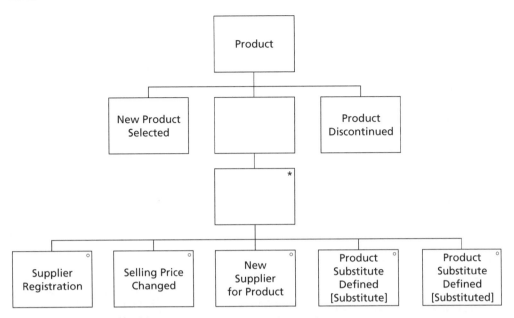

Fig 11.5 Product Entity Life History showing entity roles

is a very common one. In fact, it is so common that where we have a mid-life iteration of effects and no detail entity, we should ask whether a detail entity needs to be created.

The keen-eyed reader will have spotted something unfamiliar in the mid-life of Product: the event Product Substitute Defined appears twice. In each case the event name is followed by a role name in square brackets (see Figure 11.5). We will explain this in the next section.

Entity Roles

An event can appear more than once on an Entity Life History, but only in two well-defined circumstances:

◆ **Entity roles**: one occurrence of the event affects two or more occurrences of the same entity type, each in different ways.

◆ **Effect qualifiers**: one occurrence of the event can have different effects on an entity occurrence, depending on that entity's current state.

In the Product Entity Life History, we are dealing with the first of these situations.

When a Product Substitute entity is defined, it is tied to *two* Product occurrences, each via a *different* relationship. One Product is the substitute; the other Product is the one that is to be substituted for. Therefore two different gains take place, one via the Substitute relationship, and one via the Substituted relationship.

In Supplier Product we meet another example of entity roles. Supplier Product contains an attribute 'Main Supplier Y/N'. When a product is ordered it will be ordered from the main supplier, and the other suppliers are only used if the main supplier cannot meet the order. Each product has only one main supplier at any one time.

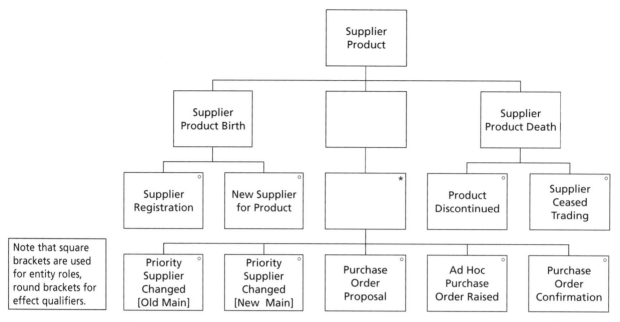

Note that square brackets are used for entity roles, round brackets for effect qualifiers.

Fig 11.6 Entity roles added to Supplier Product ELH

The preferred supplier of a product can change. When this happens, the purchaser goes to the system and updates the relevant Supplier Product record by changing the Main Suppler Y/N attribute to Y. The system must find the record of the previously preferred Supplier Product, and update its Main Supplier Y/N attribute to N.

Thus, for each occurrence of the Priority Supplier Changed event, there are two Supplier Product occurrences affected (Figure 11.6).

Effect qualifiers

The above example raises a question as to what happens when a new Supplier Product occurrence is created. If there is already a main supplier for the product, and if the new Supplier Product is to become the main source for the product, then we again have two entity roles. However, if there is no existing main supplier for the product, or if the new Supplier Product is to be a secondary source, then only one occurrence of Supplier Product is involved: the new one. It is not possible to tell which of the two alternative sets of effects is going to take place without enquiring to see if an existing main supplier for the product exists and there is no particular reason why the user of the system should be aware of the need to do this.

Thus, the processing of the event will have to take account of the two possibilities, and our Entity Life Histories will have to show that there are alternative, mutually exclusive, effects of the event.

In Figure 11.7 we show an extract of the Supplier Product Entity Life History. The Supplier Registration event appears twice as a birth effect, each time being qualified with the possible case. If the new Supplier Product is replacing another as the main

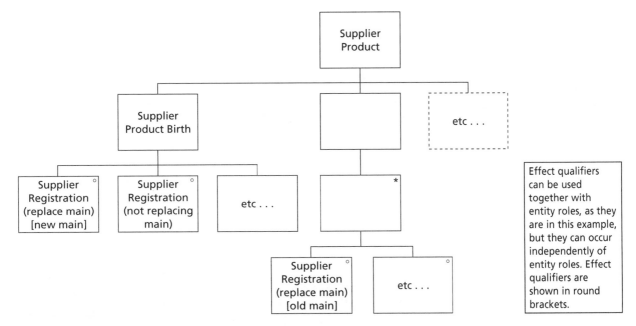

Effect qualifiers can be used together with entity roles, as they are in this example, but they can occur independently of entity roles. Effect qualifiers are shown in round brackets.

Fig 11.7 Effect qualifiers added to Supplier Product ELH

source of a product, then two occurrences of Supplier Product will be affected: one taking the role of 'new main', another taking the role of 'old main'. Thus, the event appears a third time in the mid-life of Supplier Product.

If the new Supplier Product occurrence is not replacing any other occurrence as the main source of the product, then the only effect on Supplier Product is to create a new occurrence.

It goes without saying that the same need for effect qualifiers applies to the other birth event of Supplier Product. For reasons of space, our subsequent models of Supplier Product will not show any of the qualified birth effects, but as an exercise the reader might like to redraw the Entity Life History to include them.

Mid-life sequence of events

In the two Entity Life Histories reviewed so far, the mid-life was an iterated selection of effects. But we should always examine the mid-life of an entity to see if there is any sequence to the mid-life effects.

A little care is required here, because the events that occur in a sequence for one entity may occur in a selection for a different entity.

For instance, the Purchase Order Proposal and Purchase Order Confirmation events both affect Supplier Product and Purchase Order Item. In Purchase Order Item these events are in sequence (see Figure 11.10). A real-world purchase order item cannot be confirmed before it has been placed. However, the same two events appear in a selection in the Entity Life History for Supplier Product (Figure 11.6). For Supplier Product the occurrence of these events is not ordered. Many purchase orders can be placed before one is confirmed.

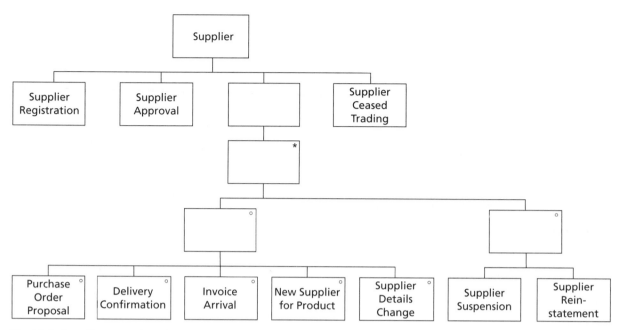

Fig 11.8 Supplier ELH showing 'suspensions'

When a mid-life sequence of events is discovered, we should consider other mid-life effects for that entity: do they fit into the sequence, are they independent of it, or are they constrained by it?

ZigZag monitors the performance of its suppliers, and if it falls below acceptable standards the supplier may be suspended. This means that purchasers are restricted from placing any further business with the supplier, until such time as the supplier is reinstated. Our first attempt at showing this is given in Figure 11.8. This figure says that once the Supplier Suspension event has happened to an occurrence of Supplier, no other events will be allowed to affect that occurrence until the Supplier Reinstatement event has happened.

For the last two entities in our discussion, initial Entity Life Histories are shown in Figures 11.9 and 11.10.

A structure of the type shown in Figure 11.9 (a sequence of two selections) will always raise a question as to its correctness: is it really the case that either of the first two events can be followed by either of the second two events?

Often the answer is 'no'. We will deal with this shortly.

Up-pass analysis of behaviour

We will now work our way back up our section of the Logical Data Model, from detail to master, looking for mutually exclusive behaviour patterns, and asking the question 'How does an iteration end?'.

When a purchase order is first raised, ZigZag waits for confirmation from the supplier that they can supply the quantities requested. If the supplier cannot supply the requested amount, then an alternative supplier will be contacted to make up the shortfall. These *ad hoc* purchase orders are confirmed, over the phone, before they

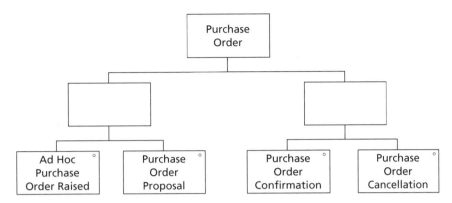

Fig 11.9 First-cut Entity Life History for Purchase Order

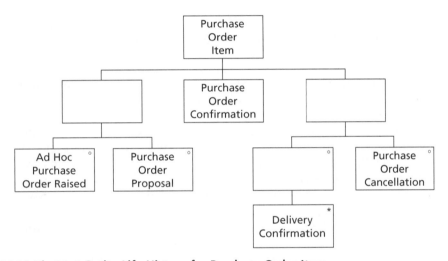

Fig 11.10 First-cut Entity Life History for Purchase Order Item

are actually raised. There is no separate Purchase Order confirmation event for an *ad hoc* purchase order. This is why the Entity Life Histories of Figures 11.9 and 11.10 did not show a true picture of events.

A better representation of the life history of Purchase Order Item is shown in Figure 11.11, where the Purchase Order Confirmation event has been placed in a sequence with the Purchase Order Proposed event. A similar technique is used on the Purchase Order structure as well.

Note that in solving one problem, we have introduced another. The structure now suggests that a purchase order that has been proposed, but not confirmed, cannot be cancelled. But of course it can. To deal with this problem we need to use a 'Quit and Resume'.

Quit and Resume notation

The *Quit and Resume* is a useful device for separating out exceptional events, or events that can occur in more than one place in an entity's life.

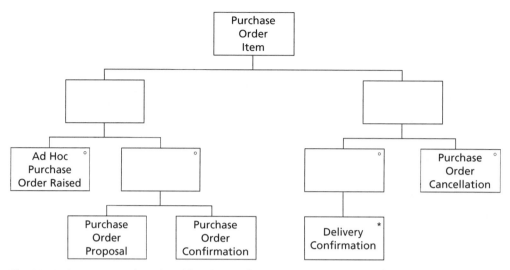

Fig 11.11 Restructured Entity Life History for Purchase Order Item

Readers who are familiar with previous versions of SSADM should note that Quits and Resumes are now used in the more disciplined way described in this section.

In Figure 11.12 the life of Purchase Order Item is modelled as a high-level selection. The left-hand option of the selection shows the normal sequence of events: the assumed case.

We assume that events are going to occur in the way that the assumed case describes unless an event occurs that proves otherwise: we expect that a purchase order proposal will be followed by a confirmation. We know that sometimes it will be followed by a purchase order cancellation instead, but, at the time of proposal, we do not know which of these two events will take place, so we assume that confirmation will take place.

If the cancellation event occurs, then it becomes clear that the confirmation event will not happen, that the alternative life has happened instead.

The 'Q' on the Purchase Order Confirmation effect indicates that it is possible that this effect can be replaced by the Purchase Order Cancellation event marked 'R' in the alternative life.

The iteration of 'events' that starts the alternative life is a convention: it represents the events in the assumed case life that might happen before the events in the alternative life occur.

When using Quits and Resumes, SSADM follows the convention that the Resume event *replaces* the Quit event. Furthermore, Quits and Resumes are only used to jump from one side of a selection to another, or to jump out of an iterated sequence of events – always from an assumed case to an alternative case.

Sharp-eyed readers will notice that there is no sequence of events in the ELH of Figure 11.12 that allows for the cancellation of an *ad hoc* purchase order. Yet, in ZigZag, purchasers can cancel any order, provided that no delivery has been arranged for that order. But if we try to fix this by putting a Quit on the iterated effect Delivery Confirmation, then the structure will permit a sequence of events that includes the cancellation of a purchase order after one or more deliveries has already been arranged.

A purchase order item can be met by more than one delivery (which is why the Delivery Confirmation event appears as an iteration in the life of Purchase Order

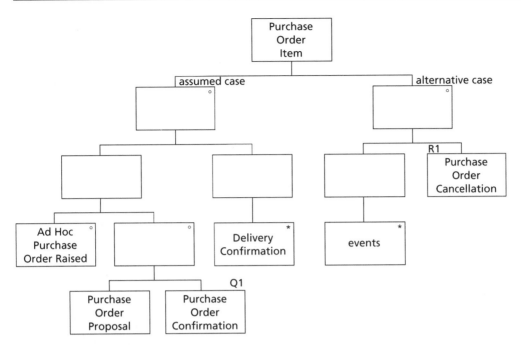

Fig 11.12 Quit and Resume notation

Item). In order to describe *exactly* the sequence of permissible events for Purchase Order Item we have to distinguish between the first delivery and any subsequent delivery, for an order item. To do this, we use effect qualifiers (see Figure 11.13). Having done this, we can put a Quit on 'Delivery Confirmation (first)', to allow for the fact that a cancellation can take place any time up to the point at which a delivery has been arranged for a purchase order (or part of a purchase order).

The Entity Life History for Purchase Order Item has now become rather elaborate. Readers who find this approach to Entity Life History analysis unnecessarily complicated might like to consider ways of 'cheating'. In the example just discussed we could attach an operation to the Purchase Order Cancellation event that says 'Fail if Delivery Items exist'. Then we would not need to distinguish between first and subsequent delivery confirmation events on the Purchase Order Item Entity Life History.

Before we finish looking at the Purchase Order Item Entity Life History, we should ask a further question of it: how does the iteration of delivery confirmations end? Put another way, how do Purchase Order Items die, assuming that they do not get cancelled?

There are two possibilities:

◆ The last delivery of goods required by a Purchase Order Item is confirmed (most Purchase Order Items will be met by a single delivery, but some will have two or more deliveries; the last delivery is the one that brings the quantity of goods delivered equal to the quantity of goods promised when the order was confirmed).

◆ Alternatively, it is possible that the time limit for delivery expires before delivery is arranged.

These two events are reflected in the Entity Life History of Figure 11.14.

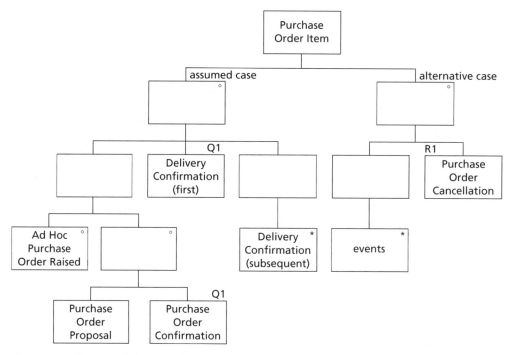

Fig 11.13 Effect qualifiers used in combination with Quits and Resumes

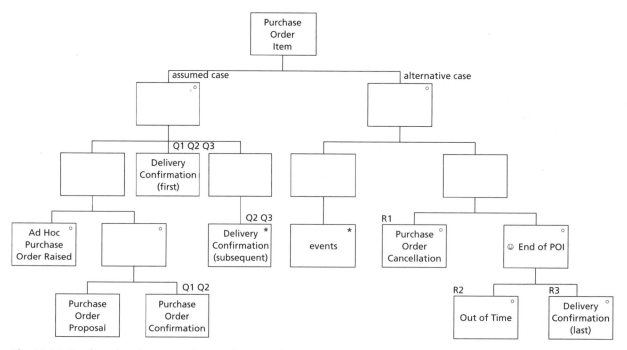

Fig 11.14 Further death events for Purchase Order Item

A new function will have to be defined for the Out of Time event. This will be an off-line function that will run overnight, when the system is not busy. It will identify all Purchase Order Items for which the required-by date has expired; a report of the products and quantities on these Purchase Order Items will be printed for the purchasers' information.

As with the cancellation event, we do not know at the outset how the life of any given Purchase Order Item will end, so the two extra death events have also been placed in the alternative life, with appropriate Quits and Resumes.

Note that it is possible to Quit from more than one place in the assumed case to the same Resume point, and that the Quits and Resumes are numbered to distinguish between the different possible resume events.

The reader is invited to return to Figure 11.9 and consider how, in the light of the treatment of Purchase Order Item, the Purchase Order Entity Life History should be developed. Our suggested solution to this exercise is shown in Figure 11.15.

Super-events

When two or more effects appear at the same point in an Entity Life History, and have the same effects, we can identify a super-event.

Super-events simplify Entity Life Histories by reducing the number of boxes appearing on them. They also identify common processing, thereby promoting re-use and reducing the amount of effort involved in specifying the update processing.

In the Entity Life History for Purchase Order Item (Figure 11.14) the super-event ☺End of POI is identified. We use a ☺ to denote a super-event, but any special symbol will do.

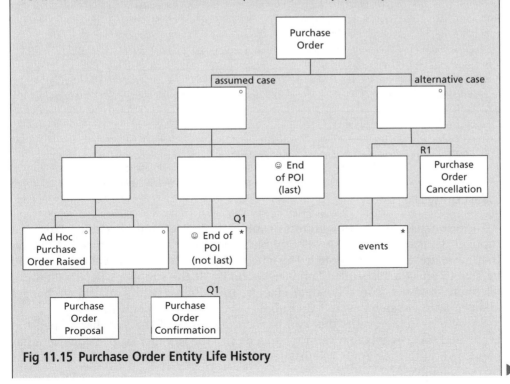

Fig 11.15 Purchase Order Entity Life History

In the Entity Life History for Purchase Order (Figure 11.15) the normal death of Purchase Order is the death of its last Purchase Order Item. Instead of showing both possible ways in which a Purchase Order can be killed by the death of its last detail, the super-event is used instead.

Note that the Purchase Order Cancellation event kills a Purchase Order and all its Purchase Order Items in one go, whereas the normal death of Purchase Order is a lingering one, with its Purchase Order Items being killed off one by one, till none remain.

Continuing our up-pass analysis of behaviour, we have only one change to make to Supplier Product – the addition of a third death event. The Product Discontinued event takes care of the situation where ZigZag stops stocking a product, but we need an event to cover the case where a supplier withdraws a product from its range (see the final Entity Life History for Supplier Product in Figure 11.16).

Moving on to consider the Supplier Entity Life History, we find that we have a problem. In the structure of Figure 11.16 we have a mid-life sequence that says that for a given occurrence no events can intervene between Supplier Suspension and Supplier Reinstatement.

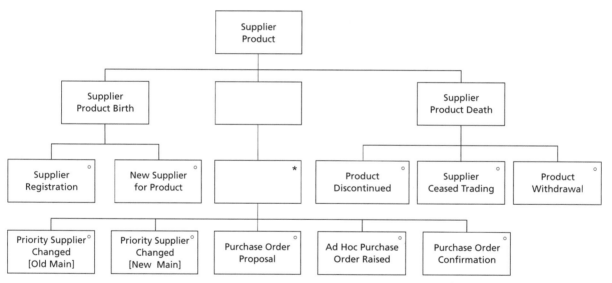

Fig 11.16 Supplier Product Entity Life History

Unfortunately, this is not quite in tune with ZigZag's business rules. First, if a supplier is suspended, ZigZag will still have to honour payment on any deliveries that have already been accepted. This means that the Invoice Arrival event should be able to affect a suspended Supplier. Secondly, if the supplier's details (address, telephone number, contact name) change, then ZigZag will want these changes updated on the system, even if the supplier is currently suspended.

So we have some events that can happen regardless of the Supplier Suspension event, and some events that cannot happen once the Supplier Suspension event has happened (until the Supplier Reinstated event occurs). In other words Supplier has two parallel lives.

Parallel lives

In Entity Life History Analysis we use a parallel life to separate groups of unrelated effects in an entity's life history. In Figure 11.17 the parallel life is denoted using a double line that straddles two sub-structures. On the left-hand side of the parallel life we have the suspension–reinstatement sequence. The other events on that side of the parallel life cannot occur if the Supplier Suspension event has just occurred.

The events on the right-hand side of the parallel life continue without regard for the supplier suspension–reinstatement sequence.

Note that the Supplier Ceased Trading event is one that is outside of ZigZag's control. It is quite possible that this event could happen whilst a supplier is suspended.

Unfortunately, the Entity Life History for Supplier does not allow for this because, following the rules of the notation, the only event that is allowed to follow Supplier Suspension is Supplier Reinstatement (ignoring events in the parallel life). This is a problem that we can again deal with using the Quit and Resume notation and the reader might like to try this as an exercise.

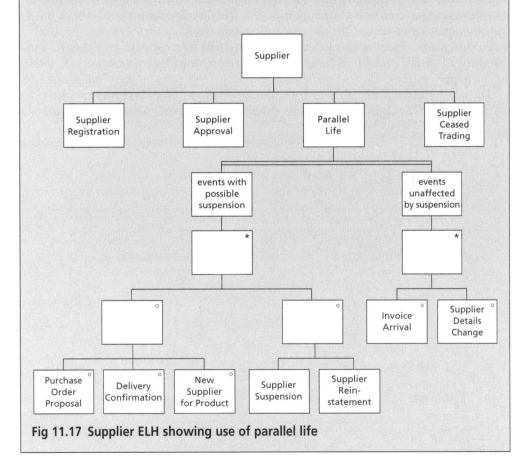

Fig 11.17 Supplier ELH showing use of parallel life

One final observation: we have forgotten to include the second of Purchase Order's birth events as a gain effect in the Supplier Entity Life History.

Down-pass analysis of behaviour

We should complete our Entity Life History Analysis by working from master to detail, looking in particular for extra death events, and at how death events cascade to detail entities. Finally, we might consider the strategy for deleting entities.

As far as the Product and Supplier Entity Life Histories are concerned we have nothing else to add.

When it comes to the Supplier Product Entity Life History, we have already shown the effects of the cascading death events of its masters. There are no further death events to add. We must, however, consider how the death of a Supplier Product might affect its details.

In the case of Product Withdrawal, we assume that any existing orders for this product will either be honoured by the supplier or lapse when the required-by date expires.

In the case of Product Discontinued we find that this only constrains the placing of future purchase orders: existing orders for the discontinued product will not be automatically cancelled.

In the case of Supplier Ceased Trading, there will be a cascading of this event: it will result in the death of all Purchase Orders, Purchase Order Items and Deliveries of the Supplier in question. The effect on Purchase Order can be seen in Figure 11.18.

The box named 'Death of Purchase Order' in Figure 11.18 looks like a candidate for a super-event. We might hope to replace the two events Purchase Order Cancellation

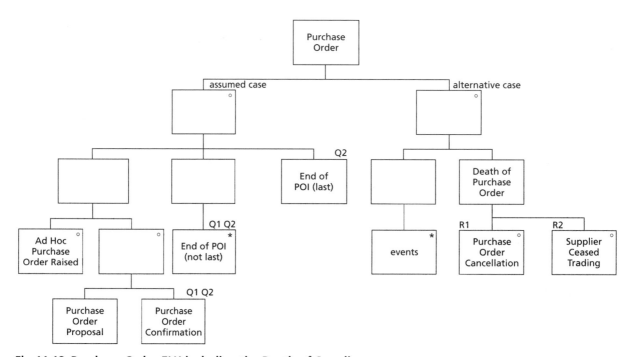

Fig 11.18 Purchase Order ELH including the Death of Supplier

and Supplier Ceased Trading on the Purchase Order Item Entity Life History of Figure 11.8 with this super-event.

Unfortunately, the quit points for these two events in the life of Purchase Order Item are not the same, so we cannot replace them with a single super-event.

The astute reader will notice that there are several common patterns in the deaths of entities. The one we have just been considering is the cascading death (death of master is also the death of its details). Another common one is the death of last detail causing the death of the master (considered earlier).

Dealing with sub-types and aspects

In dealing with mutually exclusive behaviour patterns we have used the Quit and Resume notation. This enabled us to model as an assumed case the behaviour that was common to several lives, with the alternative cases dealing with the differences in those lives. Where there are several complex and mutually exclusive behaviour patterns to be dealt with the analyst may find it helpful to model these as separate structures by defining sub-types or aspects.

When dealing with sub-types there will be two structures:

◆ a super-type structure consists of super-events, which represent the processing common to all the sub-types;

◆ a sub-type structure, which has a high-level selection and an option for each sub-type life history.

For entity aspects, each aspect is modelled as a separate Entity Life History.

Deletion strategy

A death event does not remove an entity occurrence from the system – the entity remains available for enquiry purposes. Often users will require historical information based on dead entities. If this is not the case then we might consider deleting an entity upon its death.

If historical information is to be kept on the system, then we need to identify the event that will eventually cause its deletion (although some entities may remain on the system for the life of that system).

In a system like this one, supplier and product information might never get deleted. Purchase order information might be kept for a number of years before being archived to some other medium and deleted from the live system. At a summary level purchase order information might also be periodically copied to a data warehouse. A similar pattern would apply to the sales side of the system.

We are likely to get the same patterns in deletion events as in death events, e.g. deletion of master causes deletion of details.

In ZigZag, we have to keep an audit trail, meaning that each stock of an item must be traceable to the delivery that gave rise to it. This means that the Delivery and Purchase Order entities and their details cannot be deleted until all related Stock records have been deleted. Stock records die when the quantity in Stock reaches zero. They are deleted six months later (we have to introduce a new attribute to record the date of death in order to allow for this).

As an exercise the reader might like to complete the Entity Life Histories by adding deletion events.

▶ ELH OPERATIONS AND STATE INDICATORS

▶ Operations

Each end-leaf on an ELH represents an effect on an entity. We have to describe each effect in detail, and to do this we add operations to the effects. SSADM provides a standard set of logical operations to be used with Entity Life Histories (see Table 11.2). Figure 11.19 shows part of the Purchase Order Item entity with operations added.

Not every effect carries an operation. Most of the death events of Purchase Order Item affect only its state (see next section). Operations that reflect changes in state will be built into the Effect Correspondence Diagrams, as will operations to read, write and delete entity occurrences. These are not usually added to the Entity Life History.

For every Gain in a master entity's Entity Life History there should be a Tie in the detail entity's Entity Life History and for every Lose in a master there should be a Cut in the detail.

In a relational environment a relationship is implemented by the placement of a foreign key in the dependent table (just as we have them in our detail entities on

Table 11.2 Standard set of logical operations to be used with Entity Life Histories	
Create <*Entity*>	Creates a new occurrence of the entity. In this book we will omit the create operation from our Entity Life Histories and add it when we develop our Effect Correspondence Diagrams. The need for a create operation on an Entity Life History is implicit: it applies to every birth effect
Set <*Attribute*>	Changes the value of the named attribute to the new value input with the event
Set <*Attribute*> using <*Expression*>	Sets *attribute* to the value resulting from the application of *expression* (e.g. a numeric calculation of the type 'Today's date plus seven days')
Tie to <*Entity*>	Establishes a relationship with the master *entity*
Cut from <*Entity*>	Removes the relationship with the master *entity*
Gain <*Entity*>	Establishes a relationship with the detail *entity*. Where an occurrence of the event may cause more than one detail to be gained, we can use '*gain set of <entity>*'
Lose <*Entity*>	Removes the relationship with the detail *entity*. Again, we can use '*lose set of <entity>*' where more than one detail is being lost
Invoke <*Process*>	Invokes a named process. The process could be a super-event, common process or enquiry

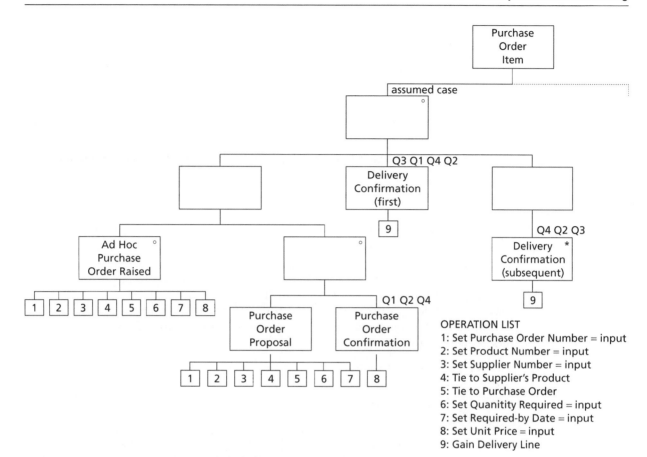

Fig 11.19 Part of the Purchase Order Item ELH showing operations

the Logical Data Model) and so you can think of a 'Tie' operation as assigning a value to a foreign key and therefore as having a very tangible effect on the entity. Gain and Lose operations represent the creation and destruction of relationships from the master's point of view and have no tangible effect. Nevertheless they help to validate the Entity Life Histories by forcing the analyst to consider the interaction between events in a detail's life and those in its master. Gains and Losses can be omitted if they are found to be unhelpful.

Organisations may wish to define their own local standards for operations rather than using the default set suggested by SSADM.

▶ State indicators

In this chapter we have shown how you can use Entity Life Histories to model the required or allowed sequence of events for each entity. You will recall how important this is from our earlier of discussion of states. The aim is to build rules into our update processing that will prevent certain sorts of errors getting into the data in our database.

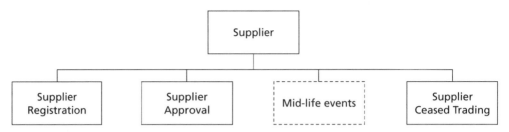

Fig 11.20 Mid-life events of Supplier cannot take place unless Supplier Approval has taken place

In the fragment of the Supplier Entity Life History in Figure 11.20 we see that the mid-life events of Supplier (omitted here for reasons of space) cannot take place *unless and until* the event 'Supplier Approval' has taken place. This means that when the processing of any of these mid-life events takes place there has to be a way of checking that the event 'Supplier Approval' has taken place.

Our way of doing this is to make the state of an entity explicit by adding an extra attribute called a *state indicator*. When the processing of any effect takes place the process will do two things:

◆ It will check that the current state indicator value is one that is allowable for the event which is being processed.

◆ It will set the state indicator value to a value that reflects that a new event has taken place.

Of course, if the current state of an entity is not one that is allowable for the event in question then the processing of the event will not continue and no changes will be made to the persistent data of our system, including its state indicator value.

Two questions arise:

◆ How do we know what the states of an entity are?

◆ How do we know what the valid sequence of events is?

Fortunately, this is exactly what the Entity Life History tells us. All we have to do is to add the following information to each effect using the following notation:

<Valid Previous Values>/<Value Set by Event>

Valid Previous Values represents a list of the values that the entity's state indicator can have for the event to be allowed. They are determined by examining the structure of the Entity Life History.

Value Set by Event specifies the value the state indicator will be set to on completion of the event.

The values can be numbers, with a unique number for each state within an entity, or they can be words, reflecting the names of states.

To give a simple example, looking at the section of Entity Life History in Figure 11.21, we can assume that event A can only take place if the state indicator of the entity currently has a value of 3 or 4. Once event A is complete the state indicator will be set to 5. Because event A and event B are in sequence, event B can only

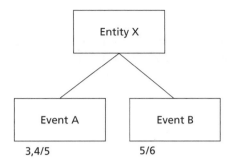

Fig 11.21 ELH section showing state indicators

take place if event A has happened last. Thus, for event B to be allowed the current state of entity X must be 5. If event B is allowed to take place, then the state of entity will be set to 6.

The values set by each event are only meaningful within the context of a single Entity Life History. The convention is for the first birth event in an Entity Life History to set the value to 1. As an entity occurrence will have no state indicator prior to its birth, the valid previous value for a birth event is null. Similarly, the value set by a deletion event would also be null.

▶ Adding state indicators

State indicators are usually added to Entity Life Histories in two passes:

◆ The values set by each effect are added.

◆ By examining the allowed sequences of events the valid previous values are added for each event

Sequence

If the Entity Life History contains a sequence of events the value set by one event is the only valid previous value of the next event (as no other events can occur in-between) as in Figure 11.22.

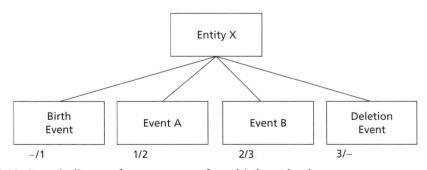

Fig 11.22 State indicators for a sequence from birth to death

Selection

The valid previous values for all events in a selection will be the same. The values set by each event in the selection will be unique.

The event that immediately follows the selection must include the 'set to' values of all selection events in its list of valid previous values (as any of them can occur), as shown in Figure 11.23.

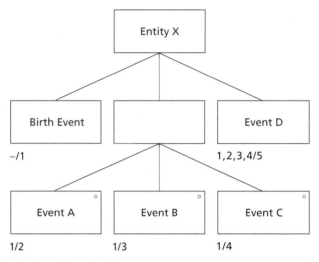

Fig 11.23 State indicators for selections

Iteration

As an iterated event must include the possibility of being preceded by itself, the list of valid previous values must include its own 'set to' value.

The event that follows an iteration will include the valid previous values of the iteration in its own list of valid previous values, as an iteration can occur '*zero or many*' times, i.e. need not occur at all (Figure 11.24).

Quits and Resumes

The valid previous values of a Resume effect are the same as the valid previous values of the corresponding Quit effects.

Parallel structures

The 'main leg' of a parallel structure will update the state indicator as normal. However, any parallel legs will leave it unchanged. This is because events in a parallel life have no effect on the position of the entity occurrence within its main life.

As a parallel life can occur at any point in the main life, the valid previous values for all parallel-life effects will be made up of all valid previous values *plus* all set to values from the main life. The set-to value for all events in the parallel life will be *, as the indicator is unchanged.

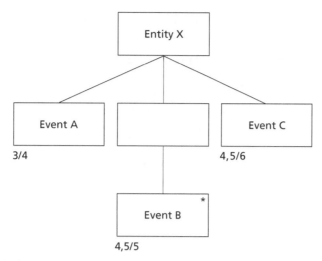

Fig 11.24 State indicators for iterations

If a parallel life itself contains a sequence of events then we will need to specify this in the same way as the main life sequence. This means introducing secondary state indicators that apply only to the parallel-life structures. The notation for a parallel-life state indicator is the same as for the main-life indicator, but is differentiated from it by the use of brackets.

Adding state indicators to an Entity Life History requires a little practice at first, but rapidly becomes a straightforward mechanical process. No knowledge of the underlying effects or meaning of events is required, just an understanding of how to navigate through Entity Life History structures. For this reason CASE tools will often generate state indicators automatically.

Optimised state indicators

The optimisation of state indicators helps to make the update processes that check them easier to maintain; this will become clearer when we have developed Effect Correspondence Diagrams in the next chapter. Optimisation also facilitates the use of super-events. Furthermore, it helps us to identify named states, which are probably more meaningful to the user.
   State indicators are optimised in the following way:

◆ The 'set to' values for each option of a selection are made the same.

◆ The 'set to' values of an iterated component can be made the same as the state that precedes the iteration.

As an exercise the reader is invited to add state indicators to the Entity Life History of Purchase Order Item. Having done this, try to optimise them. Finally, consider using names for the states instead of numbers.

▶ SUMMARY

1. The understanding of events and their effects is an important part of the specification of update processing, and the techniques employed in this chapter all contribute towards that understanding.

2. The Entity Access Matrix can provide an extremely useful view of the system processing and is concise and quite precise without being unduly burdensome to develop.

3. Entity Life History analysis is a powerful tool for systems analysis. It forces the analyst to raise important and relevant questions, many of which might otherwise not be raised.

4. Entity Life Histories are diagrammatic representations of how entity types are or can be affected by events. They detail the allowable sequence, iterations and optionality of events for each entity, and thus further capture and record business rules.

5. Used rigorously, Entity Life Histories can be arduous to complete. On the other hand, they will make the subsequent definition of update processing much more straightforward and they will ensure that the update processing incorporates many business rules that otherwise might not be enforced.

▶ EXERCISES

11.1. Using the information below produce an ELH for the Treebanks entity 'Session':

Every Friday the manager of Treebanks arranges the playing sessions for the week beginning in four weeks time. At the same time the records of playing sessions from four weeks ago are deleted.

Each session is booked, by either a team or a member. The booking may subsequently be cancelled, in which case it will become available for re-booking. When the team or member arrives to play on the booked court the session is marked as played.

The Entity Access Matrix for Treebanks records the following events for the entity Session:

Make Team Booking	(M)
Make Member Booking	(M)
Cancel Booking	(M)
Play Session	(M)
Arrange Sessions	(C/D)

11.2. Using the write-up for Exercise 3.15 and the information below, produce an ELH for the Bodgett & Son entity Estimate:

Estimates are always booked to be carried out either by a surveyor or by an employee right from the start. However, it will sometimes be necessary to change this booking before the estimate is completed.

Customers may respond to a completed estimate in several stages: accepting or rejecting some of the constituent jobs on each occasion.

If no response is received from customers after five weeks a reminder letter is to be sent. If there is still no response after a further four weeks the estimate will be marked as 'expired'.

Records of estimates will be kept for twelve months following their expiry or completion of the last job detailed on them.

Some of the more important attributes of Estimate are:

Estimate No.
Estimate Request Date
Estimate Booking Date
Estimate Completion Date
No. of Jobs
No. of Jobs Responded To
Reminder Letter Date
Expiry Date

Hint: Begin the exercise by listing the events that affect Estimate, as in the row of an Entity Access Matrix.

11.3. Add operations to the ELH for Estimate.

11.4. Using the information given below complete the Entity Access Matrix for the video rental system and develop Entity Life Histories for each entity.

The logical data model in Figure 11.25 describes a video rental system. An (incomplete) Entity Access Matrix for the system is given in Figure 11.26. The system keeps information about the hire and return of videos by customers of a chain of video stores. Customers register at a specific video store. The chain maintains a list of video titles, and copies of each title are available for hire at each store. Customers hire one or more videos at a time, and although each copy of a title can only be on loan to one member at a time, there is a need to keep a historical record of who has borrowed what.

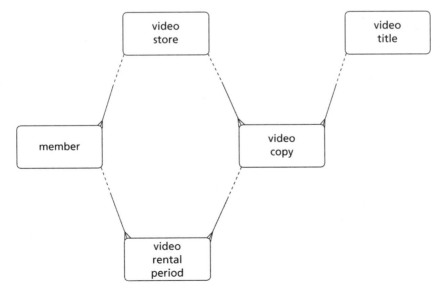

Fig 11.25 Video store LDS

ENTITY EVENT/ENQUIRY	Title	Video Store	Member	Video Copy	Video Rental Period
Copy Purchase					
Copy Sale or Destruction					
Membership Approval					
Membership Removal					
Store Closure					
Store Opening					
Title Release					
Title Withdrawal					
Video Rental					
Video Return					

Fig 11.26 Incomplete Entity Access Matrix

Video titles can be withdrawn at any time, after which the video stores will purchase no more copies of the title. A record of a video title will not be deleted from the system until the last copy of that title has been sold or destroyed.

When a video store closes, all its video copies and all its members are transferred to a replacement store (i.e. they are all transferred to one store, as designated by the company's head office).

Sometimes a member of a video store will lose or damage a video that they have rented. This is treated as a 'copy sale or destruction'.

A member cannot be removed from the system whilst he or she still has videos out on rent.

Video Copy
(Title Code)
(Copy Number)
*Store Number
Date of Purchase
Daily Rate
On Rental Indicator

Video Rental Period
(Title Code)
(Copy Number)
Member Number
Video Rental Date
Return Date
Amount Charged

Video Title
Title Code
Title Name
Title Description
Number of Copies Held

Video Store
Store Number
Store Name
Store Address
Store Manager
Total Video Stock

Member
Member Number
Member Name
Member Address
*Store Number
Copies Currently On Rental

NB. The 'on rental indicator' attribute of video copy is set to 'yes' when the copy is on loan, and to 'no' when it is available for hire.

11.5. Produce an ELH for an entity Student of a simple student administration system. This system keeps a record of a student's progress. In this college, a student normally applies to a course, is sent an unconditional offer by the college, accepts it, and then may decide to enrol, complete the course, and go. But, some students never send an acceptance after receiving an offer or never enrol after accepting an offer. It is therefore necessary to allow the deletion of a particular student from the system without having to take that student through the normal 'life' of a student. In short, we wish to be able to 'jump' from *receipt of acceptance* or from *course enrolment* directly to *student deletion*.

11.6. Suppose it is possible that after accepting an offer or enrolling on the course, a student of the previous exercise suspends his or her studies for health reasons. Adjust your ELH to accommodate this eventuality.

11.7. While SSADM is a structured method, the identification of system events is actually iterative. Provide a diagram of SSADM products to illustrate the repercussions of identifying a system event during Entity Life History Analysis.

11.8. The identification of system events lies at the heart of any systems analysis cycle. Explain how the main system events are identified when a structured approach to systems analysis is followed and show what action should be taken when a new system event is identified during Entity Life History Analysis.

11.9. State what SSADM products you would use under normal circumstances to produce an Entity Access Matrix.

11.10. Place optimised and un-optimised state indicators on the generic Entity Life History in Figure 11.27.

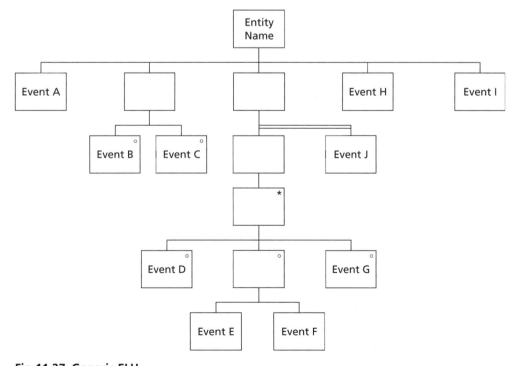

Fig 11.27 Generic ELH

11.11. Produce an Entity Life History for the entity Room Booking of the West Munster system of Exercise 8.2. The events affecting Room Booking are: Booking Requested, which may be for a provisional booking, Provisional Booking Confirmed, Three Weeks from Provisional Booking Past, Guest Arrival, Credit Check Result, which may be successful or unsuccessful, Alternative Method of Payment Accepted, Hotel Service Used, Bill Produced and End of Year, when the room bookings are all archived.

11.12. Draw a section of an Entity Access Matrix showing the effects contained in Entity Life History of Exercise 11.11. (You are expected to tally all your ties and gains but not to show other effects your events may have on the neighbours of Room Booking.)

Conceptual Process Modelling

INTRODUCTION

In Function Definition we discussed the need to specify the database processing both for updates and for enquiries. In this chapter we offer several ways of doing this.

For updates we discuss the Effect Correspondence Diagram and its close relation the Update Process Model. For enquiries we examine Enquiry Access Paths and the related Enquiry Process Models. 'Conceptual Process Modelling' is an umbrella term that covers all of these.

 Learning outcomes

By the end of this chapter you will be able to:

- ◆ use Enquiry Access Paths to validate the Logical Data Model against the users' information requirements;
- ◆ produce logical, implementation-independent specifications of the database processing triggered by events and enquiries;
- ◆ use state indicator checking to build integrity constraints into the database processing.

 Link to other chapters

- ◆ Chapter 8   The products of Conceptual Process Modelling provide the specifications for each function
- ◆ Chapter 11   The Entity Behaviour Model and the Conceptual Process Model provide cross-checking views of the effects of events on the Logical Data Model
- ◆ Chapter 14   The specifications of Conceptual Process Modelling are transferred into physical designs

▶ EFFECT CORRESPONDENCE DIAGRAMS

Once we have completed Entity Life Histories for all entities in the Required System Logical Data Model we turn our attention to analysing effects from the perspective of the event, using Effect Correspondence Diagrams (ECDs).

We develop Effect Correspondence Diagrams in order to illustrate which entities are affected by a given event. We also use them to define how the effects on these different entities correspond with each other, and ultimately to provide us with a specification of *update* processing. The development of Effect Correspondence Diagrams helps to clarify the work done during Entity Life History analysis, and often the two techniques are used in tandem in order to validate each other.

Additionally, the identification of values for state indicators for each effect on the Entity Life Histories allows us to build some integrity checking into the Effect Correspondence Diagrams.

Note

Readers with an interest in object-oriented implementations should note that the Entity Life Histories show the operations (or methods) of each business class. Attributes of each business class are described by the Logical Data Model. Each Effect Correspondence Diagram shows how different classes collaborate to support a particular Use Case. Thus, the products that we have discussed so far are quite well suited to an object-oriented implementation *without being tied to one*.

▶ Drawing Effect Correspondence Diagrams

The simplest and most effective way of drawing Effect Correspondence Diagrams is to follow the seven steps given below. A separate Effect Correspondence Diagram is developed for each event.

Draw a box for each entity

The Entity Access Matrix will show which entities have been affected by the event. Alternatively these could be identified from the Entity Life Histories directly. A good CASE tool will perform this step automatically.

If Entity Life History analysis has not been performed then the analyst will have to develop Effect Correspondence Diagrams by reference to the Event and Enquiry Catalogue, Function Definitions and Logical Data Model, trying to gain as full an understanding of the effects of the event as possible.

In any case, the analyst should refer to the Event and Enquiry Catalogue to gain an overview of the processing that is triggered by the event.

Each entity affected is represented as a round-cornered (or 'soft') box containing the entity name. A separate box is drawn for each aspect, super-type and sub-types, where these have been used.

> ## Tip
>
> There are many ways of identifying the entities that are affected by an impending (system) event. If a CRUD matrix (see Figure 6.45) has been created, then the chances are that we will have identified the necessary entities. Alternatively, we can use the Entity Access Matrix (see Figure 11.4) as a more comprehensive guide.
>
> Another way is to focus on the actual data items whose values are coming in as a result of the event. We should be able to pluck those from the screen designs or the I/O Structures that may have been produced. With the data items identified, the Logical Data Model should suffice to pinpoint the relevant entities.
>
> Also, if the function that represents the system's response to the event has been submitted to Relational Data Analysis, the precise chunk of data model needed by the event will already have been found.
>
> In the absence of all of the above, which would be surprising if a structured approach has been followed, the data model and an understanding of each function are required in order to proceed.

For the Purchase Order Proposal event, boxes will be drawn as in Figure 12.1.

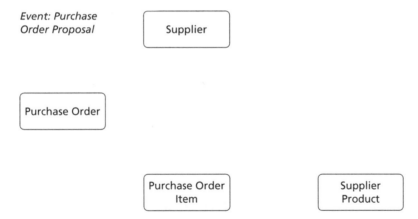

Fig 12.1 Entities affected by the Purchase Order Proposal event

Add all effects of events

Where entity roles (simultaneous effects) and effect qualifiers (alternative effects) have been shown on Entity Life Histories, these have to be added to the Effect Correspondence Diagram:

◆ alternative effects are represented by making the entity a selection, with each option showing one of the possible effects (each box will contain the entity name with an effect qualifier in round brackets);

◆ simultaneous effects are represented by drawing a separate box for each effect (each box will contain the entity name, with the role name in square brackets).

Again, a good CASE tool will assist in this step.

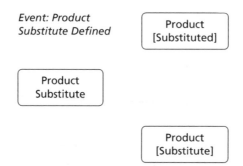

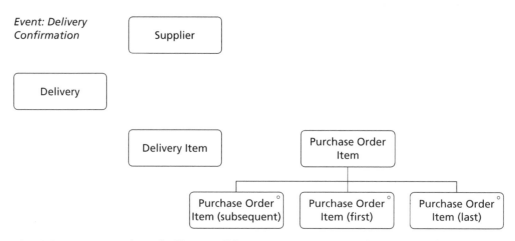

Fig 12.2 Representation of entity roles on an ECD

Fig 12.3 Representation of effect qualifiers on an ECD – note the 'o' identifying various options

Readers will remember from Figure 11.5 that the Product Substitute Defined event affects two occurrences of Product (Figure 12.2). And if we refer back to the Entity Life History for Purchase Order Item (see Figure 11.14), we will see that the Delivery Confirmation event can have one of three different effects on Purchase Order Item (Figure 12.3).

Identify entry point

The entry point is the first entity to be processed. The input data must contain enough information to identify the required occurrence(s) of the entry point entity if it is being read, or to create it if it is being created.

For the Purchase Order Proposal event we will choose Purchase Order as the entry point; for Product Substitute Defined we will choose Product Substitute; and for Delivery Confirmation the entry point will be Delivery. (Discussion point: could you choose any other entry points for any of these events?)

There may be more than one potential entry point entity, in which case several entry points can be defined, and the order of processing can be deferred as an implementation decision.

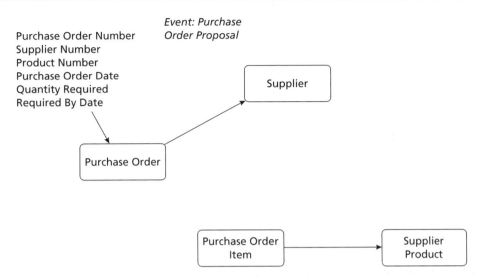

Purchase Order Number
Supplier Number
Product Number
Purchase Order Date
Quantity Required
Required By Date

Event: Purchase Order Proposal

Supplier

Purchase Order

Purchase Order Item

Supplier Product

Fig 12.4 Correspondence of effects

The entry point is shown on the Effect Correspondence Diagram using an arrow, against which is listed the data which is required as input to the event (see Figure 12.4).

Define correspondences

A correspondence shows an access to a non-entry-point entity. A correspondence is shown with an arrow. We have to be sure that for each entity there is a way of identifying which occurrences are to be affected. Usually a correspondence arrow will trace a relationship between two entities, but there are two exceptions:

♦ a detail is not directly related to a master, but contains its key as part of a compound foreign key;

♦ there is enough data in the event data to pinpoint the correct occurrence(s).

Sometimes it will be necessary to navigate via another entity to identify an affected entity. This access should be shown as a *read* on the Entity Access Matrix.

Returning to the Purchase Order Proposal event, for each Purchase Order that is affected there will be one Supplier affected, and for each Purchase Order Item affected there will be one Supplier Product affected (see Figure 12.4).

Deal with iterations

Where there is navigation from a master to a detail, it is necessary to decide how many occurrences of the detail are affected per occurrence of the event. Where more than one detail occurrence can be affected, we use a 'Set of' box, and connect it to the iterated effect, which is marked with an asterisk.

The creation of a Purchase Order will result in the creation of one or more Purchase Order Items. Thus the effect on Purchase Order Item is iterated, as shown

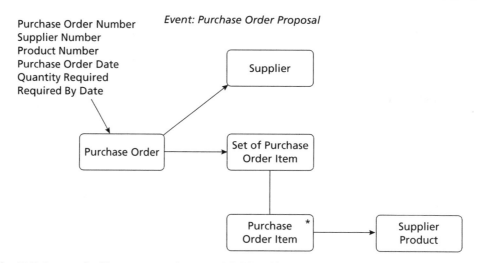

Fig 12.5 Iterated effects – note the asterisk identifying the iteration

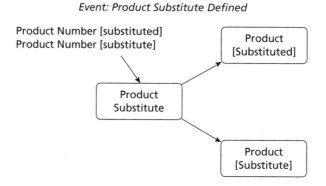

Fig 12.6 Effect Correspondence Diagram for Product Substitute Defined

in Figure 12.5. Note that many occurrences of Supplier Product will be affected too – one for each Purchase Order Item. This correspondence was identified in the previous step.

The Effect Correspondence Diagrams for Delivery Confirmation and Product Substitute Defined are similarly developed by defining entry points, identifying correspondences and considering iterations (see Figures 12.6 and 12.7).

Add conditions

A condition can be added to each option and each iteration. The condition must state the circumstances under which an option is to occur, or an iteration is to continue.

Typically a condition that is attached to an option will evaluate an attribute value (this might include state indicator values).

See Figure 12.7 for an example of using conditions.

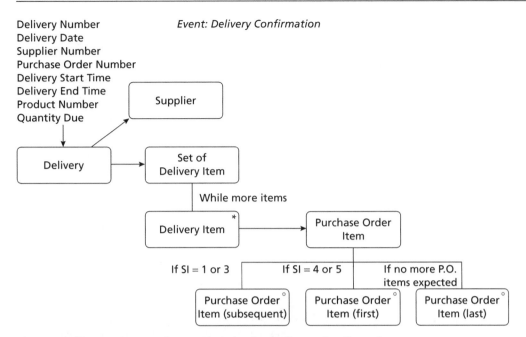

Fig 12.7 Effect Correspondence Diagram for Delivery Confirmation

Add operations

The effects that appear on Effect Correspondence Diagrams are the same effects that appear on Entity Life Histories. On the Effect Correspondence Diagram the effect carries the name of the entity affected, whilst on the Entity Life History the effect carries the name of the event that causes the effect. There are also some navigational operations that define the movement between entities. A project team may wish to produce its own set of customised operations. Table 12.1 is based on the full list recommended in the CCTA volume on Behaviour and Process Modelling (2000).

Each effect on an Effect Correspondence Diagram may have already been described with operations on an Entity Life History. These operations can be carried forward to the Effect Correspondence Diagram.

In addition to the ELH operations that we need to add, there are database operations to create/read and write/delete entity occurrences, and integrity operations to check and set state indicator values. Table 12.2 contains the expected sequencing of operations according to different effects.

The fail operations are used whenever an entity is read. The format of a fail operation is *Fail if SI value of entity outside 'range'* where *'range'* is the valid previous value(s) for the effect, as shown on the Entity Life History. Similarly, the set SI operation is of the form *Set SI of entity = 'value'* where *'value'* is the set to value for the effect, as shown on the Entity Life History.

The Effect Correspondence Diagram for Purchase Order Proposal, complete with operations and conditions, is shown in Figure 12.8.

Table 12.1 Process modelling operations

Operation Type	Logical Processing Description
Read <entity> by key	Read from the database the entity using the entity's key value which is being input
Read next <detail> of <master> [via <relationship>]	Read from the database the next entity of type <detail> related to the current occurrence of <master>. If there are more than one possible relationship paths between the two entities then we can specify via which relationship we wish the read to take place
Read <master> of <detail> [via <relationship>]	Read from the database the type <master> related to the current occurrence of <detail> (using a specified relationship if required)
Define set of <entity> matching input data	Define a set of <entity> occurrences, the members of which match the criteria in the input data. Once a set is defined it can be treated as a transient entity type to be ordered, read, output in any way desired.
Read next <entity> in set	Read the next entity of type <entity> from the currently defined set. Clearly, this operation must be preceded by a 'define set' operation.
Fail if <statement>	Abort the entire process if a pre-defined circumstance is met.
Fail if SI of <entity> outside <value range>	Usually placed after reading an entity to start an abort process when the value of a state indicator is found to be outside a prescribed range of values – which we usually read off an ELH. (If SIs are not defined then the 'fail if' statement above can be used instead)
Get <data items>	Obtain data items from the function that is invoking the event or enquiry. (Set Input is a frequent alternative to using Get)
Output <data items>	Output data items to the function invoking the event or enquiry.
Set <data item> = <value>	Set a given value to an item. This operation can also be used to input data – cf. Figure 12.8
Set <entity> SI = <value>	Set the value of the state indicator of an occurrence to a specified value.
Invoke <process>	Invoke a separately defined process. The process may be another function or a function component with its own input and output.
Write <entity>	Commit to the database all changes made to an entity.
Delete <entity>	Delete an entity occurrence from the database completely

Table 12.2 Recommended sequencing of Effect Correspondence Diagram operations

For a birth effect:	For a deletion effect:	For other effects:
Create <entity>	Read <entity>	Read <entity>
ELH operations	Fail if SI of <entity> <>	Fail if SI of <entity> <>
Set SI =	ELH operations	ELH operations
Write <entity>	Delete <entity>	Set SI =
		Write <entity>

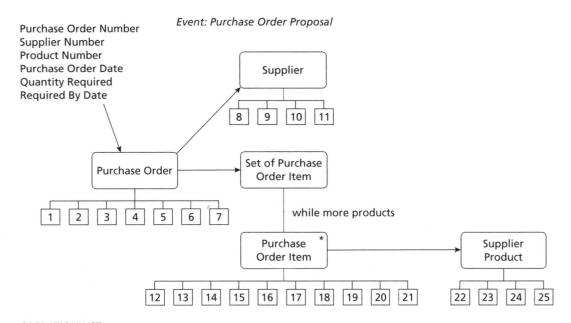

Event: Purchase Order Proposal

Purchase Order Number
Supplier Number
Product Number
Purchase Order Date
Quantity Required
Required By Date

while more products

OPERATION LIST
1: Create Purchase Order
2: Set Purchase Order Number = Input
3: Set Purchase Order Date = Input
4: Get Supplier Number
5: Tie Purchase Order to Supplier
6: Set SI of Purchase Order = 2
7: Write Purchase Order
8: Read Supplier of Purchase Order
9: Fail if SI of Supplier <> 2
10: Set SI of Supplier = 2
11: Write Supplier
12: Create Purchase Order Item

13: Set Product Number = Input
14: Set Purchase Order Number = Input
15: Set Quantity Required = Input
16: Set Required-by Date = Input
17: Set Supplier Number = Input
18: Tie Purchase Order Item to Purchase Order
19: Tie Purchase Order Item to Supplier Product
20: Set SI of Purchase Order Item = 2
21: Write Purchase Order Item
22: Read Supplier Product of Purchase Order Item
23: Fail if SI of Supplier Product <> 1
24: Set SI of Supplier Product = 1
25: Write Supplier Product

Fig 12.8 Final Effect Correspondence Diagram for Purchase Order Proposal

▶ UPDATE PROCESS MODELS

> ### Important note
>
> Through the use of Entity Life Histories and Effect Correspondence Diagrams during Specification we have painted a picture of the impact of events on the system's logical data model.
>
> For many projects the Effect Correspondence Diagram will suffice as the model of update processing to be input to Physical Design. However, the Effect Correspondence Diagram is not a proper Jackson structure, and for more complex processes it can be slightly ambiguous as to the precise order of processing.
>
> Where the system is to be built using a more traditional procedural, structured and modular approach to coding (e.g. using a language such as COBOL), this ambiguity can be usefully reduced by converting Effect Correspondence Diagrams into Update Process Models, which will represent exactly the same set of effects using Jackson notation.

Update Process Models are developed using a sub-technique of Conceptual Process Modelling with the unsurprising name of 'Update Process Modelling'.

As with all Conceptual Process Models, the Update Process Model assumes that the Required System Logical Data Model exists as an implemented database, i.e. a logical database. In environments where the final system is to be built using non-procedural code the process models can act as actual program specifications, since the translation from logical processing and databases to physical processing and databases is handled by the development software itself. Where procedural coding is to be used the process models will be translated into environment-specific program specifications during Physical Design.

As well as serving as the basis for Physical Design and implementation of the new system, Conceptual Process Models act as effective maintenance documentation once the system is running live. Any future changes to the system can be evaluated against, and applied to, the underlying logical (or business-oriented) model before any physical amendments are made.

▶ Update Process Modelling

The notation for process structures is another variation on the now-familiar Jackson-like structures of Entity Life Histories and I/O Structures. Each update process structure will describe the processing of a single event, and each end-leaf will represent a discrete module within that overall processing. To create an Update Process Model we go through the following steps.

Carry forward the ECD

In the last section we showed how to develop Effect Correspondence Diagrams to model the effects of a single event on the entities in our Required System Logical Data Model.

Effect Correspondence Diagrams provide a picture of how changes triggered in response to event (i.e. input) data are applied to and permeate through the Logical Data Model.

The purpose of each Update Process Model is to describe the processing which is to take place in response to an event. Because the two types of model share the same purpose – they actually describe the same thing – we can use Effect Correspondence Diagrams as the basis for developing Update Process Models.

We will use the Effect Correspondence Diagram for the Purchase Order Proposal event of Figure 12.8 to illustrate the creation of an Update Process Model.

Group effects in one-to-one correspondence

All effects linked on an Effect Correspondence Diagrams by arrows are in one-to-one correspondence. This means that the affected entities can be accessed and processed unconditionally as a group.

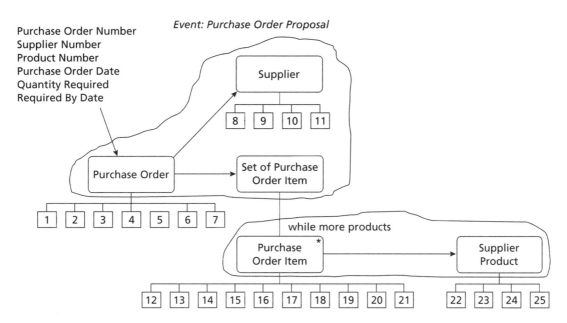

OPERATION LIST
1: Create Purchase Order
2: Set Purchase Order Number = Input
3: Set Purchase Order Date = Input
4: Get Supplier Number
5: Tie Purchase Order to Supplier
6: Set SI of Purchase Order = 2
7: Write Purchase Order
8: Read Supplier of Purchase Order
9: Fail if SI of Supplier <> 2
10: Set SI of Supplier = 2
11: Write Supplier
12: Create Purchase Order Item

13: Set Product Number = Input
14: Set Purchase Order Number = Input
15: Set Quantity Required = Input
16: Set Required-by Date = Input
17: Set Supplier Number = Input
18: Tie Purchase Order Item to Purchase Order
19: Tie Purchase Order Item to Supplier Product
20: Set SI of Purchase Order Item = 2
21: Write Purchase Order Item
22: Read Supplier Product of Purchase Order Item
23: Fail if SI of Supplier Product <> 1
24: Set SI of Supplier Product = 1
25: Write Supplier Product

Fig 12.9 Grouped Effect Correspondence Diagram

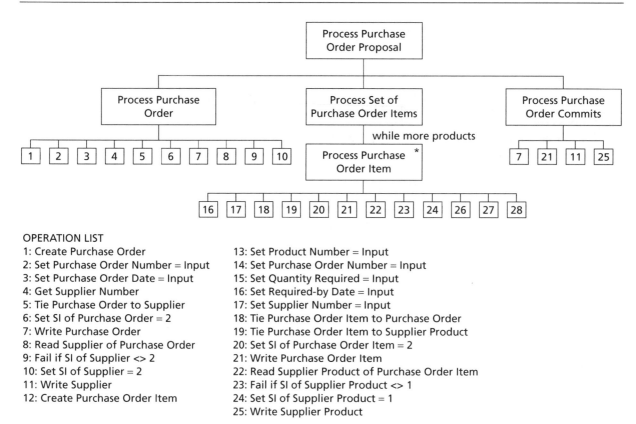

Fig 12.10 Update Process Model

In specifying the processing associated with each event we are interested in creating units of processing that can be designed and implemented in reasonably self-contained modules or blocks. By grouping together effects on each Effect Correspondence Diagram that are in one-to-one correspondence we have the basis for such modules.

Groups are added to an Effect Correspondence Diagram by drawing a boundary around all effects linked by correspondence arrows, and giving them a name that reflects the processing carried out for them. Any effects that stand alone, i.e. are not linked by arrows, automatically become single-element groups.

Figure 12.9 illustrates the Effect Correspondence Diagram for Purchase Order Proposal with processing access boundaries added.

Convert to Jackson-like structure

This is a fairly simple and somewhat mechanical procedure. First, we draw a root node containing the name of the event whose processing is being described. The wording is of the form 'Process <event name>'.

The entry-point processing group becomes a child of this root node. Then we add any subsequent processing groups. The resulting structure must adhere to the usual rules regarding combinations of box types, so a few new structure boxes may need to be added: in this case the processing of Purchase Order Items is iterated, so this group will need a structure box above it.

The structure now represents all of the processing required to apply the updates generated by the event, and to provide the required output.

Allocate operations and conditions to processing structure

In Figure 12.10 the Effect Correspondence Diagram's operations have been carried across just as they were, as have the conditions. Where several effects have been merged into one processing group, all the operations of those effects have been placed together, in the order in which they are to be processed. Thus, the processing that finally commits the new values to the database (using the 'write' operation) are placed under a node on the right-hand side of the structure.

Once again, a good CASE tool should be able to perform an automatic transformation from Effect Correspondence Diagram to Update Process Model.

▶ ENQUIRY ACCESS PATHS

We considered informal validation of access paths earlier in the text. During Specification we will build formal access models for all enquiries (other than *ad hoc* enquiries). Each Enquiry Access Path (EAP) will document the required data model accesses resulting from an enquiry trigger (the non-update equivalent of an event), using the same notation as that used for Effect Correspondence Diagrams.

> Figure 5.34 contains an informal access path produced during analysis.

Enquiry Access Paths serve two very important purposes. First, they validate the Logical Data Model against the users' information requirements. In other words, they demonstrate that the database design is capable of providing the reports and answering the queries that users have identified. Second, they provide an unambiguous specification of each enquiry process. The database programmer can therefore use them as the basis for writing code.

▶ Understanding Enquiry Access Paths

Before we look at the technique for drawing them we will try to get an understanding of Enquiry Access Paths by walking through a typical diagram. In Figure 12.11 we have an Enquiry Access Path for the Purchase Order Query. The Function Definition for this enquiry reads:

> For a given supplier return details of all confirmed purchase orders, giving a full description of each purchase order item, including the name and quantity of the product ordered.

We will start our walkthrough at the top left-hand point of the diagram where there is an arrow labelled 'Supplier Number'.

Supplier Number is part of the enquiry trigger; it is the data item entered by the user in order for this enquiry to take place. You can think of the arrow as saying 'Find the Supplier with the Supplier Number that matches the one input by the user'. This arrow is known as the *entry point arrow*.

The soft box marked 'Supplier' represents an access to the Supplier entity: the reading of one instance of Supplier. On an Enquiry Access Path the entity access point, which is annotated with the enquiry trigger, is known as the *entry point of the enquiry*. So, for this enquiry Supplier is the entry point entity.

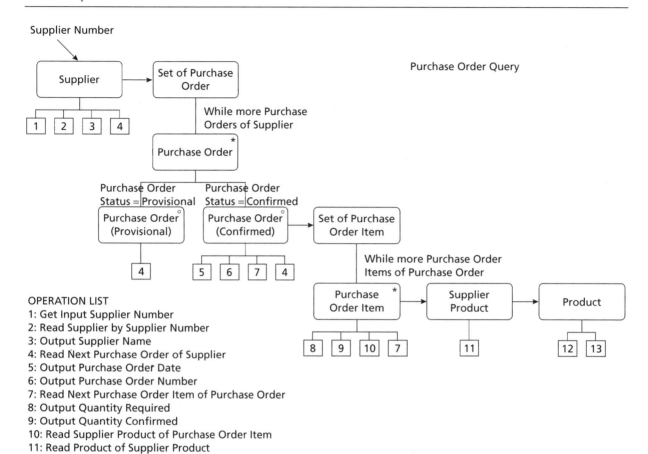

Fig 12.11 EAP for Purchase Order Query

OPERATION LIST
1: Get Input Supplier Number
2: Read Supplier by Supplier Number
3: Output Supplier Name
4: Read Next Purchase Order of Supplier
5: Output Purchase Order Date
6: Output Purchase Order Number
7: Read Next Purchase Order Item of Purchase Order
8: Output Quantity Required
9: Output Quantity Confirmed
10: Read Supplier Product of Purchase Order Item
11: Read Product of Supplier Product
12: Output Product Number
13: Output Product Name

The access to Supplier is described in detail by the operations that are attached to it:

◆ Get Input Supplier Number accepts the enquiry trigger that is input by the user via the function that invokes the enquiry.

◆ Read Supplier by Supplier Number is an instruction to find the one entity occurrence that matches the input Supplier Number.

◆ Having read an occurrence of Supplier it is possible to output to the user (via the function that invoked the enquiry) the required attribute instances of Supplier. Hence the operation: Output Supplier Name.

The arrow from the box named 'Supplier', to the one named 'Set of Purchase Orders' says 'Find all the Purchase Orders where the Supplier Number foreign key in Purchase Order matches the Supplier Number of the Supplier just found', i.e. find the Purchase Orders of the Supplier chosen by the user.

Because there is a one-to-many relationship between Supplier and Purchase Order there is a good chance that the enquiry will find more than one matching Purchase

Order, hence the use of the 'Set of Purchase Orders' node that launches us into the iteration. The box labelled Purchase Order (with an asterisk in the top right-hand corner) represents each matching Purchase Order found. Of course, there is the possibility that no Purchase Order will be found for the selected Supplier.

Each arrow on the diagram is described by a read operation. The arrow from Supplier to Set of Purchase Orders is described by the operation Read Next Purchase Order of Supplier. This operation appears three times. An iterated access on an Enquiry Access Path represents the one-by-one processing of a set of entity occurrences. To perform this processing the first occurrence has to be read, the processing of that occurrence carried out, and then the next occurrence read and processed. This loop continues until there are no more occurrences to be processed.

This enquiry only requires details of confirmed Purchase Orders. For this reason the Purchase Order access is a selection, with two options below it labelled 'Purchase Order (Confirmed)' and 'Purchase Order (Provisional)'. The Purchase Order (Provisional) access has no further arrows emanating from it, and only one operation: Read Next Purchase Order of Supplier. This means that if the enquiry finds a provisional Purchase Order it will do nothing except go on to read the next Purchase Order of the selected Supplier and see whether that one is provisional or confirmed (this cycle will continue until all the selected supplier's purchase orders have been processed).

When the enquiry finds a confirmed Purchase Order there is further processing to be done. The arrow from confirmed Purchase Order to set of Purchase Order Items says 'For each confirmed Purchase Order find all the Purchase Order Items which belong to it'; the iterated Purchase Order Item represents a one-by-one access to each Purchase Order Item; the arrow to Supplier Product says 'For each Purchase Order Item read the one Supplier Product which matches it'; and the arrow to Product says 'For each Supplier Product read the one Product which matches it'. When we arrive at Product we output the product's number and name, as required by the business rule.

▶ Drawing Enquiry Access Paths

SSADM suggests the following steps for drawing Enquiry Access Paths.

Define enquiries

Enquiries should already be documented in the Requirements Catalogue and each one should be named in at least one Function Definition. An enquiry should have a unique name and because we develop one Enquiry Access Path for each enquiry, we use the enquiry name as the title of the Enquiry Access Path.

Each enquiry should be entered on to the Entity Access Matrix and documented in the Event and Enquiry Catalogue, as discussed earlier.

We will take the Purchase Order Query as our initial example.

Note that the enquiry requests details 'for a given supplier'. In order that this enquiry can take place, the user will have to supply the system with the Supplier Number. If we looked at the I/O Structure or the screen design for the Purchase Order Query we would see that the Supplier Number appears as an input.

Data items that are input to an enquiry process so that the required information can be produced are known as the *enquiry trigger*. We should cross-check every

The EAP and ECD triggers will be used in Physical Design to provide the database entry points (see Figure 14.4).

Enquiry Access Path with the Function Definitions to ensure that any data that is required by an enquiry as a trigger is supplied by its related function.

Identify the entities to which we require access

We do this by looking at the required output of the enquiry, and identifying the entities that have the output data items as their attributes.

For the Purchase Order Query we will need access to Supplier, Purchase Order, Purchase Order Item and Product.

If the only information we required on Product was Product Number, it would be available as a foreign key in the Purchase Order Item entity and so we would not need to access the Product entity itself. However, the function's definition stipulates that Product Name is also output, so it will need to read Product too.

Draw the required view of the Logical Data Structure

Having identified the required entities we can draw an extract of the Logical Data Model containing only those entities. A good CASE tool will remove the need for this step by allowing the user to select the required entities from the Logical Data Structure and then automatically generating a first-cut Enquiry Access Path.

The required view of the Logical Data Structure includes the Supplier Product entity (Figure 12.12). None of the attributes of Supplier Product are required by the enquiry, but it provides an access path to Product.

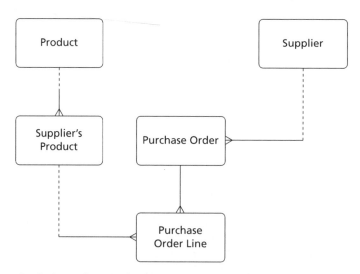

Fig 12.12 Required view of Logical Data Structure for the Purchase Order Query EAP

Develop an initial Enquiry Access Path

Enquiry Access Path diagrams use the same notation as Effect Correspondence Diagrams.

Accesses to detail entities from a master entity are shown as iterations. The entity name is placed in the iterated box and a parent box is placed above with the words 'Set of' followed by the entity name (Figure 12.13).

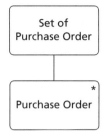

Fig 12.13 Accesses to detail entities are shown as iterations

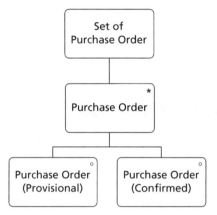

Fig 12.14 Adding selections to EAPs

Selections are added where necessary. Selections are used where the processing will follow a different path, depending on the state of the entity occurrence accessed. Each option of a selection will contain the relevant entity name, qualified in some way (see Figure 12.14). Note that a condition will be attached to each selection, and that the condition must have a way of being evaluated. In the example below, there must be an attribute of Purchase Order that tells us whether a purchase order is confirmed or provisional.

Where access is required from one entity to another we connect the relevant boxes with a single-headed arrow, indicating the direction of the access (Figure 12.15). When the access is from master to detail there will always be a 'set of detail entity' box at the head of the arrow. We try to have the arrows pointing horizontally, which has the effect of showing master-to-detail accesses cascading down the page. Applying these rules to Purchase Order Query results in the Enquiry Access Path shown in Figure 12.16.

Add entry points

There are two main types of entry allowed on Enquiry Access Paths: single-occurrence and multiple-occurrence.

◆ When the enquiry trigger contains the primary key of the entry point entity then only one occurrence of the entry point entity will be read.

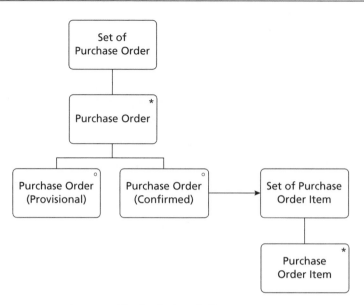

Fig 12.15 Accesses are shown with single-headed arrows

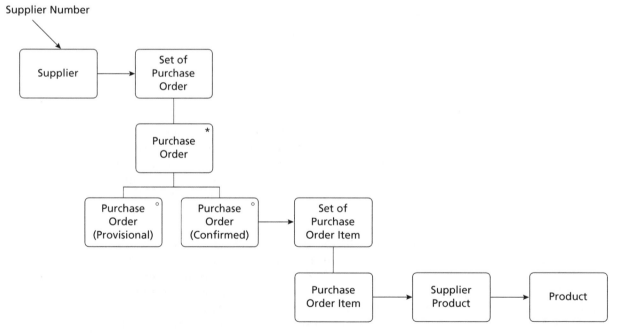

Fig 12.16 Purchase Order Query

◆ When the enquiry trigger contains non-key attributes of the entry point entity, or parameters for selecting occurrences of the entity within some range of attribute values, then many occurrences of the entry point entity will be read.

◆ A special case is where the enquiry trigger contains no selection criteria and all occurrences of the entry point entity are to be read.

In the Purchase Order Query example we are dealing with a single occurrence of Supplier, so Supplier Number is added to the Enquiry Access Path alongside the entry point arrow (see Figure 12.16).

If we wanted the enquiry to return details of confirmed purchase orders for all suppliers then the Enquiry Access Path would start like the one shown in Figure 12.17.

Entry via a foreign key is *not* allowed. If we identify a need to access an entity via a foreign key, we do so via the relevant master entity's primary key.

> If we only had the supplier's name (because we had somehow misplaced the Supplier Number) then the entry to the enquiry would look like Figure 12.18.

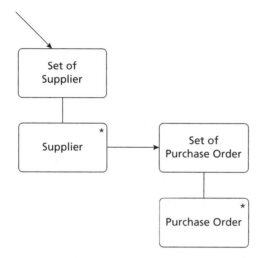

Fig 12.17 Selecting all occurrences of Supplier

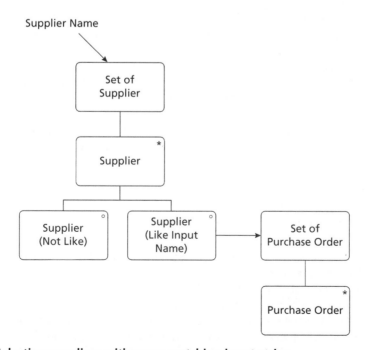

Fig 12.18 Selecting suppliers with name matching input string

Add operations

The operations that can be added to an Enquiry Access Path are a subset of those used on Effect Correspondence Diagrams. There will be one read operation for each arrow on the Enquiry Access Path. The allowable read operations are:

◆ read using the primary key (for a single-occurrence entry point);
◆ read next detail entity of a master (for an access to a 'set of');
◆ read the master entity of a detail.

If the Enquiry Access Path does not give us access to all required data using only these kinds of read, then we may need to introduce new relationships or attributes to the Required System Logical Data Model. In this way we may uncover business rules that have been missed from the Logical Data Model.

If we look again at the Enquiry Access Path in Figure 12.16 we find the following read operations taking place:

◆ direct read of Supplier using Supplier Number (the primary key);
◆ read next Purchase Order (a detail) of Supplier (the master);
◆ for each confirmed Purchase Order read the next Purchase Order Item (a detail);
◆ for each Purchase Order Item read the master entity Supplier Product;
◆ for each Supplier Product read the master entity Product.

All of these operations are of allowable types, and so the Enquiry Access Path reveals no obvious problems with the Logical Data Model.

One point to note is that the access to Supplier Product was made purely for navigation, i.e. as a way of reading the Product of the Purchase Order Item. The enquiry did not require any output from Supplier Product.

As it happens, there is a direct access path from Purchase Order Item to Product. Product Number appears in Purchase Order Item as part of the foreign key that has cascaded down from Supplier Product. This access path has not been shown as a relationship on the Logical Data Structure because it is redundant, but it would be more efficient to use it. This case is the one exception to the rule that a non-entry-point arrow on an Enquiry Access Path must be supported by a relationship on the Logical Data Structure.

As well as read operations we can also add operations to output the data items required by the enquiry. These data items will be attributes of the entities concerned, or derived from those attributes.

A note on the placement of operations

Operations should be added to each entity access, from left to right, in the order in which they are going to be executed.

To simplify the placing of operations on Enquiry Access Paths, we have adopted a convention that says that when a read operation is reached, the operations of the entity that is the subject of that read operation will be carried out. When these are complete, processing will return to the next operation after the read operation, where there is one.

Referring to Figure 12.11, this means that upon reaching the Read Supplier Product of Purchase Order Item operation, the Supplier Product access will be made, the Read Product of Supplier Product operation will take place, followed by the Output Product Number and Name operations. At this point we will return to the Read Next Purchase Order Item of Purchase Order operation and start the process all over again.

If preferred, extra structure boxes can be added to the Enquiry Access Path, and operations can be added to it just as they would on an Enquiry Process Model (but you might just as well draw an Enquiry Process Model in this case).

Strictly speaking, only the Enquiry Process Model, which is produced *after* the EAP, is a proper Jackson structure that is totally unambiguous in describing the order in which operations are to be executed. The placement of operations on Enquiry Access Paths is optional, although it is a good idea if the Enquiry Access Path is to be the final process specification, i.e. if Enquiry Process Models are not going to be developed. Since read operations are implied by arrows, the addition of read operations to an Enquiry Access Path could be considered unnecessary: a compromise might be to record only get, set and output operations on the Enquiry Access Path.

More than one access type per entity

In the above example each entity or set of entities was accessed just once. However, this is not always the case. For example, let us consider the following enquiry:

> For a given Product output the Product Number and Product Name and list all Products (Product Number, Product Name) that act as preferred substitutes.

In this case we require more than one visit to the Product entity (see Figure 12.19). The first access is a direct one using the primary key value provided. We will then need to access all of the Product Substitutes of the selected Product, via the 'For' relationship.

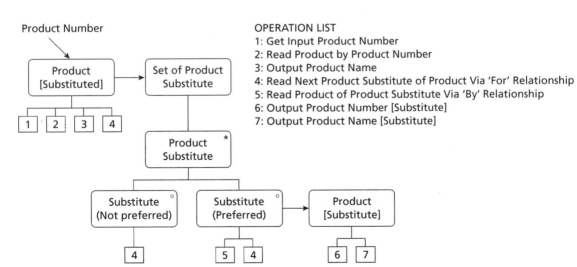

Fig 12.19 Enquiry Access Path for Product Substitute enquiry

For each 'preferred' Product Substitute we then read the Product occurrence which is related via the 'By' relationship.

We assume in this example that each Product Substitute instance has an attribute 'Preferred Substitute' which can take the value 'yes' or 'no'. There can be more than one preferred substitute for a given product.

Note that the operations for this Enquiry Access Path identify which of the two relationships between Product Substitute and Product is being used at any one time. If the read from the original Product was made along the 'By' relationship, we would find all Products for which that Product can be substituted and we would be answering a different enquiry.

Derived output data items

Our final example of an Enquiry Access Path (Figure 12.20) involves the calculation of a derived data item. A derived data item is one that is not stored as an attribute of an entity type, but is calculated from stored attributes.

In our example we suppose that a purchaser requires a report that lists the total value of all orders placed with a given supplier.

There are three operations to take note of:

◆ Operation number 4 declares a variable named 'Total Value' and sets its value to zero. This has to happen once, at the beginning of the enquiry, which is why the operation has been placed on the Supplier access.

◆ The next operation of note is operation number 6. This calculates the value of the Purchase Order Item and adds it to the Total Value. This operation has to happen once per Purchase Order Item, which is why it is placed on the Purchase Order Item access.

◆ Finally, when all the Purchase Order Items of the selected Supplier have been processed, we have to output the Total Value. This will happen once per running of the enquiry, which is why operation number 7 appears on the Supplier access.

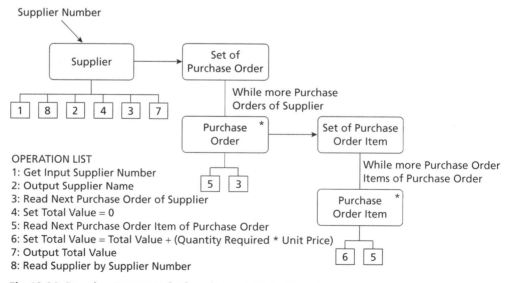

OPERATION LIST
1: Get Input Supplier Number
2: Output Supplier Name
3: Read Next Purchase Order of Supplier
4: Set Total Value = 0
5: Read Next Purchase Order Item of Purchase Order
6: Set Total Value = Total Value + (Quantity Required * Unit Price)
7: Output Total Value
8: Read Supplier by Supplier Number

Fig 12.20 Enquiry Access Path showing calculated totals

▶ ENQUIRY PROCESS MODELS

> **Important note**
> As with Update Process Models it is quite likely that the Enquiry Process Model adds no value to the specification. Once again it depends on the implementation environment that is being used.

An Enquiry Process Model is a Jackson-like structure that describes an enquiry in a procedural manner.

▶ Enquiry Process Modelling

Enquiry Process Modelling is carried out in exactly the same way as Update Process Modelling, except that the starting point is an Enquiry Access Path rather than an Effect Correspondence Diagram.

We will illustrate the conversion of an Enquiry Access Path into an Enquiry Process Model by taking as our starting point the Enquiry Access Path of Figure 12.11.

Group accesses on the EAP

As with update processes, we are interested in grouping together processing that can be carried out unconditionally as a single module. We do this by grouping

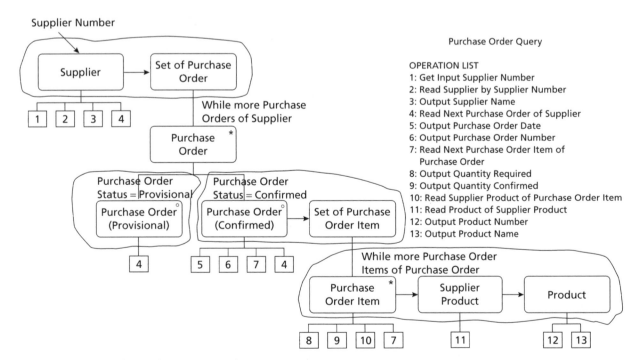

Fig 12.21 Grouped Enquiry Access Path

those accesses on the Enquiry Access Path that are in one-to-one correspondence, i.e. those that are linked by a single-headed access arrow.

All such groups should be given a meaningful name, and any accesses that stand alone will form a group of their own.

Figure 12.21 illustrates the addition of access groups to the Enquiry Access Path for Purchase Order Query.

Convert to Jackson-like structure

We now convert each grouping to a box on a Jackson-like structure. We follow the same procedure as for Effect Correspondence Diagram conversion. In this example we have added some extra structure boxes. These ensure that the structure conforms to the normal rules of Jackson structures, with which the reader will be very familiar by now.

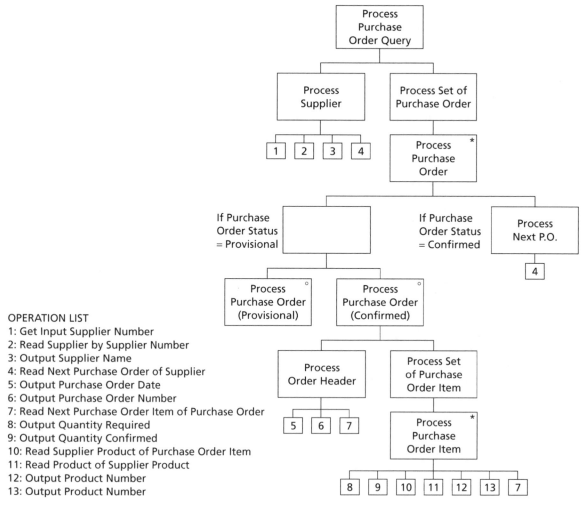

OPERATION LIST
1: Get Input Supplier Number
2: Read Supplier by Supplier Number
3: Output Supplier Name
4: Read Next Purchase Order of Supplier
5: Output Purchase Order Date
6: Output Purchase Order Number
7: Read Next Purchase Order Item of Purchase Order
8: Output Quantity Required
9: Output Quantity Confirmed
10: Read Supplier Product of Purchase Order Item
11: Read Product of Supplier Product
12: Output Product Number
13: Output Product Number

Fig 12.22 Enquiry Process Model for Purchase Order Query

Allocate operations and conditions to processing structure

Operations are carried across from the Enquiry Access Path to the Enquiry Process Model just as they were from the Effect Correspondence Diagram to the Update Process Model.

Care should be taken with the placement of *read* operations for an iteration. We need an initial read operation to access the first record. If no record exists then the condition governing the iteration evaluates to *false*, otherwise processing of the first record can take place, at the end of which the next record has to be read. Thus, each iteration is preceded by a *read next* operation, and the same *read next* operation is repeated as the last thing in the sequence under the iteration. Careful readers should note that this rule explains why operations 4 and 7 each appear twice on the Enquiry Process Model of Figure 12.22.

▶ SUMMARY

1. Effect Correspondence Diagrams illustrate which entities are affected by a given event. We also use them to define how the effects on these different entities correspond with each other, and ultimately to provide us with a specification of *update* processing. The development of Effect Correspondence Diagrams helps to clarify the work done during Entity Life History analysis, and often the two techniques are used in tandem in order to validate each other.

2. Enquiry Access Paths will document the required data model accesses resulting from an enquiry trigger (the non-update equivalent of an event), using the same notation as that used for Effect Correspondence Diagrams.

3. Enquiry Access Paths and Effect Correspondence Diagrams are logical models of the database processing. As such they are implementation-independent, are portable and can be mapped to any implementation environment.

4. Update and Enquiry Process Models are also logical models but they are more suited to a specification that is going to be implemented using a traditional procedural 3GL.

5. The act of developing Conceptual Process Models forces you to work closely with the Logical Data Model and provides some further validation and verification of the requirements.

6. The approach outlined in this book ensures that the major states of entities are defined and that these are checked before any update processing is allowed to take place. This helps ensure the integrity of the data in the information system.

▶ EXERCISES

12.1. Using the write-up for Exercises 3.15 and 11.2 produce an ECD for the event Respond to Estimate in the Bodgett & Son system. The ELHs for Bodgett & Son reveal that the entities Estimate and Job are affected by Respond to Estimate.

Assume that the Current Environment LDS from Exercise 5.10 still applies in the Required System.

12.2. Using the write-up for Exercise 11.1 produce an ECD for the event Arrange Session.

Hint: The LDS from Exercise 5.2 will provide the basis for the ECD structure but you will need to consider the effects of this event on any link entity between Session and Equipment.

12.3. Produce an EAP for the Treebanks enquiry report in Exercise 9.1, using the LDS from Exercise 5.2.

12.4. Produce EAPs for the reports in Exercise 9.5.

12.5. Use the Entity Life Histories of Exercise 11.4 to develop Effect Correspondence Diagrams for the Video Store events Copy Purchase, Video Rental, Copy Sale or Destruction, and Store Closure. For Copy Sale or Destruction the event data is Title Code, Copy Number; for Store Closure the event data is Store Number [closing store] and Store Number [opening store].

12.6. Produce Update Process Models from the ECDs of Exercise 12.5.

12.7. Draw Enquiry Access Paths for the following Video Store enquiries from Exercise 11.4: (1) List all stores. For each store give store number, store name and manager name. (2) For a given member, find the number of copies currently on rental, their title names, and rental dates. (3) For a particular store, list all its stock.

12.8. Produce Enquiry Process Models based on the EAPs of Exercise 12.7.

12.9. Produce an Update Process Model for the Bodgett & Son event 'Respond to Estimate' using the solutions to Exercises 11.2 and 12.1. Add relevant operations from the Entity Life History for Estimate, and try to make reasonable guesses as to other operations and to appropriate conditions (the Logical Data Structure from Exercise 3.15 may be helpful).

12.10. Produce an Enquiry Process Model for the Treebanks enquiry used in Exercises 9.1 and 12.3. Again make reasonable guesses about operations and conditions.

12.11. Develop an Enquiry Access Path that will produce the West Munster Hotel customer's bill shown as part of Exercise 8.2.

12.12. Turn your Enquiry Access Path from Exercise 12.11 into an Enquiry Process Model.

Physical Design considerations

An important objective of the Specification phase is to produce a Logical Design that is as far as possible implementation-independent. This provides us with a product that is portable across different hardware and software environments.

In Technical System Options we address in detail the question of how the new system is to be implemented in terms of the technical environment it will operate in, the development approach we will use to build it and the need to get the best possible value for money from the new system.

The development of detailed Technical System Options (TSOs) builds on the work carried out in defining Feasibility Options and Business System Options, and can be initiated as the Requirements Specification nears completion.

Our aim in Physical Design is to map the logical system specification onto the target environment specified in the selected TSO.

The tasks to be undertaken during Physical Design are dictated by the technical environment in which the system is to be implemented. As the tools and approaches available in each technical environment will vary greatly, SSADM (in common with other methods) can provide only generic guidelines that can be applied or adapted to a wide range of environments.

▶ Default structure

Define Technical System Options

Constraints on the choice of technical environments are established in order to provide a realistic framework for the definition of TSOs.

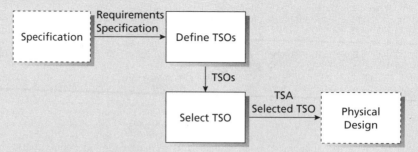

Fig P4.1 Technical System Options

A number of TSOs are outlined and discussed with users and technical experts. One or more of the TSOs will be short-listed for further expansion.

Select Technical System Options

The expanded TSOs are presented to information systems management and the Project Board who, with assistance from the project team, will select a final TSO, which may be a hybrid of more than one of the proposed TSOs.

The selected TSO will then be developed further to provide a detailed Technical Systems Architecture for use in Physical Design.

Prepare for Physical Design

The implementation environment is studied and its facilities are classified and documented (with emphasis on its strengths, weaknesses and optimisation mechanisms), Application Development Standards are drawn up, and a Physical Design Strategy agreed with management.

Create Physical Data Design

The Required System Logical Data Model is converted to a first-cut data design using techniques based on assumptions common to many database management systems. Product-specific rules are then used to produce an environment-specific data design.

Optimise Physical Data Design

The first-cut data design is tested against the performance objectives set out in the Requirements Catalogue and Function Definitions. If necessary, the design is optimised, using the facilities of the implementation environment (rather than by compromising the data structure wherever possible).

Create Function Component Implementation Map

The specification of functions is completed with the addition of physical components such as system error handling routines.

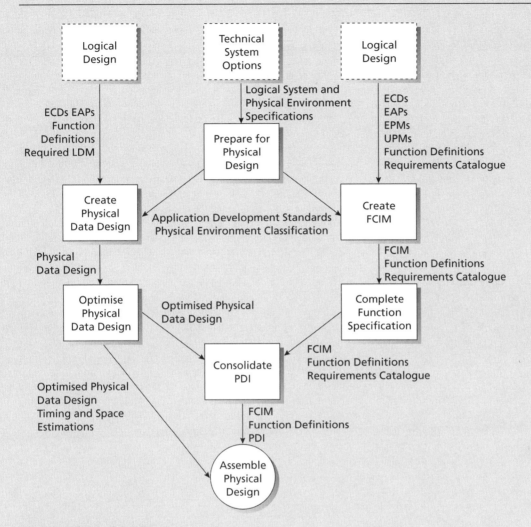

Fig P4.2 Physical Design

A Function Component Implementation Map is drawn up listing the components of each function and how they map onto physical function components. Non-procedural function components are implemented.

Complete Function Specification

Program specifications are produced for any function components that are to be implemented in procedural code.

Consolidate Process Data Interface

The optimised Physical Data Design is examined to identify where mappings are required from logical database operations to physical database operations. A Process Data Interface is developed to handle these mappings if necessary (in most relational database environments the database management system itself will fulfil the role of the PDI).

Technical System Options

INTRODUCTION

Before we embark on the physical construction of the new system, we need to address in detail the question of how the new system is to be implemented in terms of the technical environment it will operate in, the development approach and tools we will use to build it and the need to get the best possible value for money from the new system.

While defining the technical environment and standards, we may uncover information that is of such significance it requires adjustments to user requirements (and consequently to the logical design of the system). The examination of technical issues carried out previously in the project, as part of the Feasibility Study or in preparing BSOs, should limit the extent of any such adjustments. However, it is common for technical limitations to force changes to, or even the dropping of, functions that turn out to be impossible or impractical to implement using any of the available technical options.

We develop and select Technical System Options (TSOs) in much the same way as we did for Business System Options. During the Feasibility Study (if we conducted one) and the BSO definition process we will have already established a high-level view of the target technical environment. Our task now is to refine this view, and specify exactly how this environment is to be set up and operated, and to put in place plans for its purchase and construction. Most BSOs will also have left gaps in the technical environment description, in areas where the BSO decision was not affected. For example, the choice of the precise tool to be used for monitoring system performance will not affect the functionality of the required system, and is unlikely to greatly affect its overall cost. In such cases we will have noted in the BSO that a component is needed, but the actual selection of the component will be made as part of the TSO process.

> **Note**
>
> In reality, even if our BSO has identified a precise set of technical products and standards (for example where a package has been selected that specifies certain components to be used), there will still be a lot of work to do in establishing just how these products are to be used, what capacity is required, tuning the environment specification for optimum performance, and in ensuring that all components are identified.

One key difference between TSO and BSO selection is who makes the final decision. It is very unlikely that the Project Board will have sufficient technical awareness to make decisions on the technical environment or system construction methods. In these areas the management or policy-making teams with the Information Systems department will provide the decision-making body, as it is they who are responsible for the IS standards and technical strategy. The Project Board's role will be limited to assessing the impact of the technical decisions on the objectives, functionality and costs of the project.

The output from the TSO selection process is a full description of the technical environment, known as the *Technical Systems Architecture*.

> ▶ **Learning objectives**
>
> After reading this chapter, readers will be able to:
>
> ◆ understand the contents of Technical Systems Options and Technical Systems Architectures;
>
> ◆ understand the steps required to define and select the Technical Systems Option.

> ▶ **Links to other chapters**
>
> ◆ Chapters 2 and 7 The Feasibility Study and Business System Option will have defined high-level TSOs, which will act to constrain or steer the development of the chosen Technical System Architecture (TSA).
>
> ◆ Chapter 14 The TSA will specify the environment in which the Physical Design is to be implemented.

▶ DEVELOPING TECHNICAL SYSTEM OPTIONS

The Requirements Specification we have developed in such detail was completed with little consideration of technical issues, outside of basic checks of technical feasibility. An underlying assumption of SSADM is that an understanding of the user's requirements is best kept separate from a discussion of the specific hardware and software upon which the system is to be implemented.

Having defined *what* it is that we want the system to do, we can now turn our attention to *how* the system is going to do it. Thus it is the purpose of Technical

System Options to identify the possible ways of physically implementing the Requirements Specification. In so doing, we must try to ensure that each proposed technical solution is capable of meeting the service-level requirements that are spelt out in the Function Definitions.

SSADM, in common with all IS development methods, does not provide a set of specific techniques that can be applied to all possible technical and organisational scenarios. This is not really surprising, as the number of possible combinations of hardware, software and implementation strategies is almost limitless. Instead, SSADM provides guidelines on issues that should be addressed at this point, whatever the circumstances; and procedures that should be carried out as part of any selection process.

The TSOs that we develop will need to provide information and planning in five main areas:

1. **Technical environment.** It is possible that if no feasibility study has been carried out, or if the organisation does not have an overall technical strategy, this may include: consideration of the type of technical platform we require (e.g. UNIX or NT); which user interface technology will be adopted (e.g. Windows or browser); or what database will be used (e.g. Oracle or MS Access). Often, decisions of this type will already have been made, possibly as part of BSO development, in which case our concern is with the specifics of hardware and software configuration and procurement.

2. **Development strategy:** whether, for example, the system or some of its technical components are to be built in-house, or by some outside agent. Again, this type of decision will probably have been made earlier in the project (not least as part of the selected BSO). However, once the logical design is completed, and the technical environment is specified, we may find that there is a different mix of skills required for system construction from that envisaged in the BSO.

3. **Organisational impact.** This will include impact on operational working practices (e.g. the job of the picker in the ZigZag case study), other projects and even the IS department itself (e.g. new technical environment components that will need support and training). The advent of e-commerce, for example, will have a significant impact on the staff profiles of both the main business and the information systems function.

4. **System functionality.** Although this is addressed at the BSO stage, the selection of a technical environment may affect the viability of some functions, or the way in which a function is to be carried out.

5. **Procurement.** Once the technical environment has been specified, the necessary hardware and software will need to be ordered and contracts drawn up for its supply and support.

The skills needed for establishing or assessing TSOs include capacity planning and system sizing; hardware and software procurement; and the knowledge of current and anticipated technologies. These skills lie outside the scope of SSADM and it is highly unlikely that all of the skills required to carry out the required tasks will lie within the project team. We will almost certainly need to go outside the organisation to talk with suppliers of hardware and software to assess the capabilities of any proposed technical environments.

Figure 13.1 shows the process of choosing a TSO with all the likely iterations included.

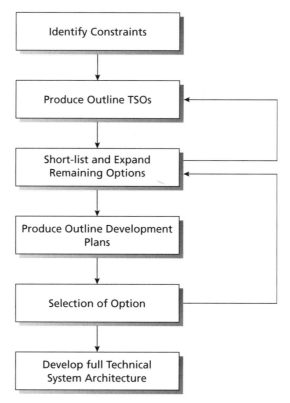

Fig 13.1 Choosing a TSO is an iterative process

We will only follow a process of this nature in cases where the PID, Feasibility Study or BSO has left the question of the technical environment or construction method sufficiently open. In many projects we will already have a pre-selected outline TSO description, in which case our task will be to expand on this pre-selected option, filling in any gaps and proposing solutions for those elements that have been left open in the outline description.

As with BSOs, the Project Board or IS management may find none of the proposed TSOs entirely acceptable, and request the development of new options, or the refinement of a presented option, before final selection is made. In this way we may get iterations of the process in Figure 13.1.

▶ Identify constraints

Before developing options for expansion and selection we need to establish the constraints on our development of possible technical environments and on how the system is to be constructed and implemented. After all, to take an extreme example, it would be pointless to waste time in developing plans to implement the system on a UNIX platform using an outside software house if the organisation has an NT technical policy requiring in-house developments using a specific 4GL.

In general constraints fall into two categories:

◆ **External:** outside or global constraints applicable to the project as a whole.

◆ **Internal:** constraints imposed by the users on specific areas within the project.

External constraints

These should have been noted in the Business Systems Options stage and should already be documented in the Requirements Catalogue (as indeed should internal constraints). The kinds of constraint commonly encountered in this area are:

◆ **Hardware and software platform.** On one project we may find that a specific language and operating system must be used, and that hardware must be of a particular type purchased from a pre-specified manufacturer. On another project we may have freedom to suggest hardware types, manufacturers and development software.

◆ **Organisational policy.** There may be policy on the use of packages, software houses, facilities management, etc.

◆ **Time.** The required delivery date for the new system will often have an enormous influence on implementation planning, possibly ruling out entire strategies.

◆ **Cost.** Most projects will have a maximum cost and/or minimum savings figure imposed on them.

External constraints may have been set up long before the TSO stage, perhaps as part of the PID. For this reason it is worthwhile confirming that they still apply and whether they are negotiable, as they are likely to dictate the nature of all the proposed TSOs.

Internal constraints

During our analysis of user requirements we will have uncovered a growing number of non-functional project-specific requirements, which will now act to constrain our choice of technical options. Possible areas of internal constraint include:

◆ **Service levels**, e.g. percentage availability of system, recovery time (in case of system failure), contingency measures.

◆ **Performance**, e.g. response time.

◆ **Capacity**, e.g. maximum number of users/transactions, data storage volumes.

◆ **Security.**

◆ **Priority.** All TSOs will inevitably lead to compromises in some areas in order to achieve targets in others. Therefore, we should identify which areas have priority or whose performance is critical.

▶ Produce outline TSOs

We now investigate possible TSOs using the identified constraints. One of the best ways of identifying potential TSOs is to use brainstorming sessions involving both the project team and outside technical experts. We will also often need to contact suppliers for information on costs, facilities and configurations.

Our aim should be to produce up to six outline TSOs, each satisfying our requirements and constraints. Generating six options at this point might appear difficult, but the main problem may actually be keeping the number that low. Even if constraints appear very rigid, there is almost always room for adjusting configurations or balancing costs in different areas. For example, we might suggest installing additional system memory, at the expense of slightly slower or poorer printing.

If there is a current computer system we might propose the use of existing hardware only, or even a *no change* option, resulting in termination of the project.

▶ Produce short-list of TSOs

A fair amount of effort can go into fully defining a TSO, so we should try to reduce the number of TSOs to around three.

Users must be involved in this short-listing process, as it is they who will eventually have to live with the implementation. To help them we should produce outline impact analyses and attempt to quantify the benefits and drawbacks of each option.

In reality we are unlikely to come out of the short-listing process with a short-list consisting of three unchanged TSOs, but rather with a number of combined options, or variations on the same basic architecture.

▶ Expand short-listed TSOs

We now develop all of the short-listed TSOs to a level that will enable them to be fully evaluated, and a final selection of TSO made. This will involve us in applying several non-SSADM techniques such as capacity planning and risk assessment in order to test the viability of the options.

Each TSO specification should contain the following components.

Outline Technical System Architecture (TSA)

As an input to the selection of TSOs the TSA will only be an outline of the proposed environment. After selection the TSA will be fully defined to provide the basis for translating the logical system design into a physical design ready for construction.

Outline TSAs give us an idea of how the system will operate, of how it will be configured, and of its likely cost. It will help to make the proposed configurations clearer if we draw diagrams illustrating how the different hardware (and software) components will interconnect. We will also need to include information on system sizing, security and back-up arrangements, and maintenance costs.

System Description

As we develop the TSOs we will need to balance the various constraints, with possible trade-offs between costs and performance, or development time and functionality (although any reduction in functionality must be considered as a final option only).

Each System Description must describe the functionality that is met by the option it describes. We develop this description by modifying existing SSADM products, such as the Logical Data Model and Function Definitions. The significance of each option can be made clearer by highlighting in the System Description any functionality that is *not* being met.

Impact analysis

We should explain the impact of the TSOs on the organisation using various non-SSADM techniques and products. Selection of the final TSO will often hinge on a comparison of the impact of each TSO, so we are likely to put a lot of work into this area.

Impact analysis should include consideration of the following:

◆ organisational and personnel changes;
◆ operating procedure changes;
◆ training requirements;
◆ system documentation (e.g. user guides);
◆ system take-on requirements. Installation (often referred to as 'take-on' or cut-over) is often more complicated for TSOs based on the existing technical environment, than for entirely new TSOs;
◆ cost savings;
◆ testing requirements;
◆ relative merits or drawbacks of each TSO (e.g. in the areas of reliability, performance, implementation time, costs and functionality).

Outline development plan

For each TSO we will need to produce an outline development plan for use by project management in estimating and scheduling the rest of the project. It is here that we will also outline the likely development costs of each TSO and propose how the option would be developed and implemented.

Issues to be covered include:

◆ overview plan for the remainder of the project;
◆ detailed plans for the Physical Design stage;
◆ human and system resource requirements;
◆ construction strategy (e.g. will we use contractors?);
◆ system testing requirements;
◆ cut-over plans.

Cost–benefit analysis

The results of CBA are frequently crucial to the adoption or total elimination of TSOs, so accurate collection and estimation of data are extremely important.

Costs will include:

◆ development and implementation costs;
◆ operating and maintenance costs.

Benefits or savings will include:

◆ displaced costs (costs of operating the current system that will not exist for the new one);
◆ avoided costs (costs of continuing with the current system, e.g. increased maintenance to deal with business expansion);
◆ tangible and intangible benefits.

▶ Select Technical System Option

The tasks and organisation of TSO selection are very similar to those of BSO selection.

For tips on presenting options, please refer to Chapter 7 – Business System Options.

We will present each TSO to IS management and/or the Project Board (supported by appropriate technical advisers) in a way that emphasises their relative advantages and disadvantages, in areas such as:

◆ development, procurement, installation and operating costs;
◆ ease and speed of development;
◆ performance;
◆ skills availability;
◆ risk;
◆ system availability, reliability and recovery.

The selected TSO may, as with BSOs, be a hybrid, in which case we must be prepared to carry out further capacity planning and impact analysis to ensure that it still lies within the identified constraints.

▶ Develop full Technical System Architecture

Once selection is complete we need to extend the outline TSA with further details of hardware, software, sizing and operating procedures. Issues that need to be addressed by the TSA are outlined as follows.

Hardware

In addition to descriptions and diagrams detailing the types, numbers and locations of hardware devices, we should cover:

◆ communications, networking and cabling;
◆ installation requirements and responsibilities;
◆ fault tolerance and back-up and disaster recovery facilities;
◆ maintenance arrangements, including upgrading and technical support;

Software

Descriptions of the following types of software are needed:

◆ database management system (DBMS);
◆ application packages;
◆ development software, including programming and interfacing tools;
◆ operating systems;
◆ systems monitoring and management facilities;
◆ communications software.

System sizing

When defining a function we have to make a statement of the service level requirement. For instance, for an on-line function, we might specify a response time of five seconds or less. At the same time we have some system-wide service level requirements, such as the hours of availability of the system.

With many users accessing the system at the same time, and with many thousands of records in our database, and with the overheads of network communications, we have to be sure that we select hardware and software that can cope with the large number of transactions taking place and that it is possible to meet all the service level requirements.

In order to reduce the risk of the new system's not performing to our expectations we employ a technique called *capacity planning*. This is not a core SSADM technique but SSADM analysts have some responsibility for it. They share this responsibility with specialists in the area of capacity planning.

The systems analyst will have to provide information on the following:

◆ target values and acceptable ranges for each functional requirement;
◆ the frequencies of each event and enquiry, highlighting any peaks or troughs in the frequency of these transactions;
◆ number of disk accesses and CPU time for each transaction;
◆ the size and number of occurrences of each entity. These are used to calculate the data storage requirements.

Capacity planners will use this data to create workload models and to evaluate the hardware configurations of any particular Technical System Architecture. The resulting predictions of system performance can be used to decide if a proposed TSA can meet the required service levels. If it cannot do this then the hardware configuration can be revised – otherwise the service level requirements must be renegotiated.

Operating procedures

With the acquisition of any new hardware or software the existing procedures for operating, monitoring and managing systems within the organisation will need reviewing. At the very least, new operating procedures will be needed to manage the new system. Frequently, a new system will introduce new components to the overall systems infrastructure that will require new skills and ways of working to be developed. The areas that need to be addressed include system performance monitoring, data back-up and restoration procedures and general housekeeping (e.g. data archiving).

▶ Completing the documentation

At this point it is not absolutely essential to commit a project to a specific supplier for every hardware and software component, as some of these components will not be required until well into the construction phase of the project. However, procurement procedures can be very drawn-out, so it will be in the project's best interest to ensure that delivery of the development environment can be made in time for the end of Physical Design by selecting suppliers as early as possible. Where the selection of technical supplier involves a formal invitation to tender (ITT) approach, the selection process may need a significant amount of time to complete. In these cases the selection would be best initiated at BSO stage if possible.

The system description and impact analysis developed as part of the selected TSO should be checked to ensure that they are complete and accurate following TSO selection, and incorporated within the TSA, to ensure that all of the documentation relating to implementation issues is in one place for input to Physical Design.

If the Project Board decides that the required system cannot be justified using any of the TSOs it may call a halt to the entire project. Otherwise it might decide to reduce service level requirements or drop some areas of functionality. In this case we would probably need to substantially update some SSADM products, such as the Requirements Catalogue, Function Definitions or the Required System LDM.

In any event, the decisions made by the Project Board must be recorded for future reference, along with the TSA.

As soon as the final implementation environment is known, we can begin work on producing an application-specific style guide based on the organisation's standard style guide.

Application Style Guide

Anyone familiar with the Microsoft Windows interface will understand the benefits of having a consistent look and feel across different applications. The user of a Windows word-processing application should find that they already know how to save a file, copy, paste or format text, or print a document in a Windows application that they have not used before. These same benefits will apply to any system.

There are benefits for developers too: design effort is reduced, and component libraries reduce the amount of code that has to be written.

Now that the target software platform has been established, we can ensure that these benefits are achieved by creating an Application Style Guide that defines standards that developers should adopt for things like:

◆ menus, check boxes, forms and dialogue boxes;

◆ user help;

◆ use of colour;

◆ tailoring of the interface (for or by the different users);

◆ use of keyboard short-cuts and mouse actions;

◆ font size, headers and footers on reports.

An Application Style Guide is of particular importance when new web pages are launched, since a best-practice look-and-feel has yet to emerge for the World Wide Web. For example, the number of frames to be used, the positioning of advertising and the manner in which an item is added to the shopping basket are issues that need to be clarified well in advance of implementation.

▶ SUMMARY

1. Technical System Options (TSOs) describe potential technical environments and construction approaches for the implementation of the system. TSOs will typically cover:

    ◆ technical environment

    ◆ development strategy

    ◆ organisational impact

    ◆ system functionality

    ◆ procurement.

2. A number of TSOs will be developed and presented to the Project Board and information systems management for selection. TSO development will be carried out in parallel with Logical Design and the selected option formally signed off before Physical Design begins.

3. The output from the TSO process is the Technical System Architecture (TSA), which contains a system description, impact analysis, development plan and cost–benefit analysis. Issues that the TSA must address include hardware and software selection, system sizing, and operating procedures.

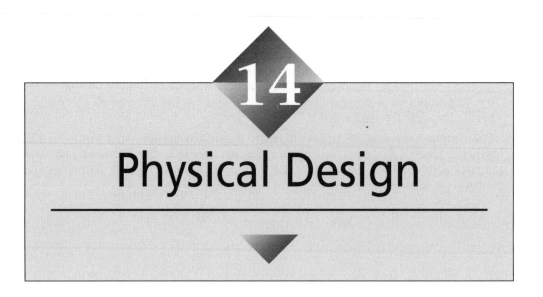

Physical Design

INTRODUCTION

The products of Logical Design provide a system specification that could be implemented in a variety of technical environments. The Technical System Architecture resulting from the TSO process will identify the chosen implementation environment. Our aim in Physical Design is to map the logical system specification onto the target environment.

By the end of Physical Design all system components will either have been specified to a level of detail that will enable programmers to build the system or, in the case of some of the non-procedural components, will actually have been implemented. In transforming the logical design into a physical design we will need to balance various competing factors such as performance, space usage and development time in order to achieve the optimum implementation. Throughout this process we will aim to preserve as direct a mapping as possible between the logical and physical designs.

To help co-ordinate and document the mapping of logical functions onto physical function components we will develop a Function Component Implementation Map (FCIM), while on the data side any mismatches between the logical database and its final physical implementation will be handled by a translation component or tool called a *Process Data Interface* (PDI).

We will then be able to design physical processing that acts as if the logical and physical databases are one and the same, with any mappings between the two being handled by the PDI. In most relational database environments the database management system itself (using query languages such as SQL) will fulfil the role of the PDI.

As the specific tasks to be undertaken during Physical Design are dictated by the chosen technical environment, SSADM (in common with other methods) can provide only generic guidelines on Physical Design, which can be applied or adapted to a range of environments. These guidelines are separated into data and process design activities, although in many environments the two are interdependent. In some 4GL or application generator environments the developers may have little or no control over how the system is physically built or tuned (particularly in the area of databases); in these cases the guidelines will be largely redundant.

There are a few physical design tasks that are applicable to a reasonable range of technical environments. Where this is the case, SSADM offers a set of more rigorous techniques (notably those of first-cut data design), and these will be discussed in greater detail during this chapter.

Physical Design can be highly technical and the involvement of database and programming specialists is likely to be essential (particularly in 3GL environments). However, the continuing presence of analysts and users from earlier in the project is also crucial to ensure that the final design satisfies user requirements.

▶ Learning objectives

After reading this chapter, readers will be able to:

◆ prepare for Physical Design by examining and classifying the facilities of the physical environment;

◆ produce a first-cut data design;

◆ develop a Function Component Implementation Map;

◆ understand the concept of a Process Data Interface;

◆ identify what non-SSADM techniques may be needed to complete the specification of the Physical Design.

▶ Links to other chapters

◆ Chapter 4   The Requirements Catalogue provides information on volumes and performance criteria that the Physical Design must meet.

◆ Chapter 5   The Logical Data Model forms the basis for the Physical Data Design, and the model against which the processing components will be specified.

◆ Chapters 8 and 9   During Physical Design specifications are developed for each Function Definition and User Interface Design.

◆ Chapter 12   The Conceptual Process Models provide the logic for program specifications.

◆ Chapter 13   The Technical System Architecture defines the physical environment that the Physical Design will be developed for, and in which the system will be implemented.

▶ PHYSICAL DESIGN PROCESS

The Physical Design process is described in the sections that follow.

As illustrated in Figure 14.1, we begin with a number of preparation and planning activities specific to the chosen technical environment. We then move on to two parallel streams of activity that translate the logical designs we have created for data and processing into physical specifications (for components that need to be built or programmed by hand), or actual physical system deliverables for non-procedural components (such as screens that can be generated directly from user interface designs, rather than translated into a program specification).

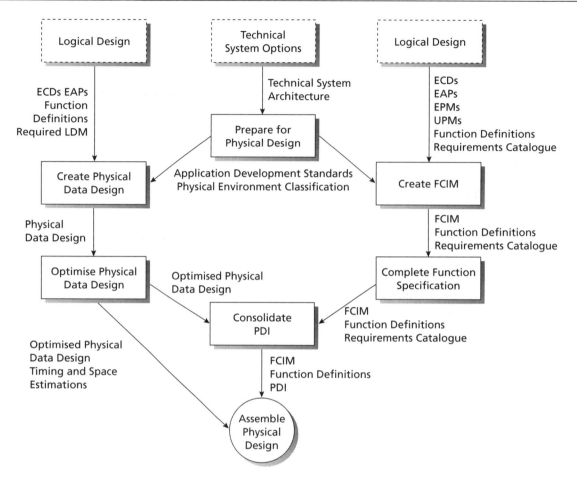

Fig 14.1 Activities of Physical Design

▶ PREPARING FOR PHYSICAL DESIGN

Before Physical Design begins in earnest there are some basic preparations to be carried out. These are intended to equip the project team with as great an understanding of the physical environment as possible, and to set a strategy for the entire Physical Design process. Where a lot of specialist help is available there may be a great deal of knowledge already within the team on how to exploit the strengths and overcome the weaknesses of the physical environment. However, it is still the job of the analysts present to gain an overall understanding of its facilities in order for them to be able to participate in Physical Design.

▶ Classifying the physical environment

The facilities offered by the physical environment will clearly affect the extent to which the Logical Design can be directly preserved in the Physical Design, and therefore determine the need for software to map between the two.

SSADM recommends that a relatively formal classification of facilities be undertaken to ensure that they are fully understood, and hence can be fully exploited. This classification is divided into three areas: DBMS Data Storage, DBMS Performance and Processing System.

In order to help classify facilities SSADM provides a checklist of mechanisms that are common to many database management systems. The idea then is to examine the physical environment to identify the way in which it supports these mechanisms.

> In practice, database and programming specialists or the suppliers of hardware and software will provide most of the information covered by these areas.

Note

In many environments (notably non-procedural or 4GL) these mechanisms will be present but may be entirely hidden from the developer. Nevertheless it is worth checking if any of them are visible and so capable of manipulation.

DBMS data storage classification

This concentrates on establishing how data is stored, accessed and updated. It is also crucially concerned with how relationships are implemented. The basic questions to be answered are:

◆ How are relationships represented?

◆ Where is relationship data held?

◆ Are relationship keys symbolic or physical, i.e. based on attribute values or physical pointers?

◆ What mechanisms are used to locate records?

◆ Can the database management system place related records near each other physically?

◆ What restrictions does the database management system place on the Physical Design? For example, on the maximum number of tables or level of the detail–master hierarchy.

DBMS performance classification

This concerns questions relating to the timing of database management system operations. In order to satisfy non-functional performance requirements we will often need to adjust our physical data design to balance the time overheads of these operations.

Time overheads may be associated with the following:

◆ transaction logging;

◆ recovery logging, i.e. the automatic backing up of the database at regular intervals, so that it can be restored to an uncorrupted state if a disaster occurs;

◆ standard operations: read, write;

◆ commit mechanisms;

◆ space management;

◆ data sorting (on input or for output).

Once the DBMS classifications have taken place it is a good idea to design forms for recording the timings and space requirements of the Physical Design. These will then be used later in testing that design against performance objectives.

Processing system classification

The final area of classification is concerned with identifying the features and facilities of the processing environment. These may differ for development and production, e.g. if different hardware is used, or if special development 'front-ends' are provided, and so we may need to carry out two classifications.

Our major aim is to understand how each function component (dialogues, logical processes, etc.) can be implemented, with emphasis on whether the mechanism is procedural (requires formal programming) or non-procedural (can be generated using tools such as screen painters).

Points to be considered include:

◆ What tools are available and what can they generate? (For example, screen or report generators.)
◆ Can procedural and non-procedural code be mixed?
◆ Can on-line and off-line function components be mixed?
◆ What error handling facilities are available?
◆ How are success units defined?
◆ What database access facilities exist?
◆ How can data be grouped on screens and reports?
◆ What types of interface can be generated?
◆ How can command structures be implemented?
◆ How can a PDI be constructed, or can its creation be automated?
◆ To what extent is the designer aware of physical data distribution?

▶ Naming standards

In most organisations, naming standards will already exist for objects such as programs, data tables, records, etc. If the physical environment is new to an organisation, naming standards may need specifying from scratch. In this case *all* function components should be subject to review with management to decide on acceptable naming conventions.

▶ Physical Design Strategy

Once the physical environment is understood a strategy for creating the Physical Design can be developed. This will include customising the Physical Design activities described below to fit with the requirements of the project, and assigning personnel to relevant activities. Particular attention should be paid to:

◆ The implementation method of each function component, i.e. procedural or non-procedural.
◆ The mapping between logical processes and actual programs or program modules, i.e. will one logical process equal one program, or a suite of programs?

◆ What tools within the processing system will be used, and for what?

◆ The extent to which physical data design can be automated, or optimised.

◆ Relative priority of timing, space and maintainability objectives, etc.

◆ Timing and space estimation methods.

◆ The mechanism for storing business rules, i.e. with the data, or using validation processes.

The level of detail required in the Physical Design Strategy will depend largely on the experience of the design team, and the level of control over the above issues allowed by the physical environment.

▶ PHYSICAL DATA DESIGN

The Required System Logical Data Model provides a picture of how users view system data and of its underlying business meaning. It does not tell us how this system data should be physically organised or stored.

In SSADM the transformation of the Required System Logical Data Model into physical data design takes place in two steps:

◆ first-cut data design, where general and then product-specific rules are applied to create a data design which matches the Required System Logical Data Model as closely as possible;

◆ optimised data design, where the first-cut design is tested against performance objectives and tuned as necessary.

It should be noted that the Required System Logical Data Model is *not* replaced by the physical data design. Indeed, it is by far the most permanent of the two data models and could be used to generate a number of physical data designs over the lifetime of the system as the physical environment alters.

▶ First-cut Physical Data Design

The technique of first-cut data design involves applying a number of transformations based on general assumptions about the target database management system, followed by product-specific rules as embodied in the DBMS data storage classification.

It is in the first stage, when applying the general 'rules of thumb', that the analyst will typically be able to participate fully in Physical Design. The activities involved are fairly straightforward and quick to apply, and a knowledge of the system as a whole will be helpful. The application of product-specific rules is a specialist activity.

DBMS assumptions

To be a suitable candidate for the application of first-cut data design SSADM assumes a database management system will have the following properties:

◆ entities are stored as record types;

◆ records are accessed in blocks;

◆ records that are likely to be accessed together are stored as groups. In practice this means physically grouping together hierarchies of masters and details, with the record type at the top of a hierarchy being known as its *root*;

◆ relationships between masters and details within a physical group are supported;

◆ relationships between records in different groups are supported.

> ### Note
>
> Not all database management systems will make these properties visible to the developer. In many non-procedural environments (e.g. Microsoft Access) they will be hidden, and a first-cut data design generated *automatically* from the Required System Logical Data Model (using application generators or data definition screens).

Developing the first-cut Physical Data Design

The Required System Logical Data Model is transformed into the Physical Data Design by carrying out the following eight activities. The first seven produce an intermediate design known informally as a physical data model. The eighth activity (applying product-specific rules) then uses this to create the Physical Data Design.

1. Adapt Required System Logical Data Model

We begin the transformation of the Required System Logical Data Model by identifying those aspects of the Logical Data Structure that are not required in a physical data model (i.e. those that deal with business rules that are not relevant to Physical Data Design).

We can then produce a physical data model to be used as a working document throughout the step.

The features of a Required System Logical Data Model that are not needed in a physical data model are:

◆ relationship names;

◆ master-to-detail optionality (not maintainable in relational database management systems);

◆ exclusion arcs.

There are some notational differences between logical and physical data models:

◆ Round-cornered, or soft, boxes are converted to square-cornered, or hard, boxes in order to differentiate between the two types of structure.

◆ Detail-to-master optionality is denoted by an 'o' on a solid relationship line, rather than a dashed line, as shown in Figure 14.2. Master-to-detail optionality is removed.

◆ Detail-to-master exclusion arcs are replaced by an 'o' on the relationship line. Figure 14.3 illustrates this using the Collection example from Chapter 5 (Figure 5.28).

Entity and relationship volumes should be documented in the Entity Descriptions; they should now be added to the physical data model. Entity volumes show the *average* number of occurrences of each entity. This information should have been collected as part of Logical Data Modelling, and documented in the Entity Descriptions.

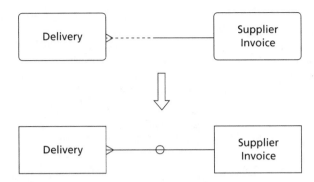

Fig 14.2 Detail-to-master optionality in Physical Design

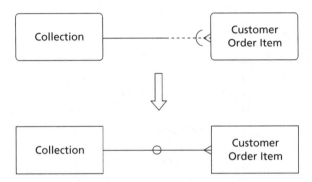

Fig 14.3 Detail-to-master exclusion arcs in Physical Design

The relationship volumes indicate how many detail entity occurrences each master entity will have on average. Usually, the ratio of detail entity volumes to master entity volumes will give the correct figure, but where optional relationships exist as a result of mutual exclusivity this will not work as the detail occurrences will not necessarily have masters, e.g. at any point in time only two-thirds of Delivery occurrences will have a related Supplier Invoice.

2. Identify required entry points

Access entry points can be identified by looking at Effect Correspondence Diagrams and Enquiry Access Paths. Each entry point is indicated on the physical data model by an arrow pointing to the relevant entity, with the input data items listed alongside. The access data items are then compared with the key of the entity, and if they do not match we have a non-key entry point. Each non-key entry point is indicated by adding a lozenge-shaped box to the entity as shown next to the Delivery entity in Figure 14.4.

3. Identify roots of physical hierarchies

Root entities lie at the top of the physical groupings based on master–detail hierarchies. There are two stages to identifying roots.

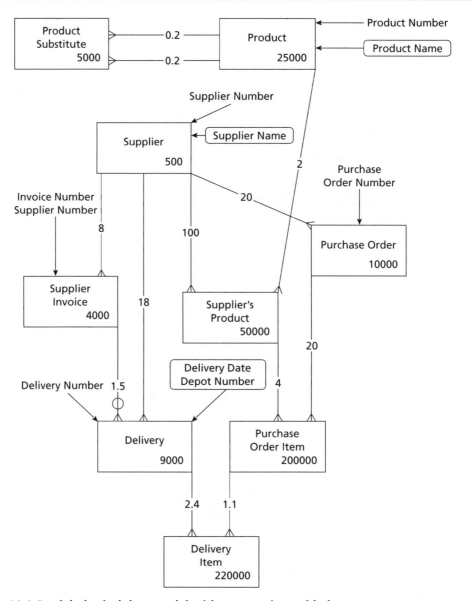

Fig 14.4 Partial physical data model with entry points added

1. An entity is a root if it has no master.
2. An entity is a root if it is a direct entry point, *unless* it has a compound key that contains the key of a root entity.

Each root entity is identified by adding a stripe to the top of its box (Figure 14.5); e.g. by applying the first rule, Supplier becomes a root.

Applying the second rule, notice that Supplier Invoice is a direct entry point, but its key contains the key of Supplier, which is already a root. Therefore Supplier Invoice is not a root. The other direct entry points become roots.

Secondary or non-key access is ignored when determining roots.

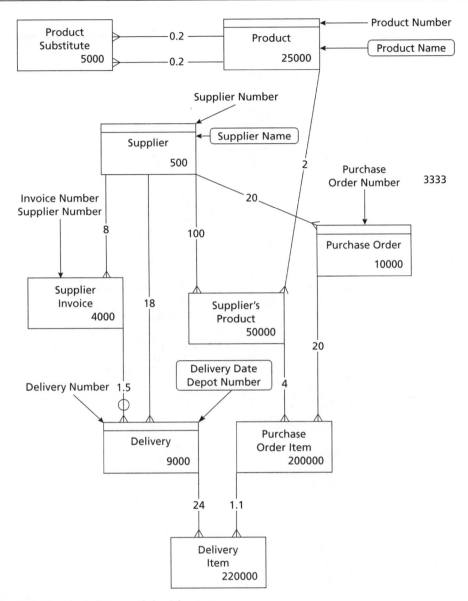

Fig 14.5 Physical data model with roots

4. Identify allowable groups for non-root entities

We now look at the physical data model to identify which root entities each non-root entity could allowably be grouped with.

A non-root entity can only belong to a group if the group contains one of its mandatory masters.

If a non-root entity has two mandatory masters and it is a direct entry point, it should be grouped with the one whose key is part of its own key.

Each allowable grouping should be drawn on the model. 'Stand-alone' root entities, i.e. those with no non-root details, will form single-entity groupings on their own. Allowable groups are shown in Figure 14.6.

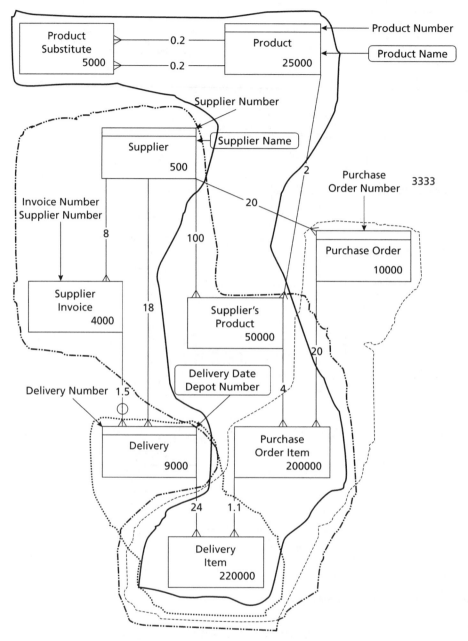

Fig 14.6 Roots and allowable groups added

5. Apply least dependent occurrence rule

If a non-root entity can allowably be placed in more than one group, we choose the one in which it occurs the least. This is done by working up from the non-root entity to the root of the group, and multiplying together the volumes of the relationships connecting all of the entities in the hierarchy. The non-root entity is then placed in the group with the lowest total.

For example, Delivery Item could be grouped with one of four groups: Product, Supplier, Purchase Order or Delivery. The relationship volume with Delivery is 24 (i.e. there are an average of 24 Delivery Item occurrences for each Delivery). For every Purchase Order there are 22 Delivery Items (each Purchase Order has 20 Purchase Order Items, and each Purchase Order Item has 1.1 Delivery Items). For every Supplier there are 440 Delivery Items. For every Product there are less than 10 Delivery Items, so Delivery Item is grouped with Product.

Continuing in this manner for the rest of the model results in the groupings shown in Figure 14.7.

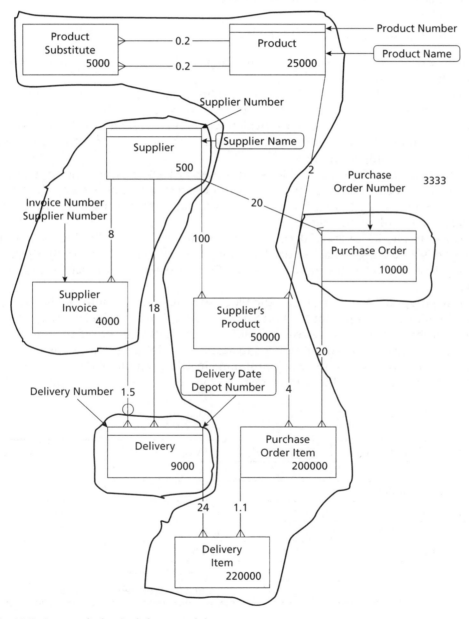

Fig 14.7 Grouped physical data model

6. Establish block size

We now need to establish the standard block size to be used in Physical Data Design. Choosing a suitable block size involves considering the sizes that can be handled by the database management system, the amount of memory available for reading blocks into (blocks must be smaller than the available memory) and the size of the most commonly used groups. In practice there may be little choice, due to installation standards or physical constraints. Ideally the block size will be large enough to accommodate even the largest groups. For ZigZag we will assume that a block size of 2000 bytes is the most suitable.

7. Split physical groups to fit block size

We now need to examine each block in the physical data model to see if it will fit into the chosen block size.

The size of each record is calculated by adding up the lengths of the attributes in the entity description. The number of records accessed in each block is calculated by starting at the root of the block (which will occur once) and working down the hierarchy, multiplying the number of occurrences of the previous entity by the connecting relationship's volume.

The total space required for the block is then arrived at by adding up the space required for each record type (the number of occurrences multiplied by the length of the record), and allowing for any space overheads involved in maintaining relationship and disk maintenance.

For example, let us look at the group containing Product, Product Substitute, Supplier's Product, Purchase Order Item and Delivery Item (Figure 14.8). Clearly a group that is 2633 bytes in size will not fit into the chosen block size of 2000 bytes.

Any group that will not fit into the block size must be split into smaller groups. We do this by working from the bottom of the grouping until the largest sub-group that will fit into the block size is reached, which we then split off to form a new group. This means that when we need to access the original grouping the necessary number of accesses is kept to a minimum. In the example above, we will split Delivery Item off into a block of its own.

Record Types in Group	Length (bytes)	Records per block	Size (bytes)
Product	200	1	200
Product Substitute	30	0.2	6
Supplier's Product	105	2	220
Purchase Order Item	100	8	800
Delivery Item	110	8.8	968
		Total size of group	2194
		Add 20% overhead	2633

Fig 14.8 Sizing of product group

8. Apply product-specific rules

The physical data model is now transformed into our first-cut Physical Data Design by applying product-specific rules. Most of these rules will have been identified as part of the DBMS classifications, and the extent of their application will be included in the Physical Design strategy.

▶ Optimised Physical Data Design

The first-cut data design is more-or-less a direct translation of the Required System Logical Data Model. If this could be implemented without amendment the resulting database should be easily understood and queried, maintainable and, above all, robust (as it would be based closely on the underlying data needs of the organisation). However, included in the non-functional requirements for the new system are preset performance objectives (relating to target data storage volumes and processing speed), and we may find that our first-cut data design would be unable to meet these objectives. In this case the design will need optimising.

Optimisation is a highly specialised activity and is well beyond the scope of both SSADM and this book; it is an activity that is really best left to database experts. The overall objective of optimisation is to meet performance objectives while preserving the one-to-one mapping between the Required System Logical Data Model and the final Physical Design as much as possible. Indeed, if we are forced to consider compromising the data structure we should try to negotiate reduced performance objectives with users first.

Clearly, the extent to which data designs can be optimised will depend heavily on the physical environment concerned. As part of the DBMS classification we will already be aware of available mechanisms, and will have developed an approach to their use in the Physical Design Strategy. It should always be borne in mind that optimisation can be an expensive activity (especially if data structure changes lead to a complex PDI), and that the law of diminishing returns will apply. If optimisation leads to too much development effort and cost, it might be cheaper to purchase more efficient or powerful data management facilities.

The brief guidelines presented below are generic in nature, and relate more to a recommended approach to optimisation than to optimisation techniques themselves. As well as trade-offs between development costs and performance improvements, there are often trade-offs to be made between meeting time and storage objectives, which may conflict. This all leads to an iterative process of optimisation as shown in Figure 14.9.

Identify storage and timing objectives

Performance objectives are recorded in the Requirements Catalogue during Investigation and confirmed during Function Definition. We now carry these forward to test against our Physical Data Design.

Estimate storage requirements

The amount of storage required by our physical data model is estimated by calculating the size of all entities, and multiplying by their estimated number of occurrences. We then add to this the storage required by data management mechanisms

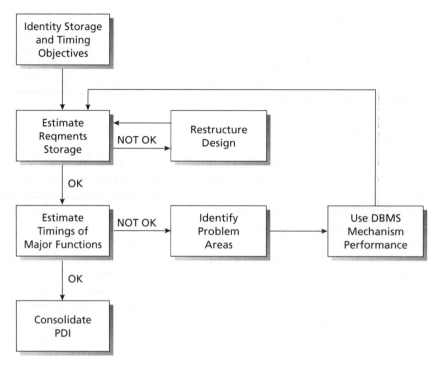

Fig 14.9 Optimisation cycle

such as indexes, pointers and security measures. Finally, we must allow some contingency for future expansion. The resulting storage requirement can then be checked against our objectives and if everything is OK we will proceed to look at timings. If the Physical Data Design does not meet our storage objectives as it stands, we will need to restructure it and carry out a re-estimation.

Restructure design

Data storage objectives are usually much less strictly defined than timing objectives (due to the relatively low cost of purchasing extra storage), so rather than restructuring the Physical Data Design we may be able to persuade users to relax the timing objectives. If not then there are two possibilities:

First, utilise data storage facilities:

◆ improve data distribution;

◆ change record types (e.g. from fixed- to variable-length);

◆ pack data (by removing blank spaces).

Secondly, compromise one-to-one mapping:

◆ reduce historical entity occurrences;

◆ introduce summary records;

◆ reduce attribute lengths;

◆ delete derivable data;

◆ eliminate 'classification' or 'type' entities.

Some of these measures (for example data packing) will have an adverse impact on timings, so may lead to problems and trade-offs with other performance objectives later on.

Estimate timings of major functions

Estimating the timings of all functions in the system may be impractical, so we must pick only the *major* functions. To do this we will look for functions that are time-critical or that are likely to involve large numbers of database accesses. The latter category is often associated with functions that create or read the most populous entities (usually those near the bottom of the Logical Data Structure hierarchy).

If the physical environment has automated database definition facilities we may be able to simulate the required accesses for each major function, and so compare actual timings with our objectives. If not, these timings will have to be estimated using specialist knowledge of the database management system's data access mechanisms (as identified in the DBMS classification). This is not an easy task and is almost certainly best left to the experts.

If timings are assessed as acceptable then we can proceed to the next step in Physical Design; if not, then we will need to identify which elements of the access mechanisms are giving us problems, so that they can be timed to achieve our preset objectives.

Improving database timings

Timings can be improved in two ways.

We can use the database management system facilities:

◆ changing access methods;
◆ placing details near most commonly associated masters;
◆ indexing;
◆ sorts;
◆ adjusting block sizes;
◆ using faster-access storage devices for critical data;
◆ holding data in memory;
◆ changing packing methods.

We can compromise the data structure:

◆ denormalising data (reduces read timings, increases update timings);
◆ postponing updates;
◆ adding redundant data (e.g. derivable data items);
◆ altering processing to fit data hierarchy.

Most of these activities, and others like them, will be carried out (if the environment allows) by database managers or data administrators, so may be beyond the abilities of the SSADM analyst.

As mentioned earlier, compromising the data structure should always be a last resort and should certainly follow attempts to renegotiate performance objectives.

Whatever the outcome, once we have completed any timing optimisation activities we will need to make another pass through the storage estimation and checking tasks, to ensure that objectives in this area can still be met.

▶ PHYSICAL PROCESS DESIGN

The main activities of Physical Process Design are:

◆ create the Function Component Implementation Map;

◆ complete the specification of each function;

◆ specify all procedural programs;

◆ specify the Process Data Interface;

◆ implement non-procedural fragments of the system.

The word *'fragment'* refers to the physical implementation of a logical function *component.*

▶ Function Component Implementation Map

The Function Component Implementation Map (FCIM) is a critically important and simple-to-operate control document that maps every logical system component to a physical system fragment. There is no standard layout within SSADM for the FCIM, which will vary in content from project to project. Therefore the Function Component Implementation Map elements and diagrams presented in this section should be taken as suggestions only. Each organisation will develop its own standards depending on the target physical environment.

The Function Component Implementation Map acts as the central reference point for tying together all of the physical fragments of the system and relating them to the logical components as identified in the Universal Function Model. As such, it underpins the entire physical process design activity.

A Function Component Implementation Map should satisfy the following three principal objectives.

(i) Map logical components to physical fragments

Each logical component of a function will be implemented as at least one physical fragment. In practice it is likely that many components will be split into several fragments, e.g. a logical dialogue may be implemented as several screens, or a logical process as several programs or modules. The FCIM will then act as a checklist that ensures that the Physical Design covers every component.

(ii) Identify common components or fragments

Ideally we should be aiming to maximise re-use of fragments such as modules or screens. Common processing fragments will often (but not always) reflect common logical processing as identified in Elementary Process Descriptions and Function Definitions. It may also reflect the grouping of functions into 'super-functions', e.g. the batching together of several off-line functions into a single execution of a super-function. Duplicate I/O fragments may reflect the appearance of a single event in more than one function, or the re-use of physical screen layouts in more than one function.

(iii) Identify implementation route

It is extremely important to fully consider which of the available implementation routes are applicable to each function component and/or fragment. The Function Component Implementation Map can assist in this by supplying a matrix that lists fragments down one side and all possible implementation methods along the other. The implementation route of each fragment can be further documented by allocating personnel to its development, and by linking it to relevant program specifications.

The concept of a Function Component Implementation Map is very sound: it enforces controls and rigour over the development of the physical design. However, the layout of a Function Component Implementation Map poses quite a problem as the document can get quite large and most organisations are probably best advised to develop their own, suited to their project procedures and physical environment. Most organisations will use some form of spreadsheet format.

Figures 14.10 and 14.11 offer suggestions for a two-part Function Component Implementation Map. Part 1 deals with the mapping of logical components to physical fragments, and the make-up of super-functions. As super-functions can legitimately be made up of other super-functions, they, along with functions themselves, will appear on both axes.

Part 2 deals with mapping physical fragments back to logical components, and in doing so identifies where they will be re-used. It also documents the implementation route of each fragment. Any processing fragments marked as N (non-procedural) may

| | Logical Component | Physical Fragment | Super Function Id | | |
			SF1	SF2	SF3
	Function 1 (F1)		X		
1	Input	1.1 (I/P)			
2	Input Process	1.2 (Proc)			
3	Event 1 Process	1.3 (Proc)			
		1.4 (Proc)			
4	Event 2 Process	1.5 (Proc)			
5	Error Process	1.6 (Proc)			
6	Valid Output	1.7 (O/P)			
7	Output Process	1.8 (Proc)			
8	Error Output	1.9 (O/P)			
	Super Function 1 (SF 1)			X	
9	Function 2 Input (F 2)	1.1 (I/P)	X		

Fig 14.10 FCIM Part 1 (extract)

Physical Fragment	Logical Component	P/N	Specification No.	Screen	Report	Screen Painter	3GL	4GL
1.1 (I/P)	1, 9, 20		1	X		X		
1.2 (Proc)	2	N				X		
1.3 (Proc)	3	N						
1.4 (Proc)	3, 12	P	2					X
1.5 (Proc)	4	N					X	
1.6 (Proc)	5	N						X
1.7 (O/P)	6		3		X			X
1.8 (Proc)	7	N						
1.9 (O/P)	8		4	X				

Fig 14.11 FCIM Part 2 (extract)

be specified or actually implemented in this step; those marked as P (procedural) will be specified formally before the end of the Physical Design stage. The number of the program specification that details its implementation is shown in the fourth column.

▶ Completing Function Specifications

Each element of the FCIM must now be fully specified. Looking at the Universal Function Model (Figure 14.12) there are several elements that have not yet been specified: syntax errors, i.e. input data errors; I/O processes, including sorts; error outputs.

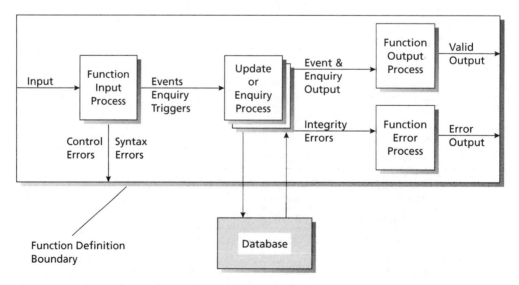

Fig 14.12 Universal Function Model

In addition, we must also decide on the physical formats of all inputs and outputs, e.g. printer, screen, file transfer system; and carry out actual physical dialogue design, i.e. the translation of User Interface Designs, Menu Structures, Windows Navigation Models and Function Navigation Models. This will involve many style issues and will usually be driven by a combination of installation standards and the capabilities of the physical environment.

In environments that contain an application generator and screen painter all of these fragments can usually be developed quickly and easily, once decisions on layout and style have been taken. However, in entirely procedural environments each screen may need building by hand and be backed by complex input and output programs, also developed with little or no automation. Every error message and all command structure translations are likely to require individual specification and construction. All of this can combine to form the largest part of the implementation phase.

Within each function we will need to decide on *success units*. A success unit is a set of inputs, outputs and processing that must succeed as a whole, i.e. that when complete will leave the system in a logically consistent state. Success units can be defined at system-event level, i.e. all effects for a single event occurrence, or at the level of groups of related effects within a function. Success units will often relate directly to programs or program modules.

Any function fragments that are entirely non-procedural in nature can be implemented at this point. Further program specification of such fragments would be a waste of effort, as part of the justification of non-procedural development tools is that we do not need to specify *how* the program should process data, but merely *what* the program should deliver.

Any function components that have procedural fragments in them now need to be fully specified using the organisation's own standards for procedural program design and specification. In largely 3GL environments this step is likely to form a substantial part of Physical Design.

It is difficult to offer detailed guidance in this area without going into details of program design methodologies such as Jackson structured programming.

One of the initial tasks, regardless of method, will be to identify how logical processes will equate with program or run-time units. Again, this is likely to be an issue of organisational standards or physical environment characteristics, as set out in the Processing System Classification.

As well as the logical processes of the Universal Function Model, we may need to specify: input or output sorts; processes to resolve any structure clashes within the Logical Design; and data optimisation processes identified during Physical Data Design.

▶ Consolidate Process Data Interface

The Process Data Interface marks the coming together of Physical Data Design and Physical Process Specification (Figure 14.13).

All of the processing elements in the Function Component Implementation Map view data as if the Required System Logical Data Model exists physically. In other words all data accesses are designed to operate on the logical database. As a result of Physical Data Design, particularly optimisation, we may not have been able to

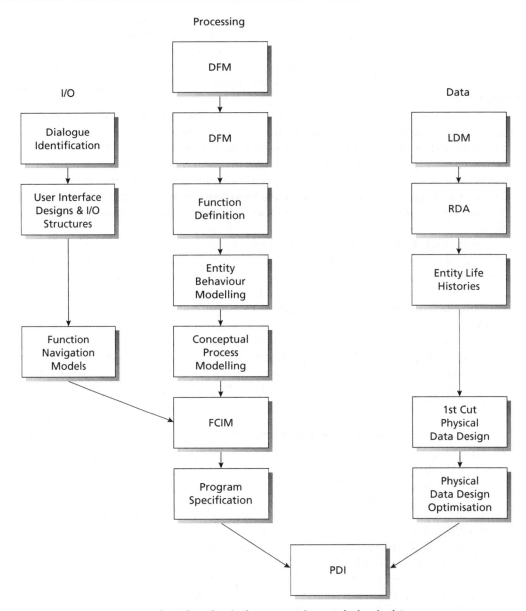

Fig 14.13 Bringing together the Physical Data Design and Physical Process Specification

implement the Logical Database directly. So we now specify a Process Data Interface (PDI) which sits between our processing fragments and the physical database and acts as a translator or mapping tool from one view to the other (Figure 14.14). In effect we are adding a further optional component to the Universal Function Model (see Figure 14.15).

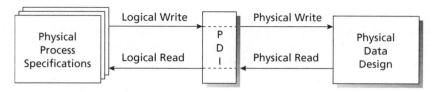

Fig 14.14 Process Data Interface (PDI)

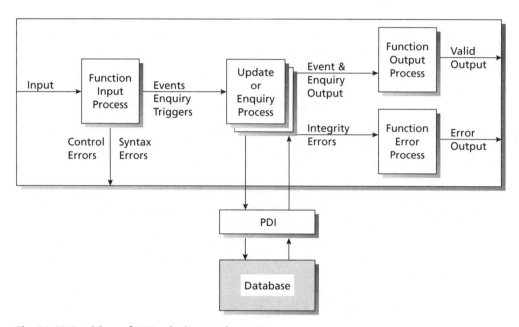

Fig 14.15 Position of PDI relative to the UFM

Note

The PDI is optional in the sense that the database management system may well provide facilities (e.g. through SQL) for defining a logical view of the data and will then handle the physical mappings itself.

In cases where the database management system does not provide these facilities we will need to develop a Process Data Interface from scratch. This will involve examining the access requirements of each function component, identifying relevant physical data items (records, keys, etc.), and designing an access path to map between the two views. In some environments this may be a very complex and time-consuming task.

However, PDI development costs should be balanced against the likely benefits:

◆ **Logical/physical data independence.** By ensuring that processing is carried out on the logical data model we build in a degree of adaptability. Changes in the physical storage of data (e.g. reorganisation of disk packs, change of database management system) will be reflected in an updated Process Data Interface, but will leave functions unchanged. Similarly, new versions of a function or entirely new functions will lead to changes in the Process Data Interface, but should minimise necessary physical data reorganisation. In addition, by shielding users from physical data structures we enable them to take as full a part as possible in the design of processing and *ad hoc* queries.

◆ **Documentation.** The Process Data Interface provides live documentation of physical design decisions.

◆ **Data design independence.** By building a buffer between data and processing, we can limit the impact of data optimisation activities. The Process Data Interface itself can then be optimised by efficient coding or choice of language.

◆ **Maintenance costs.** In most cases, despite the need to maintain the Process Data Interface itself, it will be less expensive and easier to maintain programs using a Process Data Interface than programs with hard-coded data accesses.

Process Data Interface development is yet another area of Physical Design where specialist help is almost inevitably required. The extent to which a Process Data Interface is used is likely to be based on calculations of its net benefits, and its impact on system performance.

All Process Data Interface elements should be added to the Function Component Implementation Map (in the same way as any other fragment) in order to document which components they will interface with.

Finally, once the Physical Design is complete, we must update the Requirements Catalogue with details of any design decisions that have an impact on how well user requirements have been satisfied.

▶ The products of Physical Design

The final deliverables of SSADM are the Physical Design and Application Development Standards. The following products will need to be checked for completeness and consistency:

◆ Function Component Implementation Map;

◆ Function Definitions;

◆ optimised Physical Data Design;

◆ Required System Logical Data Model;

◆ Requirements Catalogue;

◆ Process Data Interface specification;

◆ space and timing estimations.

Once these are complete, we can publish them, and move on to implementation.

▶ SUMMARY

1. As the specific tasks to be undertaken during Physical Design are dictated by the chosen technical environment, SSADM (in common with other methods) can provide only generic guidelines on Physical Design, which can be applied or adapted to a range of environments. The guidelines propose a number of steps in converting the Logical System Specification into a Physical Design:

- The implementation environment is studied and its facilities are classified and documented (with emphasis on its strengths, weaknesses and optimisation mechanisms), Application Development Standards are drawn up, and a Physical Design Strategy agreed with management.

- The Required System Logical Data Model is converted to a first-cut Physical Data Design using techniques based on assumptions common to many database management systems. Product-specific rules are then used to produce an environment-specific data design. The first-cut Physical Data Design is tested against the performance objectives set out in the Requirements Catalogue and Function Definitions. If necessary the design is optimised, using the facilities of the implementation environment rather than by compromising the data structure, wherever possible.

- A Function Component Implementation Map is drawn up, listing the components of each function and how they map onto physical function components. The specification of functions is completed with the addition of physical components such as system error handling routines. Non-procedural function components are implemented. Program specifications are produced for any function components that are to be implemented in procedural code.

- The optimised Physical Data Design is examined to identify where mappings are required from logical database operations to physical database operations. A Process Data Interface is developed to handle these mappings if necessary.

▶ EXERCISES

14.1. Carry out first-cut data design on the Natlib Logical Data Structure from Exercise 5.4, using the following volume information:

- Natlib keeps details of 500 titles.
- There are an average of 3.2 copies of each title.
- Natlib has 1000 registered readers.
- There are 2400 loan records.
- There are 250 recorded reservations.

The keys and lengths of each entity are as follows:

Title:	ISBN	400
Book Copy:	ISBN/Copy No.	100
Reservation:	Reservation No.	120
Loan:	Loan No.	60
Reader:	Reader No.	350

Direct entry is required to all entities, except Book Copy. Assume that the block size is 6000 bytes.

14.2. Turn the Logical Data Model of Exercise 8.2 into a first-cut Physical Data Model. All entities apart from Room Booking and Bill Line are 'key' entry points. *Company name* and *customer name* are also used as 'non-key' entry points.

Appendix A
Sample case study documents

Purchase Order Number: 0021113
Purchase Order Date: 4/3/01

Supplier:

2327
Bella Sonic
Lake Industrial Estate
Unit 5
NE3 7AJ

Delivery Address:

Depot 1
Harrow Way
Harrow
HA4 3NB

QTY	Your Product Ref	our Product Ref	Description	Format	Unit Price
100	BJB001	884690	The Best of Johnnie Boy	CD	6.99
500	3485VHS/3	993201	Unbranded Blank 3hr Video Tapes	BV	0.53
10	NS024	351223	Never on Sunday	CD	6.99

Fig A.1 A typical ZigZag purchase order

Delivery Note: 100213
Date: 14/3/01

To:

ZigZag
Depot 1
Harrow Way
Harrow
HA4 3NB

From:

Lake Industrial Estate
Unit 5
NE3 7AJ

P.O. No	QTY	Accepted	Ref No	Description	Format	Unit Price
0021097	10	10	BJB001	The Best of Johnnie Boy	CD	6.95
0021113	100	100	BJB001	The Best of Johnnie Boy	CD	6.99
0021113	300	300	3485VHS/3	Unbranded Blank 3hr Video Tapes	BV	0.53

Terms:
Received by:

Registered in England 0987654321
VAT Number 888888888

Fig A.2 A delivery note from a typical supplier. The delivery note contains goods ordered in two separate purchase orders. Note how two delivery note lines are used to clarify which purchase order items are being delivered with the above delivery note

Delivery Note: 100234
Date: 18/3/01

To:

ZigZag
Depot 1
Harrow Way
Harrow
HA4 3NB

From:

Lake Industrial Estate
Unit 5
NE3 7AJ

P.O. No	QTY	Accepted	Ref No	Description	Format	Unit Price
0021113	10	10	351223	Never on Sunday	CD	6.99
0021113	30	100	3485VHS/3	Unbranded Blank 3hr Video Tapes	BV	0.53

Terms:
Received by:

Registered in England 0987654321
VAT Number 888888888

Fig A.3 Another typical delivery note. Items of Purchase Order 0021113 are being delivered. Note that after the two deliveries shown here, there are still 100 outstanding Unbranded Blank 3hr Video Tapes that have not yet been delivered

Invoice Number: 22399
Invoice Date: 30/4/01

To:

ZigZag
Depot 1
Harrow Way
Harrow
HA4 3NB

From:

Lake Industrial Estate
Unit 5
NE3 7AJ

Delivery No	P.O. No	QTY	Ref No	Description	Format	Unit Price	Line Total
100213	21097	10	BJB001	The Best of Johnnie Boy	CD	6.95	69.50
	21113	100	BJB001	The Best of Johnnie Boy	CD	6.99	699.00
	21113	300	3485VHS/3	Unbranded Blank 3hr Video Tapes	BV	0.53	159.00
100234	21113	10	351223	Never on Sunday	CD	6.99	69.90
	21113	100	3485VHS/3	Unbranded Blank 3hr Video Tapes	BV	0.53	53.00

Total Excluding VAT	866.58
VAT @ 17.5%	183.82
Invoice Total	**1,050.40**

Terms: 30 days from date of invoice

Registered in England 0987654321
VAT Number 888888888

Fig A.4 A typical itemised Invoice, where each line refers to a delivery note line

Appendix B Glossary

access path A route through a data structure that is navigated to give access to all entities required by an enquiry or update process.

Application Development Standards Standards to be adopted during the Physical Design and Implementation phases of the current project.

application generator A set of software tools, usually with a user-friendly graphical interface, that automate and speed up many aspects of software construction.

Application Style Guide Standards to be adopted in the design of the human–computer interface. Although they apply to the current project they will be heavily based on the organisation-wide Installation Style Guide.

attribute A property or piece of information that describes an entity.

BAM Business Activity Model.

BSO Business System Option.

business activity An action undertaken by a person within the vicinity of the system under investigation. Business activities are studied to enhance our understanding of the business environment and to identify those activities that can be supported by a computerised information system.

Business Activity Model (BAM) A model describing business activities, business events and business rules (known in Soft System Methodology circles as a 'Conceptual Model').

Business Activity Modelling The technique within SSADM to describe business activities.

business event A happening that triggers one or more business activities. See also **event**.

Business System Option (BSO) A BSO defines the functional scope of a proposed solution. At its most basic level it consists of the set of Requirements Catalogue entries satisfied by the

solution. All BSOs must satisfy the minimum requirement as identified by users. Developed in outline during feasibility and in detail by the end of the Investigation phase.

Business System Options (BSOs) BSOs (note the plural) is the technique or set of guidelines provided by SSADM for creating a set of BSOs.

business thread A series of business activities that collectively constitute a response to a business event.

candidate key An attribute or group of attributes which together could be used to uniquely identify the occurrences of an entity.

Central Computer and Telecommunications Agency (CCTA) The UK government agency responsible for the co-ordination of government computer systems development and the promotion of effective standards in computing.

Command Structure A document that specifies where control may pass to on completion of a dialogue.

common process A piece or unit of processing that is shared or carried out by more than one function.

component A self-contained part of a logical function, usually equivalent to one or more SSADM products.

composite key A key made up of foreign keys and a qualifier.

compound key A key made up of foreign keys.

Conceptual Process Modelling (CPM) The technique of developing logical enquiry and update specifications. Its products include ECDs, EAPs, UPMs and EPMs.

Context Diagram A diagram which represents a system as a single DFD process. It is used to define the system boundary and external entities for the system.

cost–benefit analysis A financial analysis of the anticipated costs and benefits of a proposed system.

CRUD matrix A process/entity matrix

Current Environment Description A description of all aspects of a current system (manual and computerised), together with a definition of its shortfalls and problems.

Current Services Description A logicalised view of the current environment. The complete output from the Investigation phase.

database management system (DBMS) A software product for managing and controlling data within a computer system. It should govern the integrity, security and access to an organisation's data.

data catalogue A catalogue of information about the attributes and data items used or created by the system.

data flow A component of a Data Flow Diagram (DFD) illustrating where information is passed to and from in a system.

Data Flow Diagram (DFD) A diagram illustrating the flow, storage and processing of data in a system. It is essentially snapshot in nature, showing all possible movements of data. A very powerful analysis technique (being easily understood and flexible), it is less useful for design purposes.

Data Flow Model (DFM) A model consisting of a hierarchy of DFDs plus textual descriptions of their constituent processes, external entities and data flows across the system boundary.

data item The physical equivalent of a Logical Data Model's attribute. Data items are the smallest pieces of information held in data stores and passed around in data flows.

data store A place where data is held in DFDs. Data stores may be permanent or transitory (i.e. temporary), physical or logical, manual or computerised.

DBMS Database management system.

DBMS data storage classification Documentation describing the storage and retrieval mechanisms of a DBMS.

DBMS performance classification Documentation describing mechanisms and factors affecting the efficiency and performance of a DBMS.

dependent An attribute X is dependent on another attribute Y if, given the value of Y, we can always determine the value of X.

detail entity The entity at the '*m*' end of a 1:*m* relationship.

determinant An attribute Y is the determinant of X if, given the value of Y, we can always determine the value of X.

DFD Data Flow Diagram.

DFM Data Flow Model.

dialogue The on-line interaction between a user and the computer system when executing a function.

Dialogue Control Table A table detailing the allowable navigations within and through a dialogue.

Dialogue Element A logical component of a dialogue consisting of one or more attributes.

Dialogue Element Description A form detailing the attributes that make up a dialogue element.

Dialogue Structure A diagrammatic representation of the structure of a dialogue, detailing the sequence, selection and iteration of its dialogue elements.

Document Flow Diagram A working document detailing the sources, recipients and flows of actual documents around the current system, and which can be used in the development of DFDs.

domain The range of values that an attribute or data item may take.

EAP Enquiry Access Path.

ECD Effect Correspondence Diagram.

effects The set of changes to an entity caused by an event. Represented in ELHs by the bottom 'leaves' of the structure.

Effect Correspondence Diagram (ECD) A diagram illustrating all of the effects of a given event. An ECD will show all of the entities affected by an event, and the ways in which they interact. ECDs are used to define 'update access paths'.

elementary process The lowest (most detailed) level of process in a DFM. Elementary processes are those that will reveal no further understanding of the system if they are decomposed.

Elementary Process Description (EPD) A textual description of an Elementary Process.

ELH Entity Life History.

enquiry A retrieval of data that has no updating effects on the system.

Enquiry Access Path (EAP) A model of the formal navigation through the Logical Data Model to access all data required by an enquiry.

Enquiry Process Model (EPM) A model of all logical database processing required to retrieve the data for an enquiry. EPMs are derived from EAPs and are drawn using Jackson-like notation.

enquiry trigger The data that initiates or triggers an enquiry.

entity Any object or concept about which a system needs to hold information. Entities provide the core concept for Logical Data Models. Each entity must have a number of real-world occurrences.

Entity Access Matrix A matrix illustrating which entities are affected by which events or used by which enquiries. Each correspondence between an entity and an event is described as Create, Delete, Read or Modify. (Known as Event/Entity Matrix in earlier SSADM versions.)

Entity Behaviour Modelling The analysis of events and their effects on entities. Entity Behaviour Modelling anticipates each system event that can affect an entity's occurrences and provides a view of allowable sequencing of these events. (Known as Entity-Event Modelling in earlier SSADM versions.)

Entity Description A textual description of an entity, including its attributes, purpose and relationships with other entities.

Entity Life History (ELH) A Jackson-like structure diagram showing all of the events that can affect an entity, from creation to deletion. ELHs detail the allowable sequence, selection and iteration of these events, as well as detailing the operations carried out on the entity as a result.

entity role If more than one occurrence of an entity can be affected by a single event, and in different ways, then the entity is said to be adopting different Entity Roles (denoted on an ELH by the use of square brackets).

EPD Elementary Process Description.

EPM Enquiry Processing Model.

event A *business* event is a real-world happening that causes the business to perform certain actions or activities. Some of these activities will cause data within a system to be updated. An activity that causes data to be updated gives rise to a *system* event. Thus a system event will act as a trigger for update processing.

external entity A source or recipient of data for or from the system under consideration.

FCIM Function Component Implementation Map.

Feasibility Option An option for taking a project forward into full analysis and design. A Feasibility Option consists of a combination of a high-level BSO and a high-level TSO.

Feasibility Study An optional, but highly recommended stage of an SSADM project, which assesses the technical, financial and organisational feasibility of a project.

First-cut Data Design A technique for transforming the Logical Data Model into a first-cut physical data design, based on a standard set of assumptions about the facilities provided by the target DBMS.

foreign key An attribute of one entity, which is also the primary key of another, is called a foreign key. It is by maintaining foreign keys that detail-to-master relationships are established.

fragment A piece of physical processing, usually corresponding to a logical function component or operation.

function A user-defined unit of system processing. Functions represent the system's functionality from the perspective of activities carried out by users in the same time frame. Functions become the basic unit for specification of processing. First developed during Specification where they take over from Data Flow Modelling, they are constantly updated and supplemented throughout the rest of the project.

functional decomposition A way of analysing a business environment by breaking it up into functional departments.

Function Component Interaction Map (FCIM) A mapping of the logical design (represented by function components) onto the physical design (represented by fragments).

Function Definition The identification of all components of a function. The product of Function Definition (also called a Function Definition) is a package of specification products that together define a function.

Function Navigation Model A representation of the navigation through the components of a function.

functional requirement A requirement to provide a service or facility for users.

Hierarchical Task Model A breakdown of the tasks involved in performing an activity.

impact analysis An analysis of the effects of a BSO or TSO on an organisation's working practices and social structure.

Installation Style Guide Standards to be adopted in the design of the human–computer interface, throughout an organisation.

Investigation A major phase where the current system's requirements are analysed and a decision is made on the scope of the future system. Its products make up a comprehensive statement of user requirements.

I/O Description A description of all of the data items carried by a data flow (which crosses the system boundary).

I/O Structure A model of the structure and logical content of a user interface with a function. There are two components: a Jackson-like structure diagram (the I/O Structure Diagram), and descriptions of the data content of the interface (the I/O Structure Description).

IS Information system.

key An attribute or a set of attributes, the values of which uniquely identify an occurrence of an entity. The terms primary key, composite key, compound key and candidate key are used to classify keys more precisely in relational database theory.

LDM Logical Data Model.

LDPD Logical Database Process Design.

LDS Logical Data Structure.

LGDE Logical Grouping of Dialogue Elements.

Logical Data Flow Model A logical DFM representing the current system which is derived from the Current Physical DFM and the current LDM.

Logical Data Model (LDM) A rigorous model of the data requirements of an organisation, free from physical constraints and implementation considerations. It consists of a Logical Data Structure and accompanying textual descriptions. A Logical Data Model details the content, true interrelationships and business rules applicable to an organisation's data.

Logical Data Store/Entity Cross-reference A document detailing the correspondence between the entities in the Logical Data Model and the data stores used in the Logical DFM. Logical data stores must be based on a whole number of entities from the Logical Data Model.

Logical Data Structure (LDS) A diagrammatic representation of the true structure of an organisation's data.

Logical Database Process Design (LDPD) The technique of modelling the logical processing of data input to and output from the system. It is based on operations carried out on a logical database as defined in the Logical Data Model, and makes use of Jackson-like notation. Its products are Enquiry Process Models (EPMs) and Update Process Models (UPMs).

Logical Design The logical view of the final system used as input for physical design, the Logical Design consists of the Requirements Catalogue, logical process models and the Required System Logical Data Model.

Logical Grouping of Dialogue Elements (LGDE) Groupings of dialogue elements within a Dialogue Structure for use in Dialogue Design.

logicalisation The technique of transforming the Current Physical DFM into a Logical DFM.

master entity The entity at the '1' end of a 1:m relationship.

Menu Structure A hierarchical diagram illustrating how menus and dialogues will be put together in the required system.

non-functional requirement Requirements for the new system covering non-functional aspects, such as timings, performance, security and volumes. They may be used to qualify functional requirements or be applied to the system as a whole.

non-procedural Program code in which the developer states a desired result (e.g. a screen layout), without having to specify how it will be achieved.

normalisation The technique of refining unstructured tables of data into smaller tables based on the principles of Relational Data Analysis. Each refinement is known as a Normal Form.

off-line function A function that is carried out mainly 'off-line', i.e. without 'live' user interaction.

on-line function A function that is carried out mainly 'on-line', i.e. with constant 'live' user interaction.

operation Discrete units of processing carried out on the Logical Data Model, which together constitute the effects of an event or the responses to an enquiry.

optimisation The technique of tuning a Physical Data Design to meet with performance or space requirements.

Outline Current Environment Description The product of the Feasibility Study that describes an overview of current processing and data within the business area under investigation.

Outline Required Environment Description The product of the Feasibility Study that describes an overview of required processing and data for the business area under investigation.

parallel structure A structure used in ELHs to show where events can occur in parallel with the main life of an entity, without altering the course of that main life.

PDI Process Data Interface.

Physical Data Design The design of the physical database to be implemented as part of the final system.

physical data model The product of first-cut data design.

Physical Design The final phase and product of SSADM.

Physical Design Strategy The strategy for transforming the logical design into implementable physical fragments.

physical environment The implementation environment for the required system.

PID Project Initiation Document.

primary key The attribute or set of attributes chosen among an entity's candidate keys to act as the key to that entity.

Problem Definition Statement A statement of user requirements for the new system produced in the Feasibility Study. It ties together the Outline Current and Required Environment Descriptions, the Requirements Catalogue and the User Catalogue.

procedural A term applied to program code that explicitly states how every result is to be achieved.

process Activities that transform or manipulate data in the system.

Process Data Interface (PDI) A definition of how the Logical Data Model maps onto the Physical Data Design. It is then used to translate between processing that acts on the Logical Data Model, and the DBMS that controls the physical database.

process/entity matrix A matrix that illustrates which processes access which entities. It can be used to group processes during logicalisation.

Processing System Classification A description of the implementation and, if possible, the development processing environment.

program specification A document that details how a fragment of the system should be coded and implemented.

Project Initiation Document (PID) A document used to launch an IS project. It will usually include terms of reference, personnel details, high-level requirements and objectives for the project.

prototype A simulation of what the system might look like. Used to provide a demonstrable statement of requirements for verification by users.

prototype pathway A simple flowchart showing how the components of a prototype will be demonstrated.

quality assurance (QA) The process of checking products against pre-determined quality criteria.

Quit and Resume Notation used in ELHs to cater for departures from predictable patterns of events, and which cause changes to the normal life of an entity.

random event An event that can occur at any time in the life of an entity.

RDA Relational Data Analysis.

relation A collection or table of attributes, identified by a unique key.

Relational Data Analysis (RDA) The technique of transforming groups of data items or attributes into relations that obey rules of relational data design, giving rise to increased flexibility and reductions in duplication and data redundancy. In SSADM RDA involves applying the process of normalisation to produce tables in Third Normal Form (3NF), which can then be translated into small-scale Logical Data Structures for comparison with the Logical Data Model.

relationship A logical association between two entities (or between an entity and itself) on the Logical Data Structure.

relationship degree (cardinality) An indication of the number of occurrences of the entities involved in a relationship that can take part in a single occurrence of that relationship. Possible degrees are 1:*m*, *m*:*n* and 1:1.

Required System Data Flow Model A DFM illustrating the required processing for the new system.

Required System Logical Data Model A Logical Data Model illustrating the required data for the new system.

Requirements Catalogue A catalogue of all user requirements (functional and non-functional), including details of measures of success, final solutions and priority.

Requirements Definition The technique of identifying and documenting all user requirements (in the Requirements Catalogue).

Requirements Specification A detailed non-procedural specification of the new system. Made up of all the models developed during the Specification phase.

Resource Flow Diagram A variant on the DFD, illustrating the flow of physical goods or items rather than data.

Resume See **Quit and Resume**

SI State indicator.

Specification Prototyping The technique of presenting prototypes of the new system as a method of confirming or identifying requirements.

SSADM Structured Systems Analysis and Design Method. The UK government's standard IS development method.

state indicator (SI) An additional attribute added to entities, which makes explicit the position of an entity occurrence within its life. They are used to re-express the allowable sequence of events as defined in an ELH.

success unit A unit of processing which must succeed or fail as a whole.

super-function A function that consists of two or more whole functions.

system boundary The boundary of a system. In the current physical environment it will equate with the boundary of the area under investigation, i.e. will define the extent of all manual and computer systems in that area. In a logical environment it will equate with the extent of the computerised (or 'computerisable') system.

task An activity can be broken down into tasks.

Technical System Architecture (TSA) A description of the technical (or physical) environment specified and chosen as part of Technical System Options.

Technical System Option (TSO) A TSO defines a possible implementation route for the Requirements Specification.

Technical System Options (TSOs) TSOs (note the plural) is the technique or set of guidelines provided by SSADM for creating a set of TSOs.

Third Normal Form (3NF or TNF) The product of RDA is a set of tables in 3NF, in which all attributes depend on 'the key, the whole key and nothing but the key'.

transient data store Data stores on a DFD that exist to hold data temporarily, until read once. The classic example of a manual transient data store is an in-tray.

TSA Technical System Architecture.

TSO Technical System Option.

Universal Function Model An SSADM model of the standard logical components of a function.

Update Process Model (UPM) A model of logical database processing required to access and update data in response to a system event. UPMs are derived from ECDs and are drawn using Jackson-like notation.

User Catalogue A catalogue of all the prospective users of the new system, detailing their job titles and responsibilities.

User Interface Design The technique name used for all the products that lead to a more user-friendly system.

User Object Model A model following the user's perception of the system's functionality.

User Role A collective term for a set of users who share common tasks and system access requirements.

User Role/function matrix A description of which functions each User Role will need to access. Each such access gives rise to a required dialogue.

Window Specification A drawing of a function's screen, ranging in detail from a rough sketch to a more precise rendering of the final product.

Windows Navigation Model A model showing a function's navigation through a windows interface.

Work Practice Model (WPM) A mapping of business activities onto the organisation structure.

Work Practice Modelling The technique through which a user-centred view of the system is achieved.

WPM Work Practice Model.

Appendix C
Suggested solutions
to selected exercises

3.8   Fresco Organisation Chart

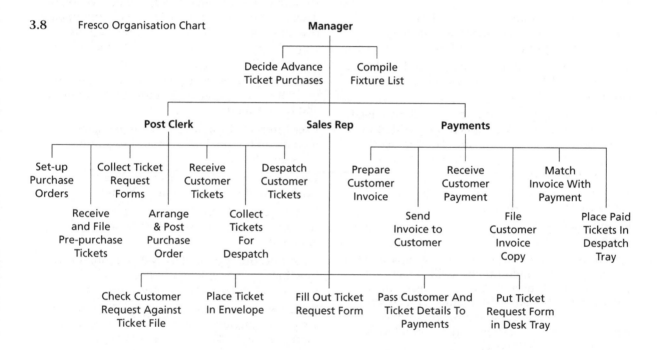

3.9

3.13

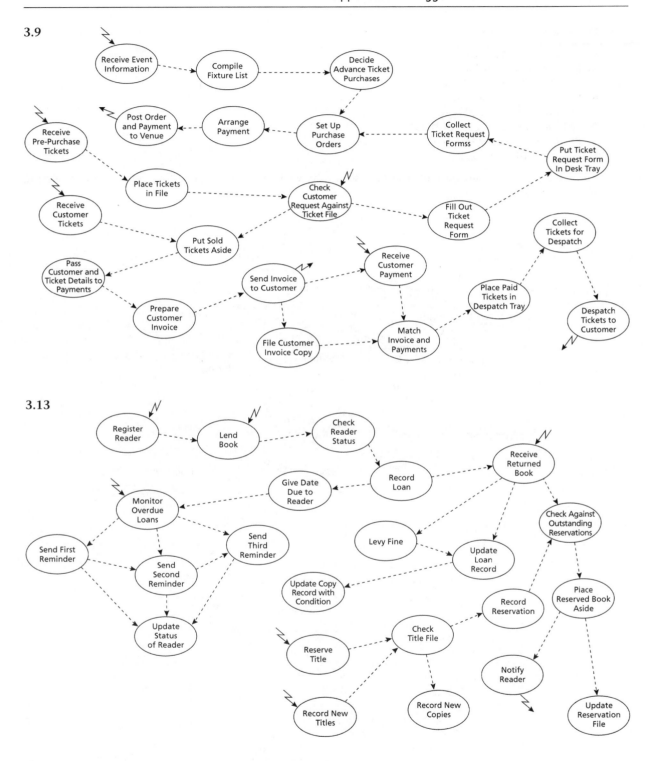

4.3 Fresco Requirement List:

1. Produce Fixtures List
2. Raise Purchase Order
3. Record New Performance
4. Record New Venue
5. Produce Customer Record
6. Sell Ticket (record sale of ticket)
7. Produce Customer Invoice
8. Record Customer Order

More requirements can be added after discussions with users, especially for enquiry and reporting requirements.

5.2

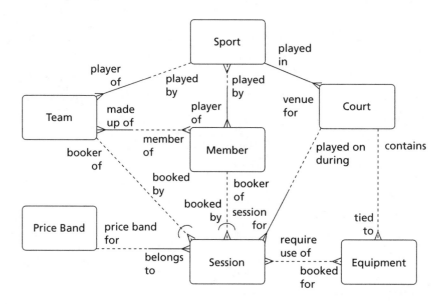

5.4

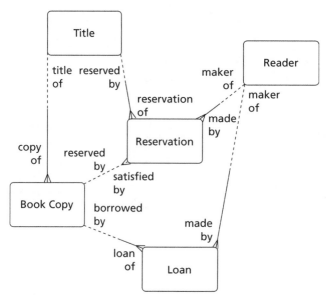

Reader

Reader Number
Reader Name
Reader Address
Reader Email
Reader Phone
Number

Title

ISBN
Title
Author (assume only the first
named author is recorded)
Publisher
Date of Publication
Classification

Book Copy

Copy Number
*ISBN
Loan Type
Condition

Reservation

Reservation Number
*ISBN
*Reader Number
Date/Time Placed
Date Notified
*Copy Number

Loan

Loan Number
*Reader Number
*Copy Number
Date Lent
Date Due
Date Returned
Date of Reminder
Number of Reminder
Amount of Fine
Amount Paid

5.10

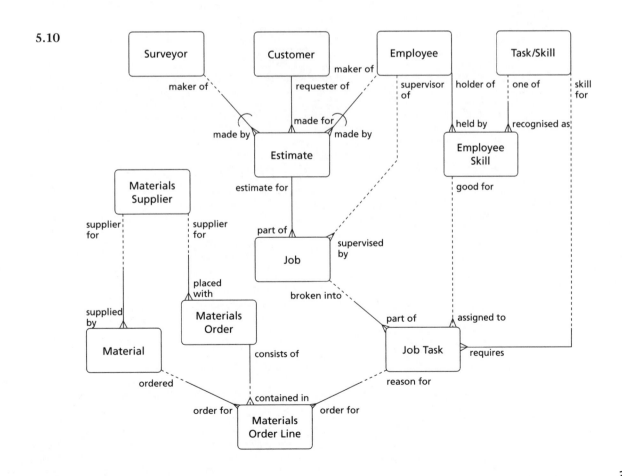

6.2

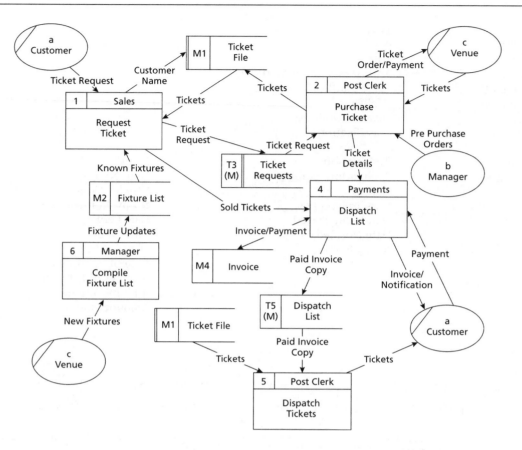

6.3

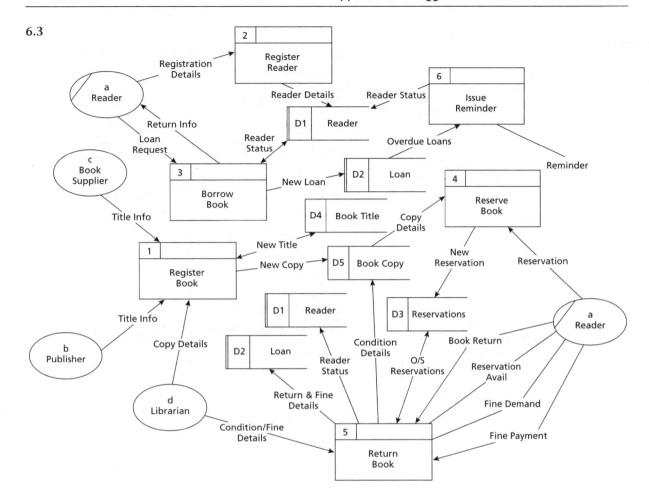

6.4

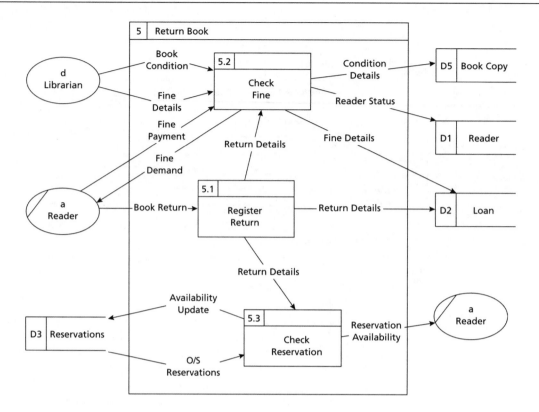

6.14

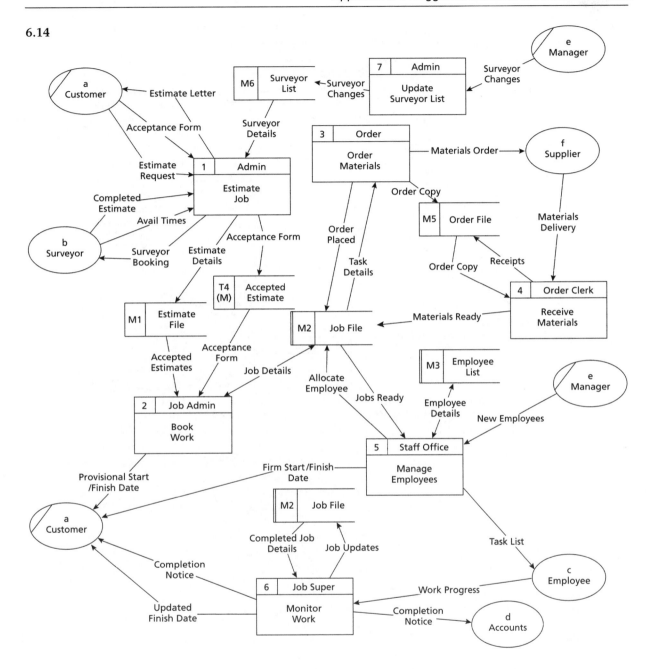

6.15

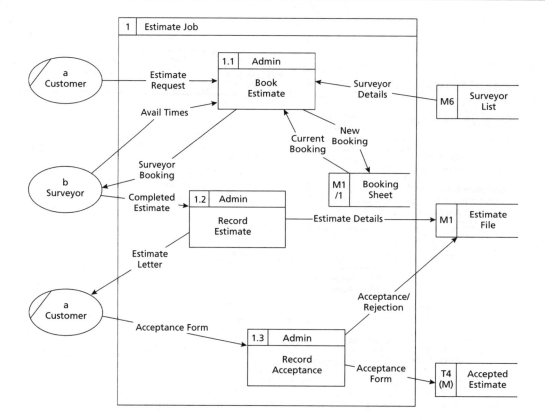

6.16

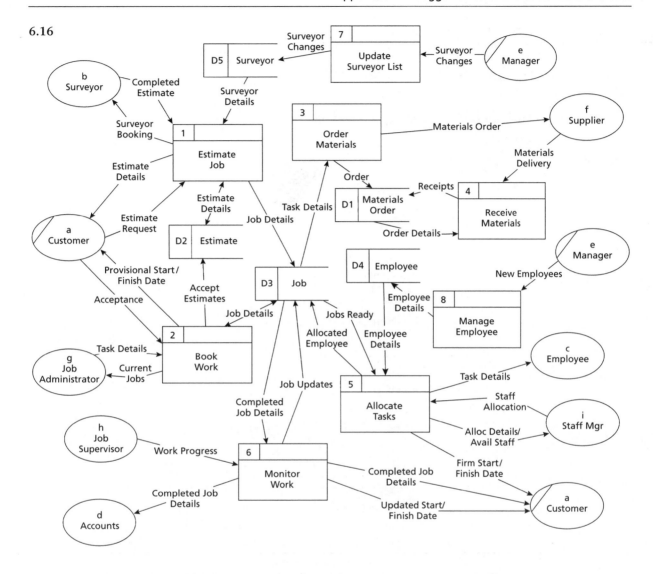

7.2 UFP = $0.58 \times 120 + 0.26 \times 120 + 1.66 \times 60 = 200.4$
$S = 130.26$
Weeks = 25.3
Effort = 1195
$H = 2.07$

7.5 UFP = 352.94
$S = 0.85 \times 352.94 = 300$
Weeks = $2.22 \times \sqrt{352.94} = 38.45$

3GL	4GL
$p = 0.119$	$p = 0.191$
∴ Effort = $S/p = 300/0.119 = 2521.01$	∴ Effort = $S/p = 300/0.191 = 1570.68$
∴ $H_{3GL} = 0.044 \times 2521.01/38.45 = 2.88$	∴ $H_{4GL} = 0.044 \times 1570.68/38.45 = 1.80$

7.6 $H_{\text{3GL coding}} = (46/25)\ H_{\text{3GL}} = (46/25) \times 2.88 = 5.3$

&

$W_{\text{3GL coding}} = 25\%\ \text{of Weeks} = 25\%\ \text{of}\ 38.45 = 9.61$

∴

3GL coding budget $= H_{\text{3GL coding}} \times W_{\text{3GL coding}} \times \pounds600.00$

$= 5.3 \times 9.61 \times \pounds600.00$

$= \pounds30,567.75$

∴ £31,000.00 is enough to cover Coding and Unit testing

8.1

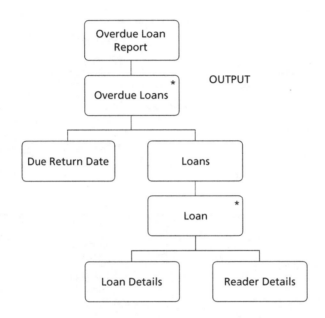

I/O Structure Description		
I/O Structure Element	**Data Item**	**Comment**
Due Return Date	Due Return Date	Latest First
Loan Details	Loan No. Book ISBN Book Title Book Copy No.	
Reader Details	Reader No. Reader Name Reader Address Reader Tel. No.	

9.2

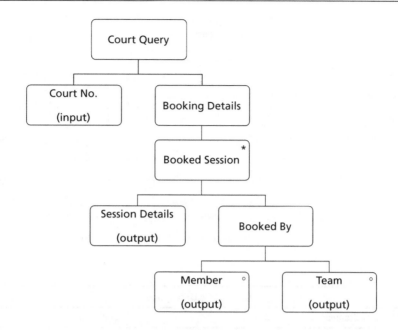

I/O Structure Description		
I/O Structure Element	**Data Item**	**Comment**
Court No.	Court No.	
Session Details	Session No. Session Start Time Session End Time	
Member	Member No. Member Name	
Team	Team No. Team Name	

9.4

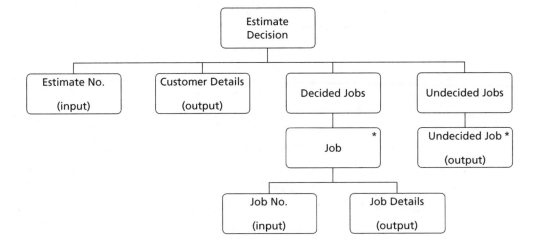

I/O Structure Description		
I/O Structure Element	**Data Item**	**Comment**
Estimate No.	Estimate No.	
Customer Details	Customer No. Customer Name Customer Address	
Job No.	Job No.	
Job Details	Job Description Job Decision	
Undecided Job	Job No. Job Description	

10.1

UNF	Lev	1NF	2NF	3NF
Court No.	1	Court No.	Court No.	Court No.
Session No.	2			
Session Start Time	2	Court No.	Court No.	Court No.
Session End Time	2	Session No.	Session No.	Session No.
Member No.	2	Session Start Time	Member No.	*Member No.
Member Name	2	Session End Time	Member Name	*Team No.
Team No.	2	Member No.	Team No.	
Team Name	2	Member Name	Team Name	Member No.
	2	Team No.		Member Name
	2	Team Name		
	2			Team No.
				Team Name
			Session No.	
			Session Start Time	Session No.
			Session End Time	Session Start Time
				Session End Time

10.2

UNF	Lev	1NF	2NF	3NF
Due Return Date	1	Due Return Date	Due Return Date	Due Return Date
Loan No.	2			
Book ISBN	2	Due Return Date	Due Return Date	Due Return Date
Book Title	2	Loan No.	Loan No.	Loan No.
Reader No.	2	Book ISBN	Book ISBN	*Book ISBN
Reader Name	2	Book Title	Book Title	*Reader No.
Reader Address	2	Reader No.	Reader No.	*Book Copy No.
Reader Tel. No.	2	Reader Name	Reader Name	
Book Copy No.	2	Reader Address	Reader Address	Book ISBN
		Reader Tel. No.	Reader Tel. No.	Book Title
		Book Copy No.	Book Copy No.	
				Book ISBN
				Book Copy No.
				Reader No.
				Reader Name
				Reader Address
				Reader Tel. No.

Note: Book ISBN and Book Copy No. constitute a composite key.

10.3

UNF	Lev	1NF	2NF	3NF
Estimate No.	1	Estimate No.	Estimate No.	Estimate No.
Customer No.	1	Customer No.	Customer No.	*Customer No.
Customer Name	1	Customer Name	Customer Name	
Customer Addr.	1	Customer Addr.	Customer Addr.	Customer No.
Job Decision	2			Customer Name
Job No.	2			Customer Addr.
Job Description	2			
		*Estimate No.	*Estimate No.	*Estimate No.
		Job No.	Job No.	Job No.
		Job Description	Job Description	Job Description
		Job Decision	Job Decision	Job Decision

10.4

Customer Request Form:

UNF	Lev	1NF	2NF	3NF	Table Name
Cust. Name	1	Cust. Name	Cust. Name	Cust. Name	1 CUSTOMER
Cust. Addr.	1	Cust. Addr.	Cust. Addr.	Cust. Addr.	
Cust. Tel.	1	Cust. Tel.	Cust. Tel.	Cust. Tel.	
Perf. Desc.	2				
Venue Name	2	Cust. Name	Cust. Name	Cust. Name	2 CUSTOMER CHOICE
Preferred Date	3	Perf. Desc.	Perf. Desc.	Perf. Desc.	
Preference Ind.	3	Venue Name	Venue Name	Venue Name	
Upper Price	2	Ticket Qty	Ticket Qty	Ticket Qty	
Lower Price	2	Upper Price	Upper Price	Upper Price	
Ticket Qty	2	Lower Price	Lower Price	Lower Price	
		Cust. Name	Cust. Name	Cust. Name	3 PREFERRED DATE
		Perf. Desc.	Perf. Desc.	Perf. Desc.	
		Venue Name	Venue Name	Venue Name	
		Pref. Date	Pref. Date	Pref. Date	
		Pref. Ind.	Pref. Ind.	Pref. Ind.	

Customer Invoice:

UNF	Lev	1NF	2NF	3NF	Table Name
Invoice No.	1	Invoice No.	Invoice No.	Invoice No.	4 CUSTOMER INVOICE
Invoice Date	1	Invoice Date	Invoice Date	Invoice Date	
Cust. Number	1	Cust. Number	Cust. Number	*Cust. Number	
Cust. Name	1	Cust. Name	Cust. Name	Total Price	
Cust. Addr.	1	Cust. Addr.	Cust. Addr.	Pay. Method	
Cust. Tel. No.	1	Cust. Tel. No.	Cust. Tel. No.	Card Number	
Perf. Number	2	Total Price	Total Price		
Perf. Name	2	Pay. Method	Pay. Method	Cust. Number	5 CUSTOMER
Venue Name	2	Card Number	Card Number	Cust. Name	
Venue Addr.	2			Cust. Addr.	
Perf. Date	2			Cust. Tel. No.	
Perf. Time	2				
Ticket Price	3	Invoice No.	Invoice No.	Invoice No.	6 INVOICED PERFORMANCE
Ticket Qty	3	Perf. Number	Perf. Number	Perf. Number	
Total Price	1	Perf. Name			
Pay. Method	1	Perf. Date	Perf. Number	Perf. Number	7 PERFORMANCE
Card Number	1	Perf. Time	Perf. Name	Perf. Name	
		Venue Name	Perf. Date	Perf. Date	
		Venue Addr.	Perf. Time	Perf. Time	
			Venue Name	*Venue Name	
			Venue Addr.		
				Venue Name	8 VENUE
				Venue Addr.	
		Invoice No.			
		Perf. Number	Perf. Number	Invoice No.	9 INVOICED ITEM
		Ticket Price	Ticket Price	Perf. Number	
		Ticket Qty	Ticket Qty	Ticket Price	
				Ticket Qty	

Rationalisation:

(a) Drop Table 1 and keep Table 5
(b) Replace Perf. Desc. and Venue Name in Tables 2 and 3 with Perf. Number.

The following Logical Data Model extract may be deduced:

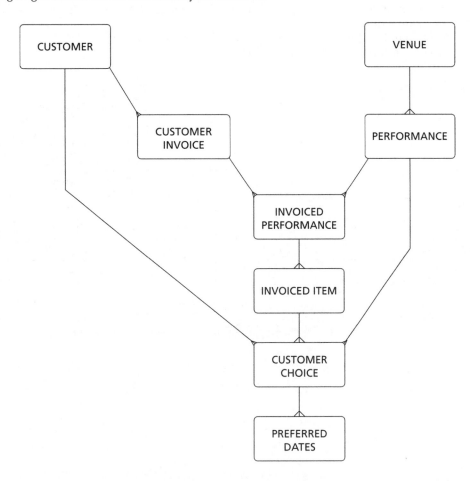

In the above data model we can debate whether to:

(a) drop Invoiced Performance and link Invoiced Item to Invoice and Performance;
(b) adjust the contents of Preferred Date and link Invoice and Performance directly with it, thus dropping Invoiced Item too.

11.2 and 11.3

Request Estimate (from Surveyor)	(C)
Request Estimate (from Employee)	(C)
Book Estimate	(M)
Complete Estimate	(M)
Receive Response	(M)
Send Reminder	(M)
Estimate Expires	(M)
Complete Last Job	(M)
Archive Estimate	(D)

OPERATION LIST
1: CREATE Estimate
2: TIE Estimate TO Surveyor
3: TIE Estimate TO Customer
4: TIE Estimate TO Employee
5: SET Estimate Booking Date = input date
6: SET No. of Jobs Respondend To = No. of Jobs
    Respondend To plus number input
7: SET Reminder Letter Date = input date
8: SET Expity Date = input date
9: SET Estimate Completion Date = input date
10: SET No. of Jobs = input

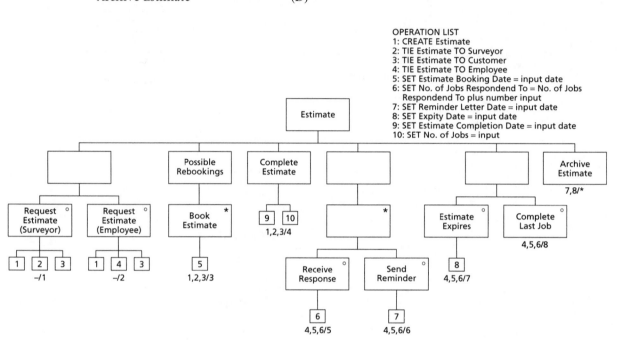

11.4 The completed entity access matrix for the Video Store looks like this:

ENTITY EVENT/ ENQUIRY	Video Title	Video Store	Member	Video Copy	Video Rental Period
Copy Purchase	M, G	M, G		I, T, T	
Copy Sale or Destruction	M	M		D	D
Membership Approval		G	I, T		
Membership Removal			D		D
Store Closure		D/M	M, S	M, S	
Store Opening		I			
Title Release	I				
Title Withdrawal	D				
Video Rental			M, G	M, G	I, T, T
Video Return			M	M	M

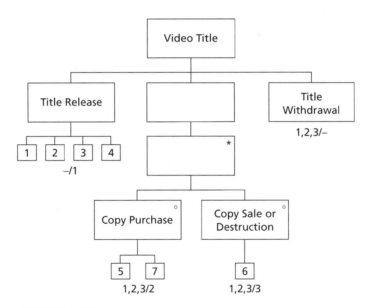

OPERATION LIST
1: CREATE Video Title
2: SET Title Name OF Video Title = input
3: SET Title Description OF Video Title = input
4: SET Number of Copies Held OF Video Title = 0
5: SET Number of Copies Held OF Video Title = Number of Copies Held + 1
6: SET Number of Copies Held OF Video Title = Number of Copies Held – 1
7: GAIN Video Copy ON Video Title

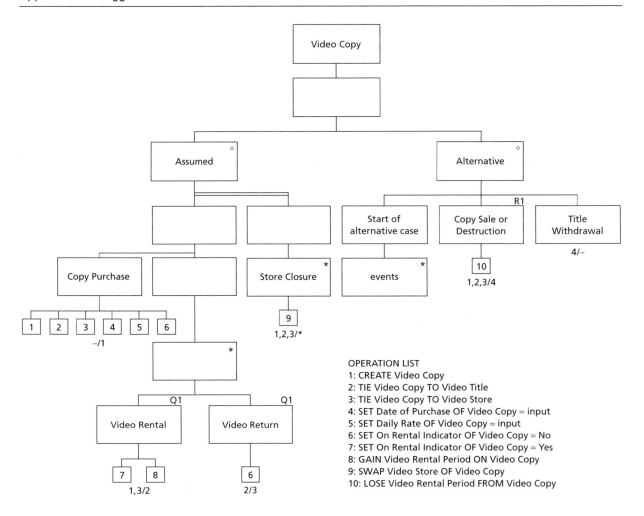

OPERATION LIST
1: CREATE Video Copy
2: TIE Video Copy TO Video Title
3: TIE Video Copy TO Video Store
4: SET Date of Purchase OF Video Copy = input
5: SET Daily Rate OF Video Copy = input
6: SET On Rental Indicator OF Video Copy = No
7: SET On Rental Indicator OF Video Copy = Yes
8: GAIN Video Rental Period ON Video Copy
9: SWAP Video Store OF Video Copy
10: LOSE Video Rental Period FROM Video Copy

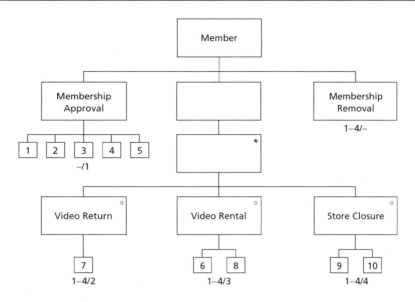

OPERATION LIST
1: CREATE Member
2: SET Member Name OF Member = input
3: SET Member Address OF Member = input
4: TIE Member TO Video Store
5: SET Copies Currently On Rental OF Member = 0
6: SET Copies Currently On Rental OF Member = Copies Currently On Rental + 1
7: SET Copies Currently On Rental OF Member = Copies Currently On Rental − 1
8: GAIN Video Rental Period ON Member
9: CUT Member FROM Video Store [closing]
10: TIE Member TO Video Store [gaining]

Alternative

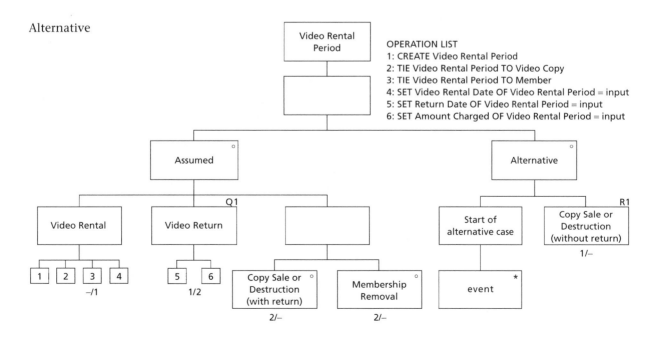

OPERATION LIST
1: CREATE Video Rental Period
2: TIE Video Rental Period TO Video Copy
3: TIE Video Rental Period TO Member
4: SET Video Rental Date OF Video Rental Period = input
5: SET Return Date OF Video Rental Period = input
6: SET Amount Charged OF Video Rental Period = input

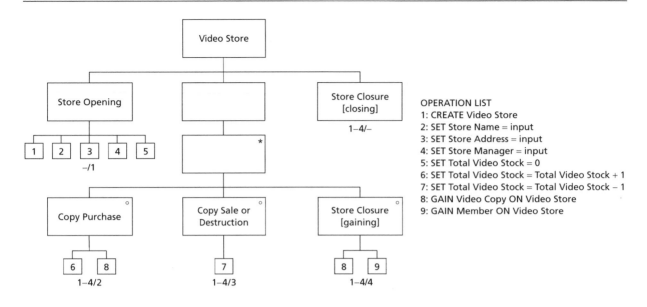

OPERATION LIST
1: CREATE Video Store
2: SET Store Name = input
3: SET Store Address = input
4: SET Store Manager = input
5: SET Total Video Stock = 0
6: SET Total Video Stock = Total Video Stock + 1
7: SET Total Video Stock = Total Video Stock − 1
8: GAIN Video Copy ON Video Store
9: GAIN Member ON Video Store

11.5

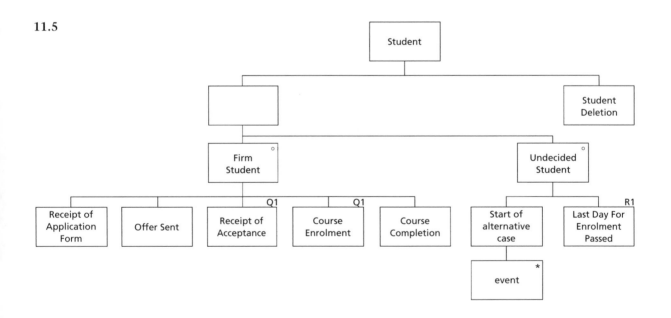

11.6

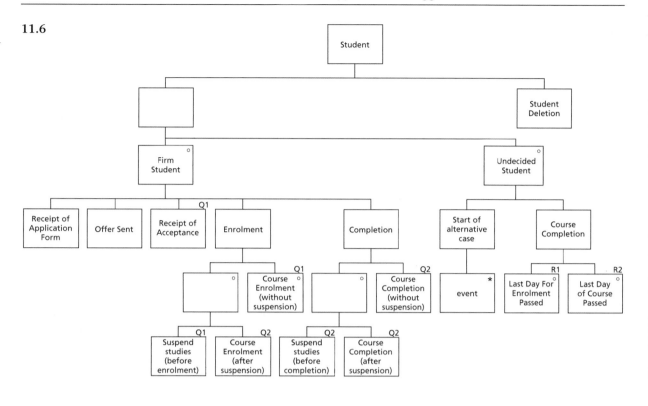

11.10

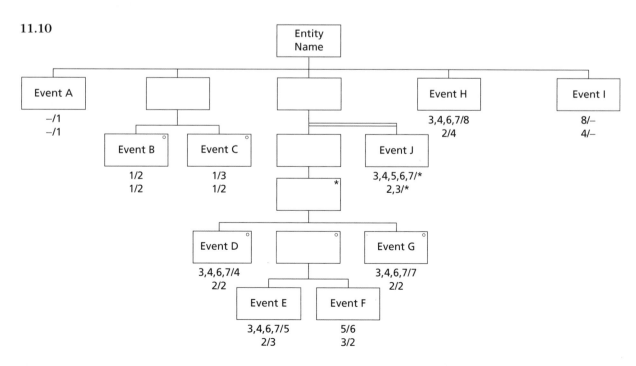

11.11

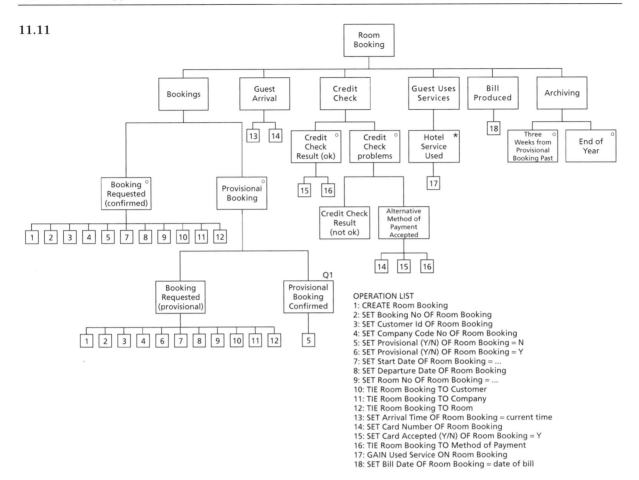

OPERATION LIST
1: CREATE Room Booking
2: SET Booking No OF Room Booking
3: SET Customer Id OF Room Booking
4: SET Company Code No OF Room Booking
5: SET Provisional (Y/N) OF Room Booking = N
6: SET Provisional (Y/N) OF Room Booking = Y
7: SET Start Date OF Room Booking = ...
8: SET Departure Date OF Room Booking
9: SET Room No OF Room Booking = ...
10: TIE Room Booking TO Customer
11: TIE Room Booking TO Company
12: TIE Room Booking TO Room
13: SET Arrival Time OF Room Booking = current time
14: SET Card Number OF Room Booking
15: SET Card Accepted (Y/N) OF Room Booking = Y
16: TIE Room Booking TO Method of Payment
17: GAIN Used Service ON Room Booking
18: SET Bill Date OF Room Booking = date of bill

11.12

	Room Booking	Room	Customer	Company	Method of Payment	Used Service
Booking Requested	I, T, T, T	G	G	G		
Provisional Booking Confirmed	M					
Three Weeks from Provisional Booking Past	B					
Guest Arrival	M					
Credit Check Result	M, T				G	
Alternative Method of Payment Accepted						
Hotel Service Used	G					T
Bill Produced	M					
End of Year	B					

12.1

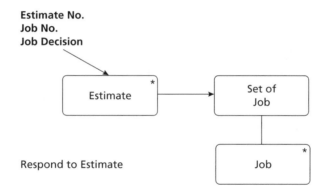

12.3

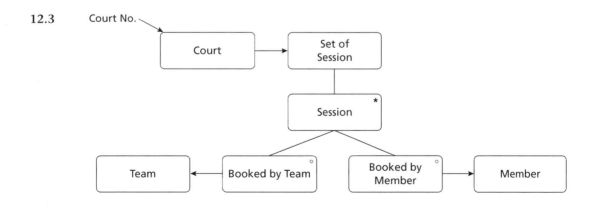

12.5

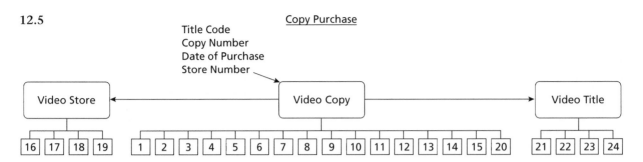

Copy Purchase

Title Code
Copy Number
Date of Purchase
Store Number

OPERATION LIST
1: CREATE Video Copy
2: GET Copy Number OF Video Copy
3: GET Title Code OF Video Copy
4: GET Date of Purchase OF Video Copy
5: GET Store Number OF Video Copy
6: SET Title Code OF Video Copy = input
7: SET Date of Purchase OF Video Copy = input
8: SET Store Number OF Video Copy = input
9: SET Copy Number OF Video Copy = input
10: TIE Video Copy TO Video Store
11: TIE Video Copy TO Video Title
12: SET On Rental Indicator OF Video Copy = N
13: SET SI OF Video Copy = 1
14: WRITE Video Copy
15: READ Video Store OF Video Copy
16: FAIL IF SI OF Video Store OUTSIDE RANGE 1–4
17: SET Total Video Stock OF Video Store = Total Video Stock + 1
18: SET SI OF Video Store = 2
19: WRITE Video Store
20: READ Video Title OF Video Copy
21: FAIL IF SI OF Video Title OUTSIDE RANGE 1–3
22: SET Number of Copies Held OF Video Title = Number of Copies Held + 1
23: SET SI OF Video Title = 2
24: WRITE Video Title

12.5

<u>Store Closure</u>

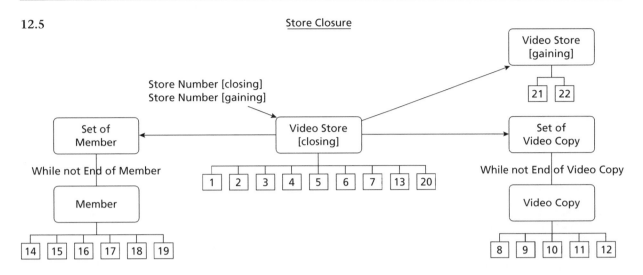

OPERATION LIST
1: GET Store Number OF Video Store [closing]
2: GET Store Number OF Video Store [gaining]
3: READ Video Store [closing] BY KEY
4: FAIL IF SI OF Video Store OUTSIDE RANGE 1–4
5: SET SI OF Video Store = Null
6: DELETE Video Store [closing]
7: READ NEXT Video Copy OF Video Store
8: FAIL IF SI OF Video Copy OUTSIDE RANGE 1–3
9: SET Store Number OF Video Copy = Store Number [gaining]
10: CUT Video Copy FROM Video Store [closing]
11: TIE Video Copy TO Video Store [gaining]
12: WRITE Video Copy
13: READ NEXT Member OF Video Store
14: FAIL IF SI OF Member OUTSIDE RANGE 1–4
15: SET Store Number OF Member = Store Number [gaining]
16: CUT Member FROM Video Store [closing]
17: TIE Member TO Video Store [gaining]
18: SET SI OF Member = 4
19: WRITE Member
20: READ Video Store BY Store Number [gaining]
21: SET Total Video Stock OF Video Store = Total Video Stock + Total Video Stock [closing]
22: WRITE Video Store [gaining]

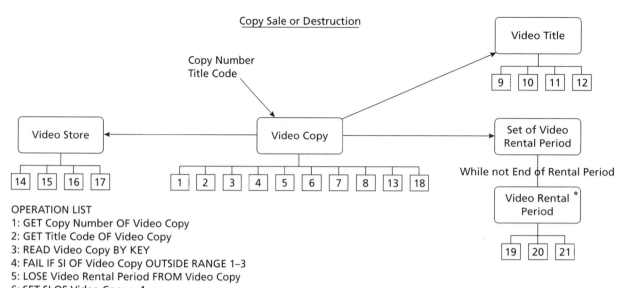

Copy Sale or Destruction

OPERATION LIST
1: GET Copy Number OF Video Copy
2: GET Title Code OF Video Copy
3: READ Video Copy BY KEY
4: FAIL IF SI OF Video Copy OUTSIDE RANGE 1–3
5: LOSE Video Rental Period FROM Video Copy
6: SET SI OF Video Copy = 4
7: DELETE Video Copy
8: READ Video Title OF Video Copy
9: FAIL IF SI OF Video Title OUTSIDE RANGE 1–3
10: SET Number of Copies Held OF Video Title = Number of Copies Held – 1
11: SET SI OF Video Title = 3
12: WRITE Video Title
13: READ Video Store OF Video Copy
14: FAIL IF SI OF Video Store OUTSIDE RANGE 1–4
15: SET Total Video Stock OF Video Store = Total Video Stock – 1
16: SET SI OF Video Store = 3
17: WRITE Video Store
18: READ NEXT Video Rental Period OF Video Copy
19: FAIL IF SI OF Video Rental Period OUTSIDE RANGE 1,2
20: SET SI OF Video Rental Period = Null
21: DELETE Video Rental Period

Video Rental

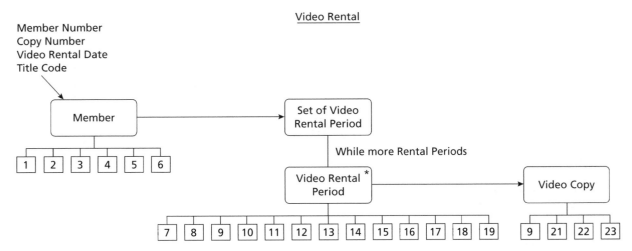

OPERATION LIST
1: GET Member Number OF Member
2: READ Member BY KEY
3: FAIL IF SI OF Member OUTSIDE RANGE 1–4
4: SET Copies Currently On Rental OF Member = Copies Currently On Rental + 1
5: SET SI OF Member = 3
6: WRITE Member
7: CREATE Video Rental Period
8: GET Copy Number OF Video Rental Period
9: GET Video Rental Date OF Video Rental Period
10: GET Title Code OF Video Rental Period
11: SET Member Number OF Video Rental Period = Input
12: SET Copy Number OF Video Rental Period = Input
13: SET Video Rental Date OF Video Rental Period = input
14: SET Title Code OF Video Rental Period = input
15: TIE Video Rental Period TO Video Copy
16: TIE Video Rental Period TO Member
17: SET SI OF Video Rental Period = 1
18: WRITE Video Rental Period
19: READ Video Copy OF Video Rental Period
20: FAIL IF SI OF Video Copy OUTSIDE RANGE 1–3
21: SET On Rental Indicator OF Video Copy = Y
22: SET SI OF Video Copy = 2
23: WRITE Video Copy

12.6   Copy Purchase

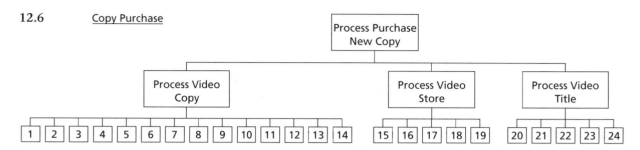

OPERATION LIST
1: CREATE Video Copy
2: GET Title Code OF Video Copy
3: GET Copy Number OF Video Copy
4: GET Store Number OF Video Copy
5: GET Date of Purchase OF Video Copy
6: SET Date of Purchase OF Video Copy = input
7: SET Store Number OF Video Copy = input
8: SET Title Code OF Video Copy = input
9: SET Copy Number OF Video Copy = input
10: TIE Video Copy TO Video Store
11: TIE Video Copy TO Video Title
12: SET On Rental Indicator OF Video Copy = N
13: SET SI OF Video Copy = 1
14: DELETE Video Copy
15: READ Video Store OF Video Copy
16: FAIL IF SI OF Video Store OUTSIDE RANGE 1–4
17: SET Total Video Stock OF Video Store = Total Video Stock + 1
18: SET SI OF Video Store = 2
19: WRITE Video Store
20: READ Video Title OF Video Copy
21: FAIL IF SI OF Video Title OUTSIDE RANGE 1–3
22: SET Number of Copies Held OF Video Title = Number of Copies Held + 1
23: SET SI OF Video Title = 2
24: WRITE Video Title

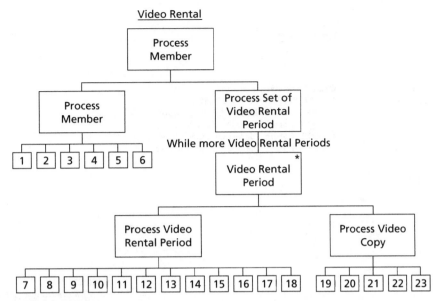

Video Rental

OPERATION LIST
1. GET Member Number OF Member
2: READ Member BY KEY
3: FAIL IF SI OF Member OUTSIDE RANGE 1–4
4: SET Copies Currently On Rental OF Member = Copies Currently On Rental + 1
5: SET SI OF Member = 3
6: WRITE Member
7: CREATE Video Rental Period
8: GET Copy Number OF Video Rental Period
9: GET Video Rental Date OF Video Rental Period
10: GET Title Code OF Video Rental Period
11: SET Member Number OF Video Rental Period = Input
12: SET Title Code OF Video Rental Period = input
13: SET Video Rental Date OF Video Rental Period = input
14: SET Copy Number OF Video Rental Period = input
15: TIE Video Rental Period TO Video Copy
16: TIE Video Rental Period TO Member
17: SET SI OF Video Rental Period = 1
18: WRITE Video Rental Period
19: READ Video Copy OF Video Rental Period
20: FAIL IF SI OF Video Copy OUTSIDE RANGE 1–3
21: SET On Rental Indicator OF Video Copy = Y
22: SET SI OF Video Copy = 2
23: WRITE Video Copy

12.7

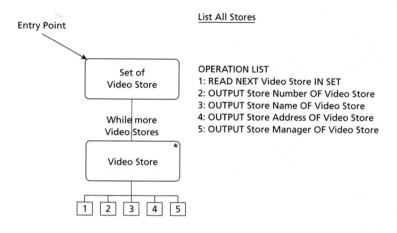

List All Stores

OPERATION LIST
1: READ NEXT Video Store IN SET
2: OUTPUT Store Number OF Video Store
3: OUTPUT Store Name OF Video Store
4: OUTPUT Store Address OF Video Store
5: OUTPUT Store Manager OF Video Store

Stock List

OPERATION LIST
1: GET Store Number OF Video Store
2: READ Video Store BY KEY
3: OUTPUT Store Name OF Video Store
4: OUTPUT Store Address OF Video Store
5: READ NEXT Video Copy OF Video Store
6: OUTPUT Copy Number OF Video Copy
7: OUTPUT Date of Purchase OF Video Copy
8: OUTPUT Title Code OF Video Copy
9: READ Video Title OF Video Copy
10: OUTPUT Title Name OF Video Title
11: OUTPUT Title Description OF Video Title

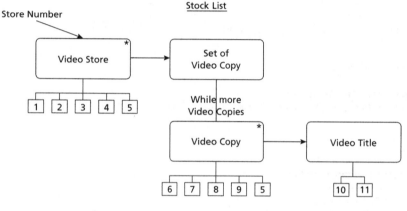

Member's Current Loans

OPERATION LIST
1: GET Member Number OF Member
2: READ Member BY KEY
3: OUTPUT Member Name OF Member
4: OUTPUT Member Address OF Member
5: READ Next Video Rental Period OF Member
6: OUTPUT Video Rental Date OF Video Rental Period
7: OUTPUT Amount Charged OF Video Rental Period
8: READ Video Copy OF Video Rental Period
9: OUTPUT Title Code OF Video Copy
10: READ Video Title OF Video Copy
11: OUTPUT Title Name OF Video Title
12: OUTPUT Title Description OF Video Title

12.8

Member's Current Loans

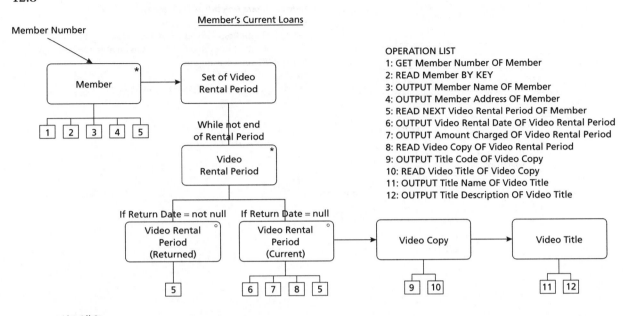

OPERATION LIST
1: GET Member Number OF Member
2: READ Member BY KEY
3: OUTPUT Member Name OF Member
4: OUTPUT Member Address OF Member
5: READ NEXT Video Rental Period OF Member
6: OUTPUT Video Rental Date OF Video Rental Period
7: OUTPUT Amount Charged OF Video Rental Period
8: READ Video Copy OF Video Rental Period
9: OUTPUT Title Code OF Video Copy
10: READ Video Title OF Video Copy
11: OUTPUT Title Name OF Video Title
12: OUTPUT Title Description OF Video Title

List All Stores

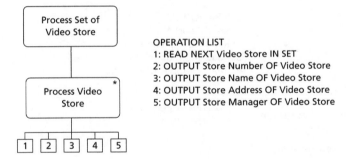

OPERATION LIST
1: READ NEXT Video Store IN SET
2: OUTPUT Store Number OF Video Store
3: OUTPUT Store Name OF Video Store
4: OUTPUT Store Address OF Video Store
5: OUTPUT Store Manager OF Video Store

Stock List

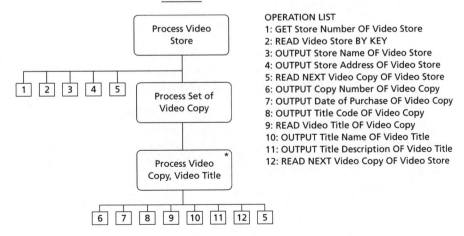

OPERATION LIST
1: GET Store Number OF Video Store
2: READ Video Store BY KEY
3: OUTPUT Store Name OF Video Store
4: OUTPUT Store Address OF Video Store
5: READ NEXT Video Copy OF Video Store
6: OUTPUT Copy Number OF Video Copy
7: OUTPUT Date of Purchase OF Video Copy
8: OUTPUT Title Code OF Video Copy
9: READ Video Title OF Video Copy
10: OUTPUT Title Name OF Video Title
11: OUTPUT Title Description OF Video Title
12: READ NEXT Video Copy OF Video Store

12.9

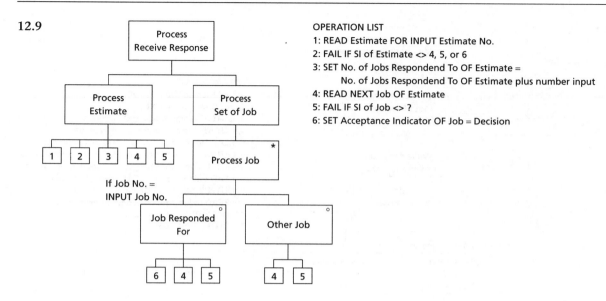

OPERATION LIST
1: READ Estimate FOR INPUT Estimate No.
2: FAIL IF SI of Estimate <> 4, 5, or 6
3: SET No. of Jobs Respondend To OF Estimate =
    No. of Jobs Respondend To OF Estimate plus number input
4: READ NEXT Job OF Estimate
5: FAIL IF SI of Job <> ?
6: SET Acceptance Indicator OF Job = Decision

12.10

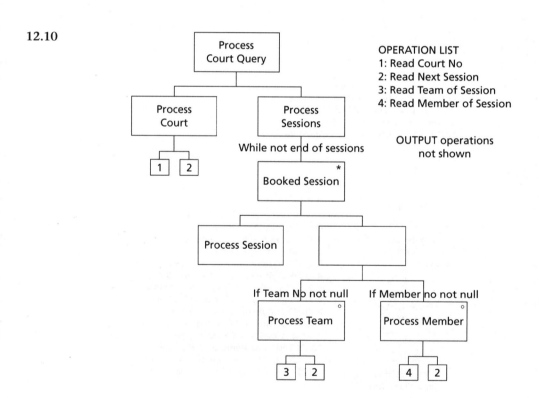

OPERATION LIST
1: Read Court No
2: Read Next Session
3: Read Team of Session
4: Read Member of Session

OUTPUT operations
not shown

12.11

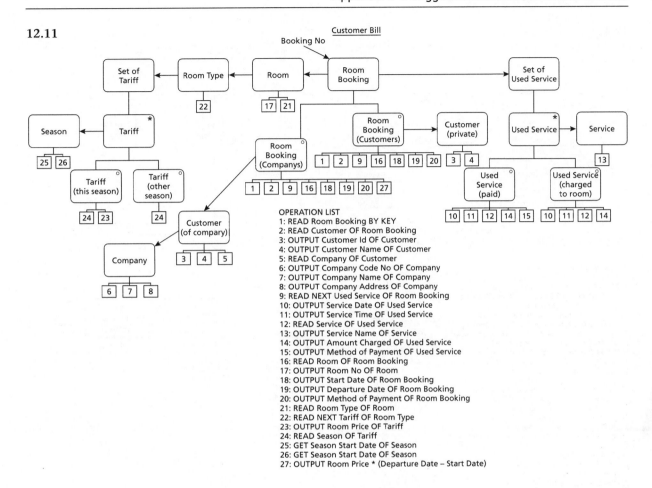

OPERATION LIST
1: READ Room Booking BY KEY
2: READ Customer OF Room Booking
3: OUTPUT Customer Id OF Customer
4: OUTPUT Customer Name OF Customer
5: READ Company OF Customer
6: OUTPUT Company Code No OF Company
7: OUTPUT Company Name OF Company
8: OUTPUT Company Address OF Company
9: READ NEXT Used Service OF Room Booking
10: OUTPUT Service Date OF Used Service
11: OUTPUT Service Time OF Used Service
12: READ Service OF Used Service
13: OUTPUT Service Name OF Service
14: OUTPUT Amount Charged OF Used Service
15: OUTPUT Method of Payment OF Used Service
16: READ Room OF Room Booking
17: OUTPUT Room No OF Room
18: OUTPUT Start Date OF Room Booking
19: OUTPUT Departure Date OF Room Booking
20: OUTPUT Method of Payment OF Room Booking
21: READ Room Type OF Room
22: READ NEXT Tariff OF Room Type
23: OUTPUT Room Price OF Tariff
24: READ Season OF Tariff
25: GET Season Start Date OF Season
26: GET Season Start Date OF Season
27: OUTPUT Room Price * (Departure Date – Start Date)

12.12

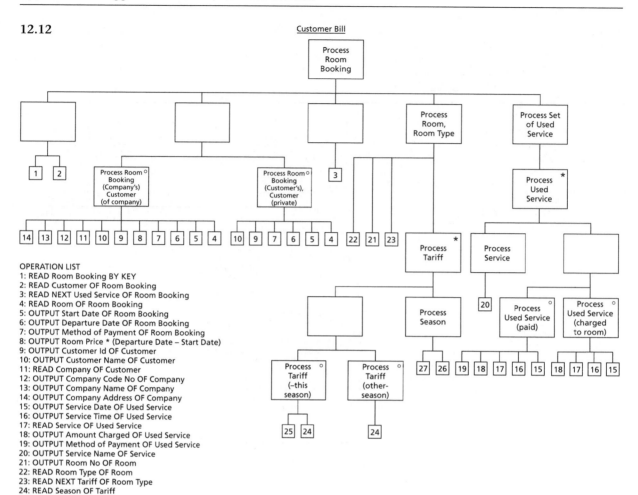

Customer Bill

OPERATION LIST
1: READ Room Booking BY KEY
2: READ Customer OF Room Booking
3: READ NEXT Used Service OF Room Booking
4: READ Room OF Room Booking
5: OUTPUT Start Date OF Room Booking
6: OUTPUT Departure Date OF Room Booking
7: OUTPUT Method of Payment OF Room Booking
8: OUTPUT Room Price * (Departure Date – Start Date)
9: OUTPUT Customer Id OF Customer
10: OUTPUT Customer Name OF Customer
11: READ Company OF Customer
12: OUTPUT Company Code No OF Company
13: OUTPUT Company Name OF Company
14: OUTPUT Company Address OF Company
15: OUTPUT Service Date OF Used Service
16: OUTPUT Service Time OF Used Service
17: READ Service OF Used Service
18: OUTPUT Amount Charged OF Used Service
19: OUTPUT Method of Payment OF Used Service
20: OUTPUT Service Name OF Service
21: OUTPUT Room No OF Room
22: READ Room Type OF Room
23: READ NEXT Tariff OF Room Type
24: READ Season OF Tariff
25: OUTPUT Room Price OF Tariff
26: GET Season Start Date OF Season
27: GET Season Start Date OF Season

14.1

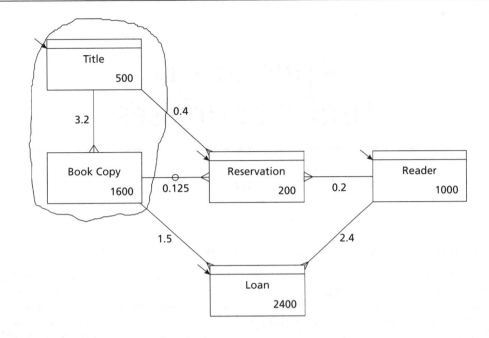

14.2

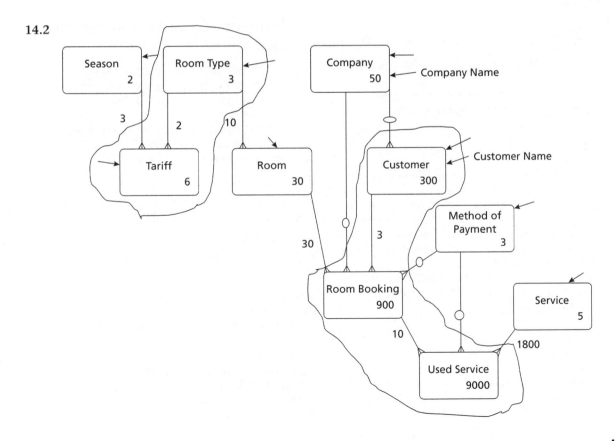

Appendix D
Useful addresses

International SSADM Users Group

The International SSADM Users Group (ISUG) is the largest representative group of SSADM users and providers and is involved in all aspects of the industry surrounding SSADM and its use in application development.

Sue McGowan
Administration Officer
Hazeldene, Halstead Lane
Knockholt
Kent, TN14 7JY
Tel: 01959 534337, email: sue@ssadm.com.uk

The University of Westminster supports the ISUG website at: www.cscs.wmin.ac.uk/~ssadm

CCTA

SSADM has been developed under the management and sponsorship of the CCTA.

CCTA, the Government Centre for Information Systems, is responsible for promoting business efficiency and effectiveness through the development and use of information systems by Government departments and agencies.

The CCTA ceased to exist in April 2001 and has been taken into the Office of Government Commerce (OGC) (www.ogc.gov.uk)

Customer Services Manager
CCTA Help Desk
Rosebery Court
St Andrew's Business Park
Norwich NR7 OHS
Tel: 01603 704567

UK Academy for Information Systems

The aims of the UK Academy for Information Systems (UKAIS) are to promote a better knowledge and understanding of information systems within the United Kingdom, and to improve the practice of information systems teaching and research.

UKAIS information can be found on www.ukais.org

Information Systems Examinations Board

The ISEB is an independent and nationally recognised body that oversees and administers accreditation and certification in SSDAM qualifications (amongst others).

The ISEB offers modular Business Systems Development Certificates and Diplomas. Diplomas are available in five specialised areas:

◆ Systems Analysis and Design
◆ Data Management
◆ SSADM 4+
◆ Business Analysis
◆ Rapid Application Development

To obtain a diploma, candidates must obtain the appropriate core modules as defined on the ISEB web site (www.bcs.org.uk/iseb).

More information can also be obtained from

Customer Services
Information Systems Examinations Board
1 Sanford Street
Swindon, SN1 1HJ
Tel: 01793 417424

Appendix E
The SSADM default structure

The SSADM default structure acts as a framework for planning an SSADM-based project. It is not intended as a project plan that any SSADM project should follow. The default structure is a set of stages and steps, with associated techniques and products, that describes all of the tasks that an SSADM project could utilise. The process of customisation will tailor this default structure to the characteristics of a project, dropping some steps and adding in other non-SSADM activities. Once tailored, the default structure *does* provide an excellent start point for the production of the project plan.

Feasibility Study		
Step	**Technique**	**Product**
Define the Problem	Data Flow Modelling	Overview Data Flow Model – Required Environment Overview Data Flow Model – Current Environment
	Logical Data Modelling	Overview Logical Data Model – Required Environment Overview Logical Data Model – Current Environment
	Requirements Definition	Requirements Catalogue
	Work Practice Modelling	User Catalogue
Select Feasibility Option	Business System Options Technical System Options Logical Data Modelling Data Flow Modelling	Feasibility Option

Investigation of Current Environment		
Step	**Technique**	**Product**
Develop Business Activity Model	Business Activity Modelling	Business Activity Model
	Work Practice Modelling	User Catalogue Work Practice Model
Investigate and Define Requirements	Requirements Definition	Requirements Catalogue
	Work Practice Modelling	User Catalogue
Investigate Current Processing	Data Flow Modelling	Current Physical Data Flow Model
	Requirements Definition	Requirements Catalogue
Investigate Current Data	Logical Data Modelling (Relational Data Analysis)	Current Environment Logical Data Model
	Requirements Definition	Requirements Catalogue
Derive Logical View of Current Services	Data Flow Modelling	Logical Data Flow Model Logical Data Store/Entity Cross Reference Requirements Catalogue Current Environment Logical Data Model
	(Logical Data Modelling)	Current Environment Logical Data Model
Assemble Investigation Results		Current Services Description Requirements Catalogue

Business System Options		
Step	**Technique**	**Product**
Define BSOs	Business System Option (Logical Data Modelling) (Data Flow Modelling) (Work Practice Modelling)	Business System Options
Select BSO	Business System Option	Selected Business System Option

Requirements Specification		
Step	**Technique**	**Product**
Define Work Practice	Work Practice Modelling	Work Practie Model User Catalogue User Roles
	Requirements Definition	Requirements Catalogue
Develop Users' Conceptual Models	User Object Modelling	User Object Model
	(Work Practice Modelling)	Work Practice Model
	Requirements Definition	Requirements Catalogue
Define Required System Processing	Data Flow Modelling	Required System Data Flow Model Logical Data Store/Entity Cross-reference
	Work Practice Modelling	User Roles
	Requirements Definition	Requirements Catalogue
Develop Required Data Model	Logical Data Model (Relational Data Analysis)	Required System Logical Data Model Data Catalogue
	Requirements Definition	Requirements Catalogue
Derive System Functions	Function Definition	Function Definitions I/O Structures User Role/Function Matrix
	Requirements Definition	Requirements Catalogue
Design, Prototype and Evaluate User Interface Design	User Interface Design	Function Navigation Models Window Navigation Models Help System Specification
	Specification Prototyping	Prototyping Report
	Dialogue Design	Menu Structures Command Structures
	Work Practice Modelling	Work Practice Model
	Requirements Definition	Requirements Catalogue
Enhance Required Data Model	Relational Data Analysis (Logical Data Modelling)	Required System Logical Data Model Data Catalogue
Develop Processing Specification	Entity Behaviour Modelling	Entity Life Histories
	Conceptual Process Modelling	Effect Correspondence Diagrams Enquiry Access Paths
	Requirements Definition	Requirements Catalogue
Assemble Requirements Specification		Requirements Specification

Logical Design		
Step	**Technique**	**Product**
Define Update Processes	Conceptual Process Modelling	Update Process Models
Define Enquiry Processes	Conceptual Process Modelling	Enquiry Process Models
Assemble Logical Design		Logical Design

421

Technical System Options		
Step	**Technique**	**Product**
Define TSOs	Technical System Option	Technical System Options
Select TSO	Technical System Option	Technical System Options Technical System Architecture Application Style Guide

Physical Design		
Step	**Technique**	**Product**
Prepare for Physical Design	Physical Data Design Physical Process Specification	Application Development Standards Physical Design Strategy
Create Physical Data Design	Physical Data Design	Physical Data Design (1st-cut) Space Estimation
Create FCIM	Physical Process Specification	FCIM Function Definitions Requirements Catalogue
Optimise Physical Data Design	Physical Data Design	Physical Data Design Function Definitions Requirements Definition Space Estimations Timing Estimations
Complete Function Specification	Physical Process Specification	FCIM Function Definitions Requirements Catalogue
Consolidate PDI	Physical Process Specification	Process Data Interface FCIM Function Definitions Requirements Catalogue
Assemble Physical Design		Physical Design

References and further reading

▶ Further reading on SSADM

Central Computer and Telecommunications Agency (CCTA) (2000) *Business Systems Development with SSADM*, in 7 volumes: *SSADM Foundation; Data Modelling; The Business Context; User-centred Design; Behaviour and Process Modelling; Function Modelling; Database and Physical Process Design*, Norwich: HMSO

Hargrave, D. (1996) *SSADM4+ for Rapid Systems Development*, Maidenhead: McGraw-Hill

Robinson, K. (1994) *Object-oriented SSADM*, Hemel Hempstead: Prentice Hall

▶ Business information systems

Chaffey, D., Bocij, P., Greasley, A. and Hickie, S. (1999) *Business Information Systems*, Harlow: Financial Times Prentice Hall

Checkland, P. (1981) *Systems Thinking, Systems Practice*, Chichester: Wiley

Checkland, P. and Scholes, J. (1990) *Soft Systems Methodology in Action*, Chichester: Wiley

Griffiths, G. (1998) *The Essence of Structured Systems Analysis*, London: Prentice Hall

Stowell, F. and West, D. (1994) *Client-led Design*, Maidenhead: McGraw-Hill

Tudor, D. J. and Tudor, I. J. (1997) *Systems Analysis and Design: A Comparison of Structured Methods*, Maidenhead: McGraw-Hill

Yeates, D., Shields, M. and Helmy, D. (1994) *Systems Analysis and Design*, London: Pitman

▶ IS development management

Robson, W. (1997) *Strategic Management and Information Systems*, Harlow: Financial Times Prentice Hall

Yeates, D. and Cadle, J. (1996) *Project Management for Information Systems*, London: Pitman

▶ Useful publications from CCTA

Drummond, I. (1992) *Estimating with Mk II Function Point Analysis*, CCTA

O'Neil, P. B. (1993) *Prototyping within an SSADM Environment*, CCTA

▶ Databases and programming

Date, C. J. (2000) *An Introduction to Database Systems*, Reading, MA: Addison-Wesley

Elmasri, R. and Navathe, S. (2000) *Fundamentals of Database Systems*, Reading, MA: Addison-Wesley

Ingevaldsson, L. (1989) *Software Engineering Fundamentals: the Jackson Approach*, Bromley: Chartwell-Bratt

Jackson, M. (1975) *Principles of Program Design*, London: Academic Press

Index